SEED SCIENCE AND TECHNOLOGY

SEED SCIENCE AND TECHNOLOGY

Editor

Stewart Arterbury

scitus academics

Seed Science and Technology

Edited by **Stewart Arterbury**

ISBN: 9781681179179
Library of Congress Control Number: 2017948946

SCITUS Academics LLC,
616, Corporate Way, Suite 2, 4766,
Valley Cottage, NY 10989

www.scitusacademics.com

Notice

Printed in United States of America.

Preface

Seeds are fundamental to agriculture. They are the starting point for the production of most crops and delivery system for advanced genetics. Seeds constitute 70% of our food and recent additional uses of seeds as stored energy has increased both seed and commodity prices worldwide. The past 50 years has seen many research-driven improvements in seed genetics and technology that have been responsible for dramatic increases in crop productivity worldwide. In broad sense, seed is a material which is used for planting or regeneration purpose. However scientifically, Seed is a fertilized matured ovule together covered with seed coat is called seed or it is a propagating material i.e., part of agriculture, sericulture, silviculture and horticultural plants used for sowing or planting purpose. In agricultural systems, the seed used for cultivation is considered one of the means of production. Moreover, the seed is the starting point, the first determinant of the future plant development and consequently the master key to success with its cultivation. The expression "seed quality" is used in practice to describe the overall value of the seed for its intended purposes. It is a multiple concept resulting from the genetic characteristics of the seed and from other factors affecting its development, maturation and storability. Seed quality is a combination of different characteristics. Within the seed, there usually is a store of nutrients for the seedling that will grow from the embryo. The form of the stored nutrition varies depending on the kind of plant. In angiosperms, the stored food begins as a tissue called the endosperm, which is derived from the parent plant via double fertilization. The usually triploid endosperm is rich in oil or starch and protein. In gymnosperms, such as conifers, the food storage tissue is part of the female gametophyte, a haploid tissue. In some species, the embryo is embedded in the endosperm or female gametophyte, which the seedling will use upon germination. In others, the endosperm is absorbed by the embryo as the latter grows within the developing seed, and the cotyledons of the embryo become filled with this stored food.

'Seed Science and Technology' focuses on seed quality and physiology as related to seed production, harvest, processing, sampling, storage, genetic conservation, habitat regeneration, distribution and testing. The contributed chapters are written by seed science experts from the field of seed physiology, ecology, molecular biology, biochemistry, and seed technology was gathered. We hope that this book will attract the attention of researchers and technologists from academia and industry, providing points for interactive and fruitful discussion on this fascinating topic. To meet future world needs for food, fiber and energy, additional research advancements in seed genetics and technology will be critical. This comprehensive guide is designed to meet the needs of researchers, advisers and all those involved in the improvement and technical control of seed quality.

Table of Contents

CHAPTER 1

Physical Properties of Seeds in Technological Processes

B. Dobrzański[1] and A. Stępniewski[1]

[1] *Institute of Agrophysics Polish Academy of Sciences, Lublin, Poland*

1. INTRODUCTION

The physical properties of seeds and grain is a wide knowledge that can be useful in the farming, harvesting and storage or in processing such as drying, freezing and other. This knowledge is important in the designing of machinery to harvest and in preparation of processing chain from grain to food. Accurate design of machines and processes in the food chain from harvest to table requires an understanding of physical properties of row material. Considering either bulk or individual units of the agricultural material, it is important to have an accurate estimation of shape, size, volume, density, specific gravity, surface area, and other mechanical characteristics, which may be considered as designing parameters for food production. The measurement techniques allow computation of these parameters, which can than provide information about the effects of processing. Some of characteristics, such as colour, mechanical parameters, rheological properties, thermal and electrical resistance, water content and other physical quantities give excellent description of product quality.

2. PHYSICAL PROPERTIES OF AGRICULTURAL MATERIALS

In a Newtonian sense, the physical properties of an object may include many varietals properties. Physical properties are determined for many raw materials and agricultural products, including seeds and grain, however, Barbosa-Cánovas et al. (2004) divided them into the following groups:

- Thermal properties such as specific heat, boiling point, conductivity, temperature, thermal transfer, diffusivity, and boiling point rise, freezing point depression.
- Optical properties, primarily colour, but also gloss and translucency.
- Electrical properties, electric charge, electric field, electric potential, primarily conductivity, resistance, dielectric, conductivity, impedance, inductance, magnetic field, magnetic flux, and permittivity.
- Structural and geometrical properties such as mass, density, particle size, shape, volume, length, location, porosity, surface roughness, and cellularity.
- Mechanical properties (including strength, ductility, tension, compressibility, pressure, deformability moment, and textural properties, malleability,) and rheological properties (such as viscosity, flow rate, fluidity,).
- Others, including mass transfer related properties (diffusivity, permeability), surface tension, cloud stability, gelling ability, and radiation absorbance.

Seeds in general can display large compositional variations, inhomogeneities, and anisotropic structures. Composition can change due to seasonal variations and environmental conditions, or in the case of processed grain, properties can be affected by process conditions and material history.

Early physical property analyses of seeds and grain required constant uniform values and were often oversimplified and inaccurate. Nowadays, computational engineering techniques, such as the finite element method, are much more sophisticated and can be used to evaluate non-uniform properties (for example, thermal properties) that change with time, temperature, and location in grain that are heated or cooled. Improvements measuring the compositions of seeds are now allowing predictions of physical properties that are more accurate than previously.

3. PHYSICAL METHODS FOR QUALITY EVALUATION OF SEEDS

It was found that radiation in the near-infrared region of the spectrum can provide information related to many quality factors of agricultural products. Short wave radiations such as X-rays and gamma rays can penetrate through most seeds. The level of transmission of these rays depends mainly on the mass density and mass absorption coefficient of the material, that it allow find internal cracks invisible because of the seed coat (Dobrzański et al, 2003).

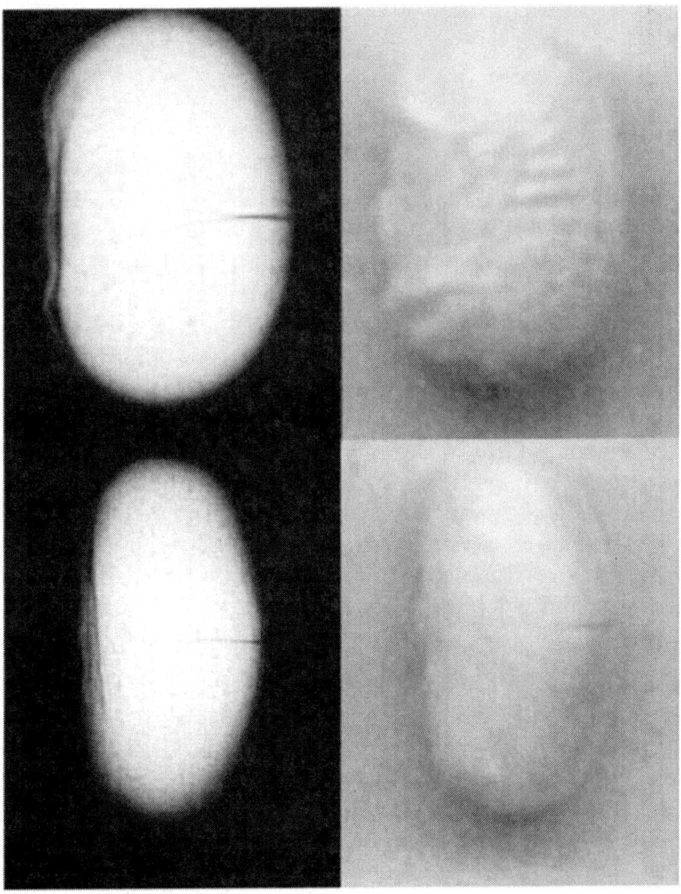

FIGURE 1. Invisible crack on the seed coat visible in X-ray image

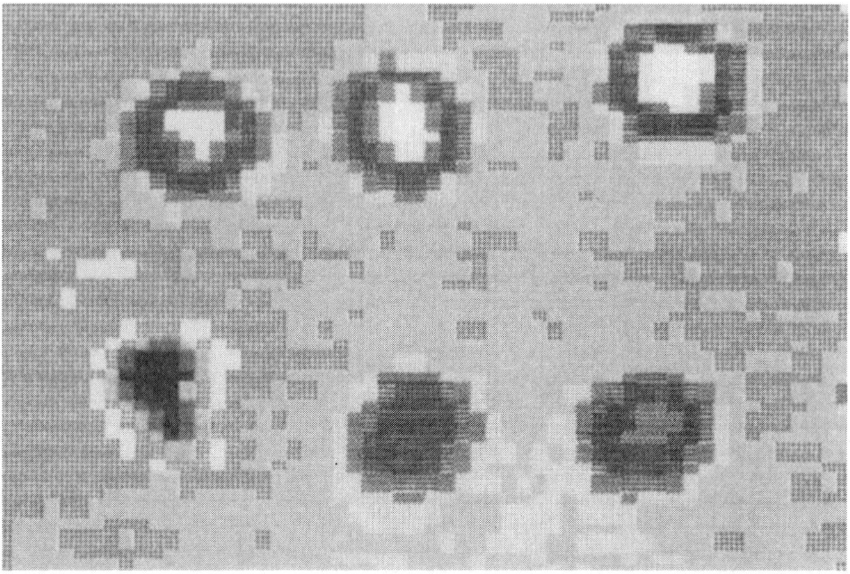

FIGURE 2. Thermal image of seeds after 3 hours of water soaking and imbibitions

Nuclear magnetic resonance (NMR) is a technique that detects the concentration of hydrogen nuclei (protons) and is sensitive to variations in the concentration of liquid in the material. Although NMR imaging (MRI) has been used frequently in the medical field and other quality factors in grain has not been fully explored. Chen (1996) presents an overview of various quality evaluation techniques that are based on one of the following physical property: density, firmness, vibration characteristics, X-ray and gamma ray transmission, optical reflectance and transmission, electrical properties, aromatic volatile emission, and nuclear magnetic response (NMR).

One of the most practical and successful techniques for nondestructive quality evaluation and sorting or separating impurities is the electro-optical technique, based on different optical properties of seed and impurities. Thus, determining such optical characteristics of seeds and grain can provide information related to quality factors of the product. Looking for more advanced method of germination capacity estimation, Dobrzański et al (2003) using thermal technique found slightly differences in temperature of dead and alive seeds, being soaked, that thermal properties should indicate live potential of seed in first hours of germination test. According to traditional germination test, that needs two weeks or more for forming germ (depends to species), thermography should be rapid technique for estimation of germination capacity.

4. INSTRUMENTAL MEASUREMENT OF MECHANICAL PROPERTIES OF SEEDS AND GRAIN

There are many types of mechanical loading: puncture, compression, shearing, twisting, extrusion, crushing, tension, bending, vibration, and impact. And there are four basic values that can be obtained from mechanical properties tests: force (load), deformation (distance, displacement, penetration), slope (ratio of force to deformation), and area under the force/deformation curve (energy). Mechanical characteristics included basic parameters (hardness, cohesiveness, viscosity, elasticity, and adhesiveness) and secondary parameters (brittleness or fracturability, chewiness, and gumminess).

The engineering terms based on these measurements are stress, strain, modulus, and energy, respectively. Stress is force per unit area, either of contact or cross-section, depending on the test. Strain is deformation as a percentage of initial height or length of the portion of sample subject to loading. Modulus of elasticity (tangent, secant, chord, or initial tangent) is a measure of stiffness based on the stress/strain ratio. Force and deformation values are more commonly used in food applications than stress and strain values, that are determined at seed loading.

Generally, determination of mechanical properties based on the high precision measurement, with expensive equipment (Figure 3 and 4) requires a complete force-deformation curve. From the force-deformation curve, stiffness; modulus of elasticity; modulus of deformability; toughness; force and deformation to point of inflection, to bioyield, and rupture, and maximum normal contact stress or stress index at low levels of deformation can be obtained. The force-deformation curve was frequently used in the study of physical properties of seeds and grain; however this measurement is not widely utilized in practice because of expensive and complicated procedures. A typical force/deformation characteristic for a cylindrical piece of tissue compressed at constant speed gives F/D curve for puncture tests look similar to compression curves. The portion of the initial slope up to point represents nondestructive elastic deformation; point is the inflection point where the curve begins to have a concave-downward shape and is called the elastic limit. The region before this point is where slope or elastic modulus should be measured. Beyond the elastic limit, permanent tissue damage begins. There may be a bioyield point where cells start to rupture or to move with respect to their neighbors, causing a noticeable

FIGURE 3. Universal testing machine Instron 1253 for fatigue test and dynamic load

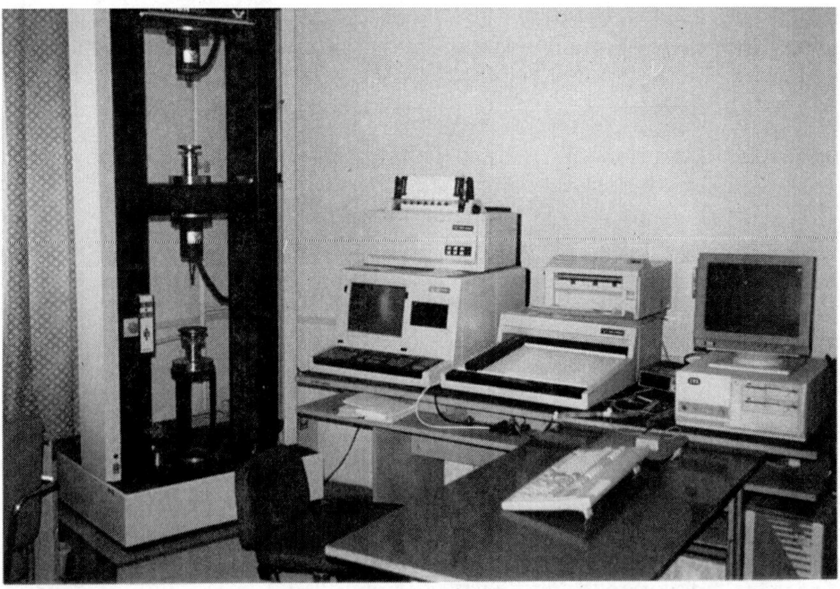

FIGURE 4. Universal testing machine Instron 6022 for quasi-static tests.

FIGURE 5. Temperature chamber mounted to 8872 Instron machine (High Wycombe)

decrease in slope (Abbott and Harker, 2003). (Dobrzański and Rybczyński, 2011 divided the measurements on direct (contact; compression, shear, impact, rebound) and indirect techniques (non-contact; vibration, sonic).

Seed hardness is related to the quality factors or behaviour of grain in relation to harvest, however, through use of simply test, only the maximum squeezing force has been frequently correlated with numerous quality factors. It is also well known, that including external and internal properties of seed, the hardness should depends to the shape and size of seed; size and contact area with plate; rate of deformation; the way of seed fixing, and other procedures influenced a final accuracy. It is well known, that all mechanical properties and behaviour of agricultural materials are influenced by moisture and temperature, but universal testing machine with temperature chamber (Figure 5) must be equipped.

5. THE CRACKING MECHANISMS OF GRAIN LEGUME IN TECHNOLOGICAL PROCESSES

The models describing some geometrical relations for different cracking mechanism were main aim of presented paper. In this study, the verify of cracking mechanism needs to determine the elasticity strain limit of cotyledon and the resistance of the seed coat to tension for various legume seeds. The change of shape as well as the mechanical values of the seed coat and cotyledon at various moisture contents, during swelling and drying were determined. This approach lead to an understanding of the possible causes and mechanisms that affect the cracking of the seed coat. Further, a major problem in the drying of legume seeds with heated air is the splitting of seed coat; affected by the temperature and moisture. The estimation of the effect of drying on the stress and strain development of the seed coat was determined at wide range of moisture.

Legume seed quality is greatly affected by methods of harvesting and handling. In these operations, seeds are subjected repeatedly to many impacts on metal surfaces and against other seeds. Impacts may be result of threshing cylinder or rotor motion, the movement of buckets in vertical elevators, centrifugal discharging, the filling and discharging of screw conveyors, spouting and free-fall dropping (Palusen *et al.*, 1981 a,b). In nearly all handling operations, seeds are accelerated to some velocity and then discharged onto stationary objects or other grains.

Physical properties of agricultural products are needed for adequate design of processing machines but their specific applications should be clearly understood before determining it experimentally (Misra *et al.*, 1981). The modulus of elasticity of seed is a physical property which has been suggested as an evaluation of firmness and hardness (Mac Donald, Jr.,1985). It is also an important property for determination of the stress cracks in seeds. However, in study of stress cracking due to drying of cereal grains from about 25 percent moisture content to about 14 percent moisture content, it is essential to understand the variation of modulus of elasticity according to wide range of moisture (Dobrzański, 1998; 2011).

The moisture content of seed coat and cotyledon influences the cracking mechanism and many researches have already published the results of mechanical properties including a modulus of elasticity for various seeds at a given moisture content and deformation rate. However, in order to clarify the mechanisms of seed cracking, it is necessary to carefully recognise the mechanical properties of cotyledon and seed coat separately.

The major problem in the drying of legume seeds with heated air is the splitting of the seed coat (Liu *et al.*, 1989). Temperature and moisture gradients can cause stresses in the seed coat during drying. Overhults *et al.* (1973) observed the physical damage in soybeans during drying, in the form of indentations and cracks. This condition makes the beans susceptible to microbial attack during storage, and also reduces their germination potential (White *et al.*, 1980). In view of these problems, there is need, specially in the leguminous seed industry, to find the optimum combination of drying parameters which will minimize seedcoat cracking (Mensah *et al.*, 1984).

The following types of cracking mechanisms were observed during mechanical loading of the legume seeds:

- tension of seed coat caused by the shape deformation at axial compression,
- tension of seed coat caused by wet cotyledons sliding,
- shearing of dry seed coat caused by a force that smashed the seed in two cotyledons,
- crack opening displacement of the cotyledon under compression.
- splitting of the seed coat; affected by the high temperature and low moisture (a major problem in drying practice of grain legume). A rapid change of seed moisture during drying caused the shrinkage of seed coat leading to its stress. The splitting of the coat, caused by shrinkage, is frequently the most important mechanism of coat cracking.

6. TENSION OF SEED COAT CAUSED BY DEFORMATION OF THE SEED COMPRESSED

With the increase of moisture content seed become very plastic and even slight forces, which do not produce negative biological results, may cause considerable deformations of shape.

The description of the shape deformations of the legume seed compressed between parallel plates accomplishment needs some simplifying assumptions:

- the seed is a sphere surrounded by elastic shell,
- the compressibility of the cotyledon is very low that the volume is steady,
- the seed coat thickness is small that in geometrical calculations could be ignored,
- The seed compressed between parallel plates is circular in initial phase, deforming further in to the barrel of parabolic shape at large deformation.

Using the assumption of constant volume, make it possible to determine the changes of surface area describing the spherical body being compressed between parallel plates. When the deformation is slight the sphere does not lose its shape and hence it was assumed that lateral surface of barrel created by compressing the opposite cups of sphere will also be spherical in its shape.

In initial phase of a shape deformation of viscoelastic seed was observed. Further cotyledons deformation causes seed coat filling, and when this process completed, an increase of the inner pressure leads to the seed coat tension; till its disruption. Assuming that during seed coat tensions the pressure necessary for shape deformation is constant, the increase of pressure $\Delta p\ w$ resulting in the seed coat cracking can be determined from the force-deformation curve.

Substituting characteristic values, taken form the force-deformation curve obtained in compression tests of legume seeds and the diameter of contact area as well as diameter of barrel calculated using geometrical formulas presented in previous papers (Dobrzański, 1998) it is possible to determine the modulus of elasticity of seed coat $E\ om$ from the following equation:

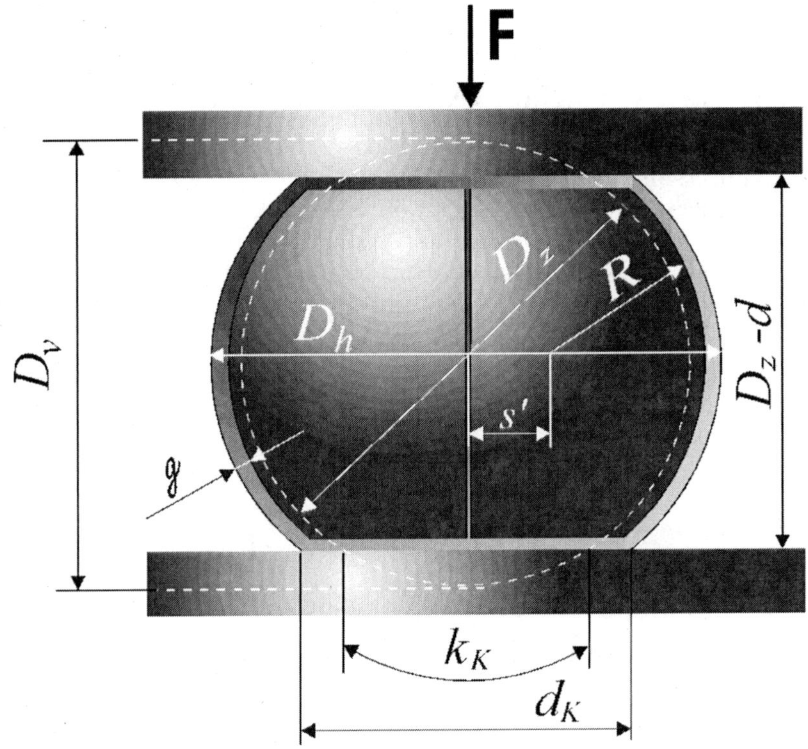

FIGURE 6. The shape deformation of seed at axial compression.

$$Eom = \frac{D_h \Delta p_w}{\Delta D_h}\left(\frac{(0.5D_h + g)^2 + (0.5D_h)^2}{(0.5D_h + g)^2 - (0.5D_h)^2} + \nu\right) \quad (1)$$

The above method enabled the interpretation of the force-deformation curve for large deformation of non-elastic wet seeds where mechanical strength is mainly connected with the resistance of the seed coat to tension.

7. TENSION OF WET SEED COAT CAUSED BY THE SLIDING DISPLACEMENT OF THE COTYLEDONS

The compression of seed leads the shape deformation, however, low increase of lateral surface at seed strain under ε<0.3 proves, that there are any stresses of it. On the other hand, sliding of wet cotyledons during compression caused larger shape deformation of seed leads to tension of the seed coat and stress.

Sliding cotyledons surrounded by the seed coat caused tension strain of the seed coat ε_o, which can be calculated from simple equation based on geometrical relations according to the figure 7:

$$\varepsilon_o = \frac{\pi r_w - k_U + d_U}{\pi r_w} \tag{2}$$

Substituting geometrical relations for above values of formula (2) it is possible to determined strain of the seed coat largest diameter in cross section of body formed by sliding of cotyledons:

$$\varepsilon_o = 1 - \frac{1}{180}\arccos\frac{D_z - 2d}{D_z} + 2\frac{\sqrt{\left(\frac{D_z}{2} + \frac{D_z d}{D_z}\right)^2 - \left(\frac{D_z}{2}\right)^2}}{\pi D_z} \tag{3}$$

The change of seed shape during compression caused increase of the surface area surrounding cotyledon. A sphere surrounded by a shell imitating a seed is deformed on both sides and change the shape into a barrel when compressed between parallel plates. When the deformations are slight the sphere does not lose its shape and hence it was assumed that the lateral surface of the barrel created by compressing opposite cups of sphere will also be spherical.

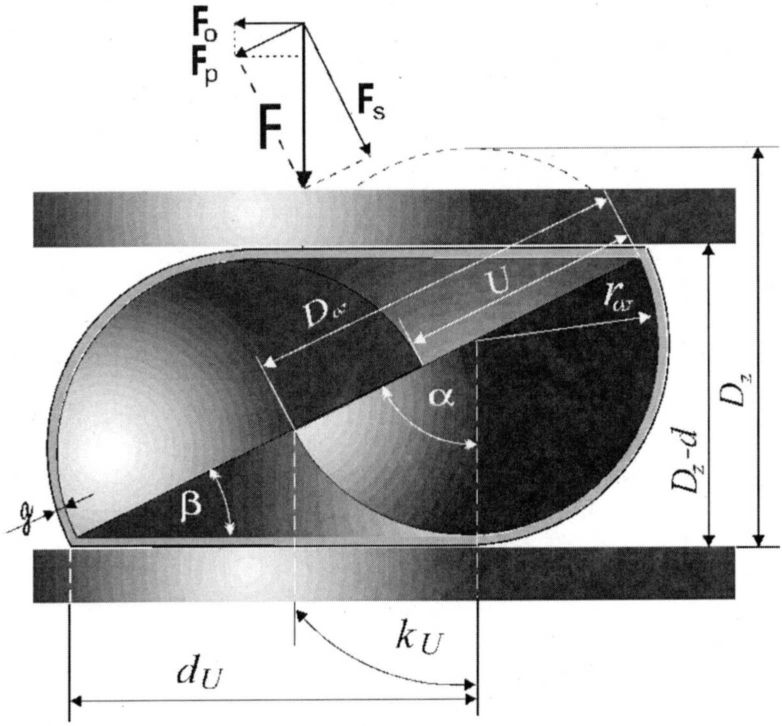

FIGURE 7. Tension of seed coat caused by the sliding displacement of the cotyledons.

Circumferential strain $\varepsilon\, o$ at tension caused the sliding displacement of the cotyledons determined using formula (3) is represented on the figure 8. The increase of shape deformation is represented by barrel diameter of parabolic $D\,P$ and circular shape $D\,K$. Both diameters increases with an increase of axial deformation, however, almost 10 % of the increase was obtained for 0.4 axial strain. In comparison, cotyledon sliding caused coat strain $\varepsilon\, o$ up to 15 % for axial strain ε close to 0.2 only. It suggests that sliding may involve rupture easier than shape deformation at compression. The range of legume seed coat elongation at rupture for various moisture are presented on Figure 8. It is easy to conclude, that fababean and bean seed coat is most influential on tension at axial compression and sliding of the cotyledons. At 0.18 of axial strain caused cracking of the seed coat of some fababean and bean.

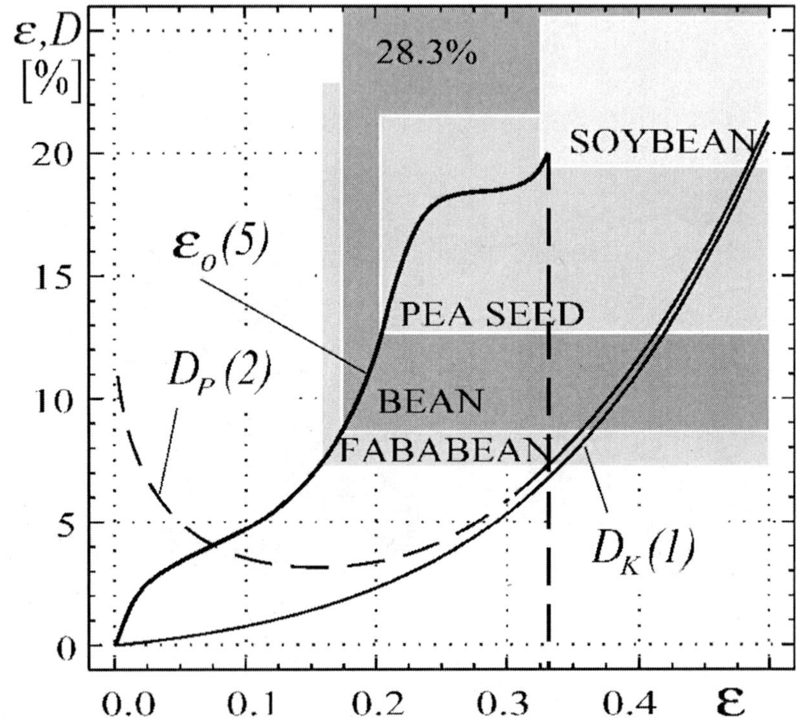

FIGURE 8. Increase of the seed diameter caused by the shape deformation D_K, D_P and cotyledons sliding ε_o.

Fababean *(Vicia faba* L.*)*, soybean *(Glicine hispida* Max.*)*, bean *(Phasoleus* L.*)*, and pea seed *(Pisum sativum* L.*)* cultivars were studied to mechanical resistance of seeds subjected to load at different rates of deformation. The resistance of legume seeds to compression was determined over a wide range of moisture contents. The diameter of seed and shape deformation under compression, as an additionally measurement, were determined with strain gauge. Most resistant species of legume seeds is soybean, for which the seeds could not be damaged in this way. The determination of the cracking mechanism of the cotyledon was carried out at low moisture only with single cotyledons -- a half of bean was compressed between parallel plates at various rates of cross-head movement.

The resistance of seed coat to tension was studied for all seeds at wide range of moisture. The drying test was conducted to evaluate the effect of drying parameters on the stress and strain development of the seed coat. To evaluate the effect of drying conditions on the splitting of seed coat, the samples were dried with air. The time-strain curves at different drying conditions were obtained with 2630-107 Instron strain gauge. Three air velocities: 4.3; 10.5; 15.6 m/s and four temperatures: 22; 42; 52; 62°C were used as parameters of the drying tests.

The results obtained from compression tests of cotyledons of fababean cultivars as an example was presented in previous paper (Dobrzański, 1998). Different values of cracking force for each cultivar indicate that the strength of cotyledons differ. However, the modulus of elasticity is almost the same and there is no significance difference among (p=0.05) the cultivars. This was also noted for other legume species.

8. SHEARING OF THE SEEDCOAT AT THRESHING OF DRY SEED

The shearing of dry seedcoat most frequently observed mechanism of bean cracking at trashing, is caused in dry seed only, by a force which is parallel to the plane of cotyledons contact. This mechanism is characteristic also for rape seed and other legume seeds such as pea seed, french bean, soybean and fababean, which covers under the seed coat two cotyledons. Some of dry seed during processing are separated two parts. It may be result of threshing cylinder or rotor motion, centrifugal discharging, the filling and discharging of screw conveyors, sprouting and free-fall dropping. Each part consists the cotyledon and half of the seed coat, which is shearing by a force parallel to the plane of contact of the cotyledons (Figure 9).

Approximately cross section area of the seed coat, for oblong seeds such as: soybean, bean and french bean, was determined according to the figure 4 from the formula:

In this mechanism of dry seed cracking the force F made work L at deformation d, leads to the seed coat shearing, which for elastic body may by calculate using simple formula:

$$L = \frac{Fd}{2} \tag{4}$$

Both of these values are determined in experimental test presented on Figure 9. According to the energy conservation law it is possible to determined permissible velocity limit Vp, for which the seed of known mass m did not cause the seed coat cracking. Equating both energies: kinetic energy of spouting seed and work of cracking at seed coat shearing following formula was obtained:

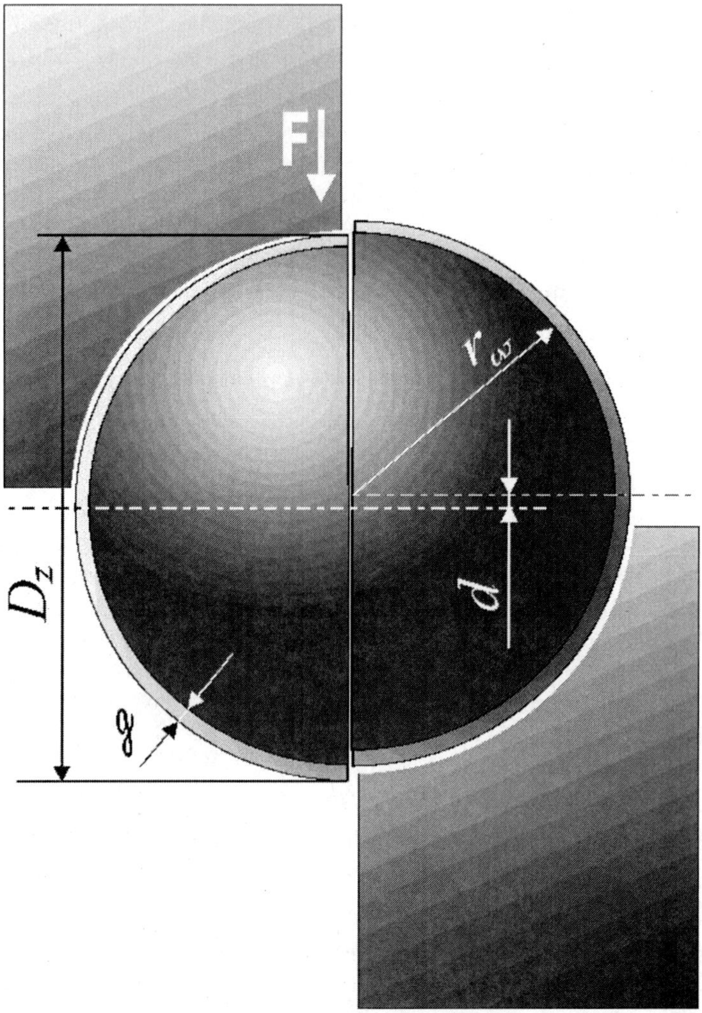

FIGURE 9. Shearing of the seed coat.

$$V_p = \sqrt{\frac{Fd}{m}} \tag{5}$$

This method of velocity limit Vp calculation covers some simplifying assumptions, however, it allows to estimate the seed cracking of dynamic processes frequently meets in practice, by simply way at quasi-static test.

9. SHRINKAGE STRESS OF THE SEED COAT AT DRYING

The effect of moisture content on expansion volume of soybean was shown in the experimental results presented in previous paper (Dobrzański, 1996). Seed dimensions increased during swelling for all cultivar. More distinctly effects of moisture content on dimensions of soybean were observed in the initial phase of swelling. Further seed's filling, leads to the seed coat tensioning. Similar behavior and significant decrease of volume in other grain legume was observed (Figure 10).

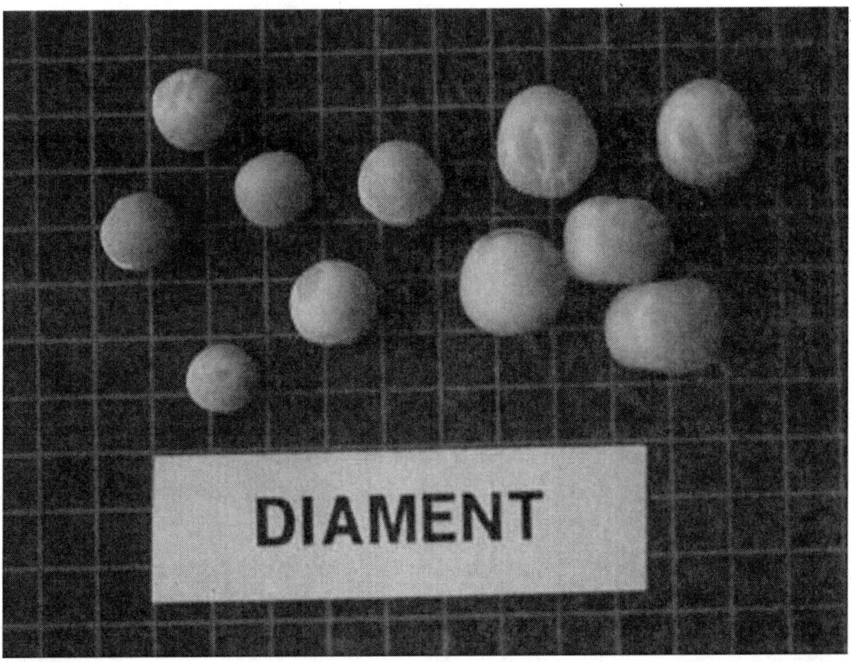

FIGURE 10. Volume expansion of pea seed (v. Diament) caused by water imbibitions during soaking.

The water vaporizing from the seed is limited by coat and the results presented in previous paper (Dobrzański, 1996) proved that this process was not rapid. On the other hand, all dimensions of seed decreased at this process and exponential relationship was observed.

Rapid change of moisture content of seed coat during drying caused the shrinkage of coat leading its stress. Highest shrinkage for rudest condition of drying was observed. When the drying conditions were milder, the shrinkage of the seed coat was lower. The Figure 11 presents a model of spherical seed in the stream of air $T\,tV$. The wet seed of volume $V\,Kw$ is surrounded by the seed coat of diameter $D\,W$. During drying the seed coat try to reduce its volume to $V\,Ks$, following its surface and diameter $D\,S$.

The shrinkage of the seed coat on wet seed of the large volume leads the stress, which involve increase of the inner pressure $p\,w$ of incompressibility wet cotyledons.

In order to determine strength criteria for tension seed coat at drying, Dobrzański (1998) found that, it is sufficient to estimate circumferential stress $\sigma\,o$ in cross section of seed. He determined the circumferential stress in the seed coat using Lamé equation concerning radial strain and stress of a pipe subjected to the inner pressure $p\,w$:

$$\sigma_o = \frac{p_w(0.5D_W^2 + D_W g + g^2)}{D_W g + g^2} \tag{6}$$

During drying, the circumferential shrinkage stress $\sigma\,o$ must meet the strength criteria:

$$\sigma_o \le k_r \tag{7}$$

Taking into account, that stress of the seed coat $\sigma\,s$ at drying depends to shrinkage strain $\varepsilon\,s$ and modulus of elasticity of the seed coat $E\,o$ it must be verify experimentally.

The single seed presented on figure 11 is in the stream of hot air, however, most seeds at real condition are in the layer being in contact to other. It limits inflow of air to some parts of seed in contact to other (Figure 11).

Hot air caused the shrinkage of part available for the stream of air. Residual part of seed coat being in the contact to other seed keeps higher moisture content. Dobrzański and Szot, 1997; Dobrzański, 2011) found that mechanical strength of seed coat decreases with the increase of moisture content. Finally, shrinkage of dry seed coat caused tension of wet part of it. Figure 12 presents the cross section of two seed being in the contact.

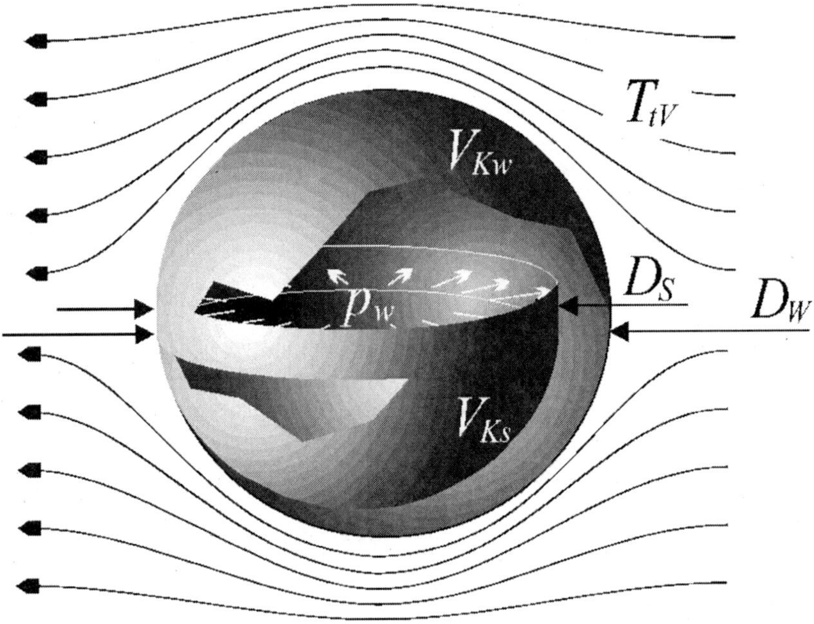

FIGURE 11. The shrinkage of the seed coat in the stream of air.

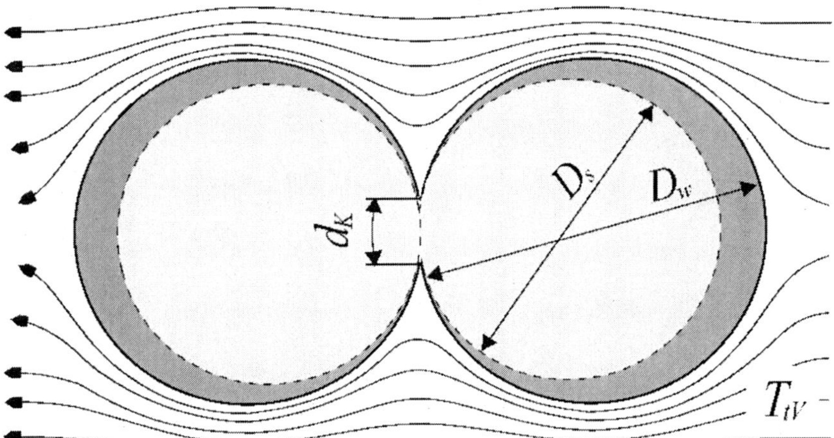

FIGURE 12. The shrinkage of part of seed coat in the stream of air.

The circle diameter $D\,W$ decrease to $D\,S$, however only the part $(D\,W - d\,K)$ of the seed coat is being subjected to the shrinkage. It allows determination of the shrinkage strain $\varepsilon\,s$ of dried part of the seed coat:

$$\varepsilon_s = \frac{\pi(D_W - D_S)}{\pi D_W - d_K} \tag{8}$$

The shrinkage leads to the stress σs, which depend to the modulus of elasticity of dry seed coat $E\ oS$:

$$\sigma_s = E_{oS}\varepsilon_s \tag{9}$$

The shrinkage of dry seed coat caused tension $\sigma\ o$ of wet part $d\ K$. Determining strain $\varepsilon\ o$ of wet part of the seed coat subjected to extension and its modulus of elasticity $E\ oW$, it is possible to estimate the tension strength limit $\sigma\ u$:

$$\sigma_u = E_{oW}\varepsilon_o \tag{10}$$

For safe condition of drying must be fulfil an inequality:

$$\sigma_u > \sigma_s, \tag{11}$$

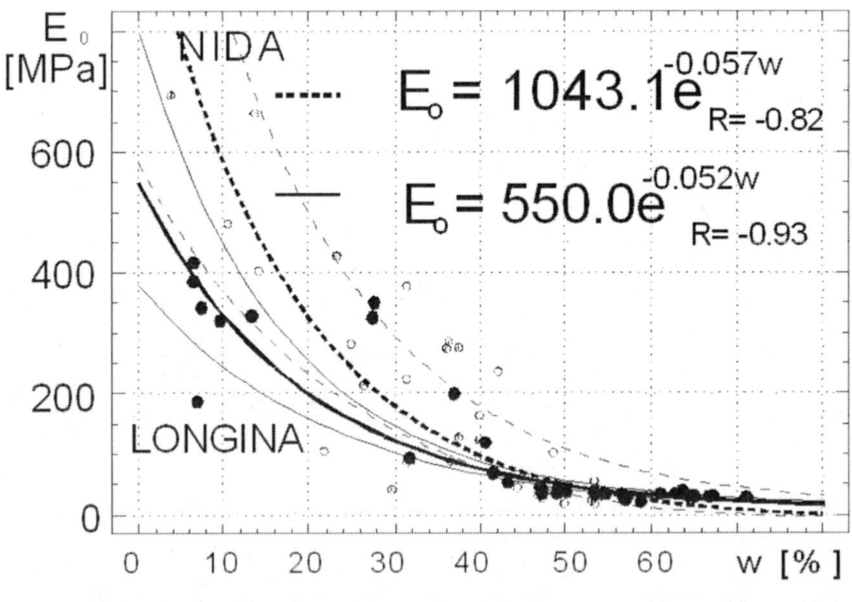

FIGURE 13. Modulus of elasticity of the seedcoat for Nida and Longina varieties in the range of moisture from 6 to 70 %.

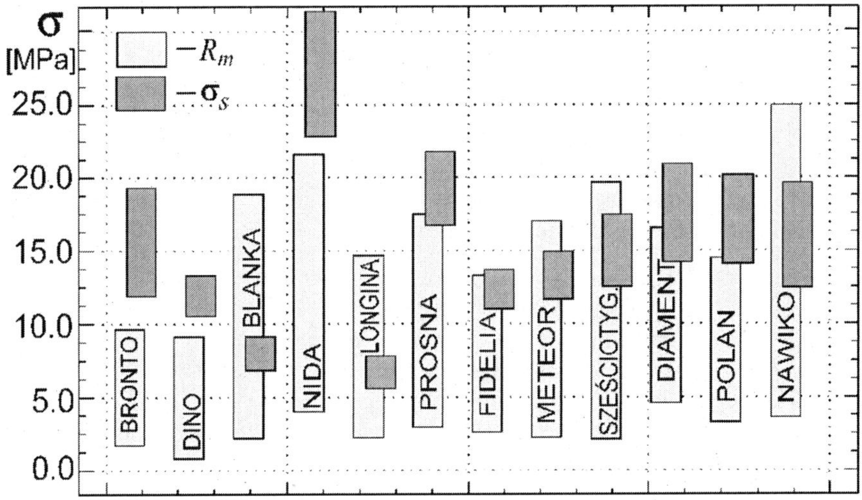

FIGURE 14. The range of seed coat strength R_m to tension and shrinkage coat stress σ_s at drying for fababean (Bronto, Dino) bean (Blanka, Nida, Igolomska, Prosna), pea seed (Fidelia, Szesciotyg, Diament, Meteor) and soybean (Polan, Nawiko).

For this purpose, the shrinkage stress σs and strength $R m$, (Figure 14) strain εs, ultimate elongation at rupture εo and the modulus of elasticity $E o$ (Figure 13) must be determine experimentally in wide range of moisture.

A preliminary study was conducted to establish the time required for strain of seed coat after being dried to final moisture content. The second series of drying tests was conducted to evaluate the effect of drying parameters on the stress and strain development of seed coat. To evaluate the effect of drying conditions on the splitting of seed coat, the samples were dried with air. The seed coat ring samples were prepared according to the method elaborated by Dobrzański (1998).

The results obtained at tension of the seed coat shoved that, the increase of moisture content involves the decrease of force at breakage. Cultivar's differentiation of seed coat strength in tension test was observed. The results obtained in all experiments concerning with strength of seedcoat proved that frequently the stress was higher than strength of seed coat (Figure 14). Probably, it is one of the most important reason of the seedcoat cracking during drying. The rapid decrease of air temperature during drying causing additional shrinkage of seed coat, clarifies the origin of the coat cracking after disconnection of the heating set in dryers, observed in practice. The highest shrinkage for all cultivar at most severe drying condition was observed. The value of force induced by the shrinkage depended also on drying conditions. Drying of the seed coat at temperature of 62 °C gave highest shrinkage stress. Further drying with the cold air caused the increase of seed coat shrinkage for 25 % more. This phenomenon was observed for other cultivars. Increase of moisture content leads to a decrease of force at breakage. The shrinkage of the seed coat during drying was more than 26 % for all seeds.

The theoretical considerations of cracking mechanisms allows to verify some models describing geometrical relations during shape deformation and drying. The modulus of elasticity of the seed coat was determined at compression and shape deformation. However, this method enabled the interpretation of the force-deformation curve for seed covering moisture over 12%. On the other hand, the seed strain (under 30%) leads shape deformation without stress in the coat. For large deformation of wet seeds mechanical strength is mainly connected with the resistance of seed coat to tension. Sliding of the cotyledons is one of most important reason of seed coat cracking for seed over 20% of moisture content.

Shrinkage of the seed coat at drying leads to its stress and splitting. The results obtained in all experiments concerning with strength of seed coat proved that frequently the stress σs was higher than strength $R m$ of seed coat. Probably, it is one of the most important reason of the seed coat cracking during drying. The additional shrinkage of the seed coat clarifies the origin of cracking, after disconnection of the heating set, frequently observed in practice.

10. POSTHARVEST RAPESEED QUALITY ACCORDING TO SEEDS PHYSICAL PROPERTIES

Physics of plant material concentrates on estimation of physical features important for material quality during vegetation, harvest, transport, storage and processing. It also studies micro-structure of plant materials and it influence on differentiation of properties. Physics of plant materials develops methods of estimation of damage in agricultural products and the mechanisms of its appearance taking into account external factors. The variations between the varieties are also under investigation. All the data collected from the above mentioned measurements allows elaborating technological improvements for preventing the unfavorable effects of damage. This in consequence increases the quality of raw materials for food production.

Agrophysical metrology allows widen and deeper analysis of processes undergo during each stage of agricultural production with a special emphasis on harvesting and postharvest operations, deals with adaptation, improvement and designing measuring methods, which can be used to explain quantitative and qualitative changes, which take place in agriculture. It develops methods as well as equipment for measuring, monitoring and analyzing the time and space variability of physical parameters of agricultural materials subjected to agrotechnical and technological processing.

The investigations were performed in field conditions and some simulations were also a subject of investigations in laboratory. Samples were collected at regular storage rooms and driers. Physical parameters of rapeseed were obtained in compression tests of single seeds. Compression curves were registered and analyzed with numeric methods to calculate apparent modulus of elasticity and strength of seeds.

Monitoring of quality of storied seeds was made in specially constructed storage silos which allowed measurement of pressure and temperature. The influence of seed moisture content on changes of quality within the storage time was described.

The investigations showed considerably differentiated quality features (percentage of broken seeds) of rapeseed within the whole typical process of their postharvest processing. The amount of broken seeds increased even two times comparing samples taken from harvester (threshed seeds) and samples taken from storage room of dry seeds.

TABLE 1. The changes of the amount of broken seeds during rapeseed postharvest processing

OPERATIONS	BROKEN SEEDS Min [%]	BROKEN SEEDS Max [%]
Unloading of threshed seeds	1,6	7,5
Storage room of wet seeds	1,6	8,0
Before drying	1,5	8,9
After drying	1,7	9,1
Storage room of dry seed	3,2	8,2

Investigations showed considerable decrease of quality features of rapeseed during harvesting and postharvest operations. The amount of broken seeds considerably depended on their moisture content. The dependence between moisture content of seeds and their resistance to external loads resulted in quality losses was proofed. The most resistant seeds were observed at 6 % of moisture content and decreased both with increased and decreased of seed moisture.

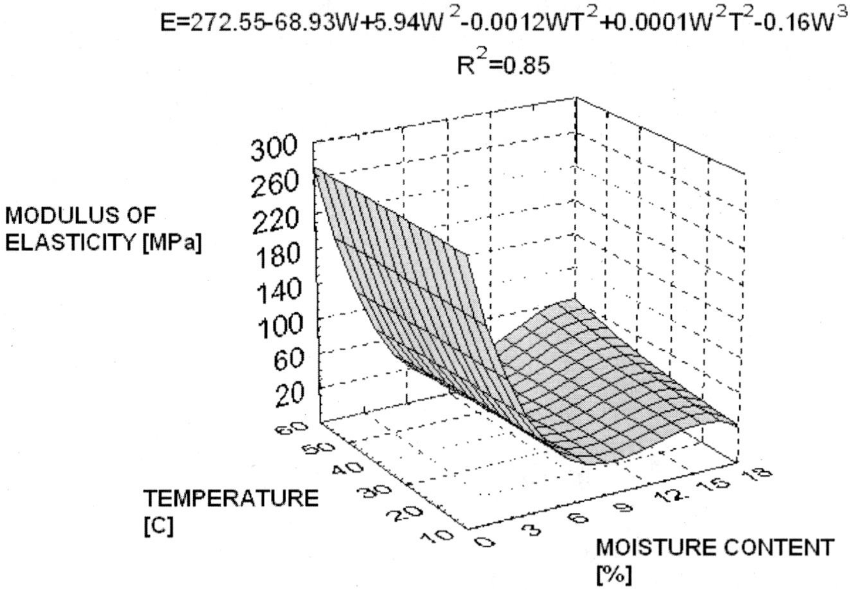

FIGURE 15. Influence of moisture content and temperature on rapeseed apparent modulus of elasticity

Quality features of stored seeds depended on their moisture content, physical state of seed cover (broken or untouched), pressure and temperature inside silo. Safe time of storage was longer for dryer seeds. In that case of unbroken seeds biological activity was lower as well as biochemical processes underwent slower. The methods could be used to evaluate physical condition of rapeseed. All the methods allowed to monitor changes of quality features of rapeseed as raw material for oil production.

According to the results obtained it can be stated that the most negative influence on quality features of rapeseed (amount of broken seeds) had harvest operation. The amount of broken seeds increase during successive postharvest operations i.e.: transport, cleaning, drying, however the initial quality parameters had considerable significance for final quality of raw material for oil production. The methods used allowed a precise monitoring of changes of rapeseed quality parameters.

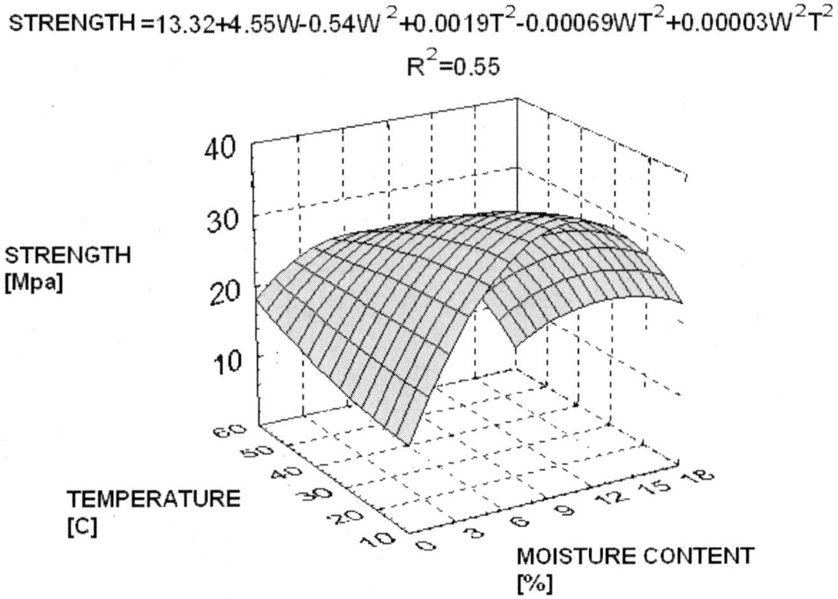

FIGURE 16. Influence of moisture content and temperature on rapeseed strength

Laboratory studies of mechanical strength of rapeseed allowed indicating range of "safety" moisture content for handling. Seed of 6 % of moisture content had the highest strength. Both the increase and the decrease of moisture content negatively influence strength features of seed. The influence of temperature was much lower than the influence of moisture content in the case of modulus of elasticity as well as in the case of strength. Some methods as well as results could be utilized also for other oilseeds i.e. soybean.

11. TECHNOLOGICAL PARAMETERS OF LENTIL SEEDS

Lentil is not a significant plant from the point of view of the economy and agronomy (crop rotation), however have unique nutritional properties and most of all is regarded as so called healthy food. Therefore lentil sees become more and more popular and are sought after and desired by consumers as an additive or raw material for the food, pharmaceutical and chemical industry. Physical parameters of such "exotic" plant are under investigation as they are important from the technological point of view.

Air-dry lentils has a mass of 1000 seeds 49.36 g for the variety Tina and 48.05 g for the variety Anita. This parameter increases with increase of seeds moisture content and at 24% seeds achieve mass of 57 g for Tina and 58 g for Anita respectively. The high correlation coefficient was noticed in case of both varieties i.e. 0.987 for Anita and 0,975 for Tina.

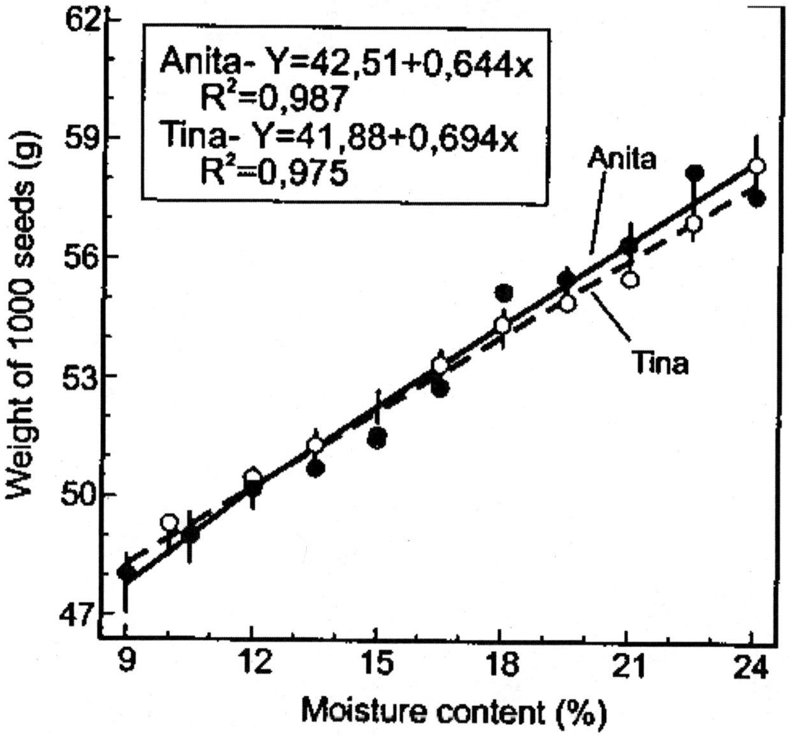

FIGURE 17. The influence of moisture content of seeds on their weight

Similarly, with an increase of moisture content of the lentils seed its porosity increase. This characteristic was observed for both tested varieties and porosity parameter ranged from 44 to 47% within moisture content range of 8 – 24%. The high correlation coefficient was noticed in case of both varieties i.e. 0.931 for Anita and 0,927 for Tina. The differences between varieties were statistically significant and higher values of this parameter was observed for Tina variety.

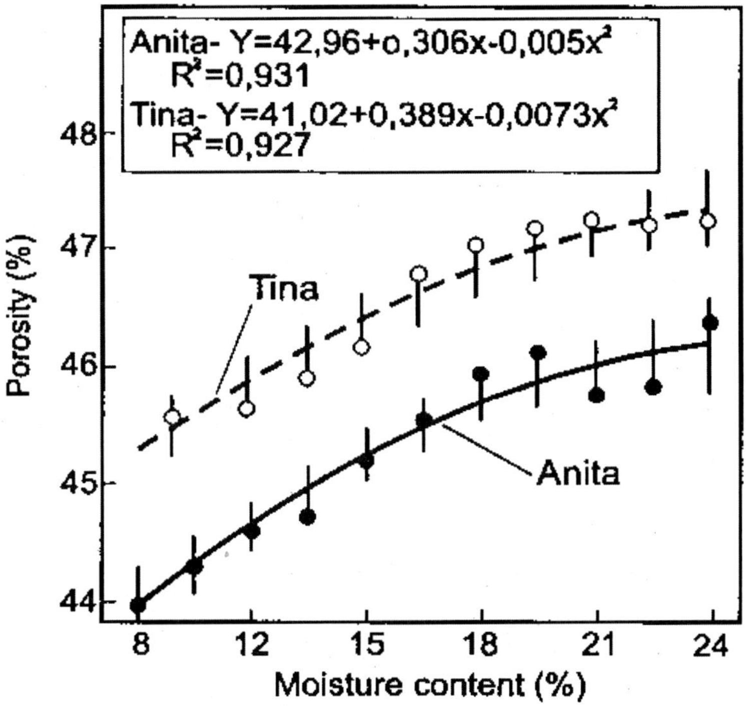

FIGURE 18. The influence of moisture content of seeds on their porosity

Density of lentil seed decreased with an increase in their moisture content. The maximum value of this parameter was observed at 8% moisture content for the variety Tina was 806 kg/m^3. The lowest recorded value is 747 kg/m^3 and this value was similar for both varieties at the humidity of 24%. Also in this case the high correlation coefficient was noticed for both varieties i.e. 0.959 for Anita and 0,913 for Tina.

With the increase of moisture content both angle of repose, and slippery increase. These values of the angle of repose was measured in the range of 27.5 ° for air-dry seeds of Tina variety to 34 º for the same variety at moisture content of 24%.

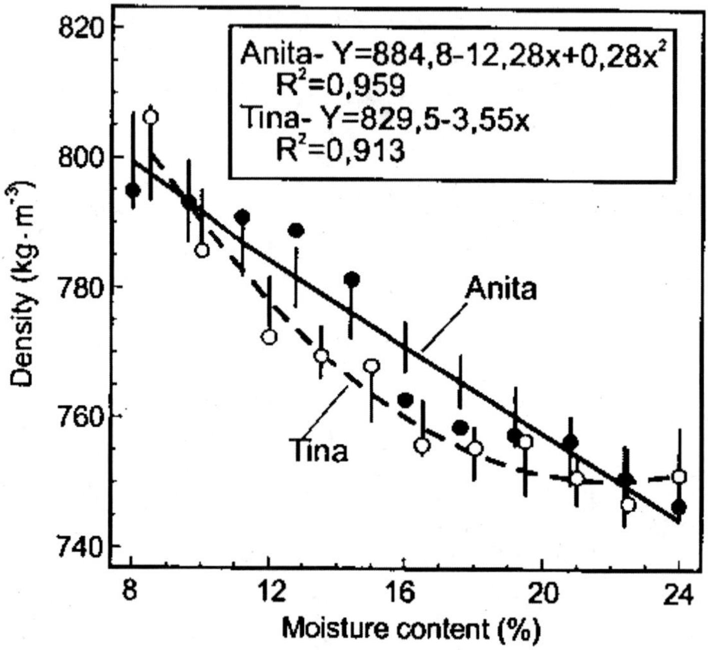

FIGURE 19. The influence of moisture content of seeds on their density

12. SWEET CORN KERNELS MECHANICAL PROPERTIES AND THE EFFECT OF HARVEST TIME, STORAGE AND BLANCHING

Sweet corn attains its consumption and processing ripeness within 24 to 28 days from blooming that is from the appearance of browning that manifests ripeness. In the phase of late milk ripeness, kernels assume yellow-gold colouring. The time of harvest not only determines the physical properties of sweet corn cobs and kernels, but also generally affects the mechanical strength of the kernels. Although the time of harvest mainly affects the mechanical properties of kernels, one should also expect chemical changes related to the ripeness stage. The differences between the mean values of the cutting force were statistically significant and varied within the range from 34.2 for harvest term I to 10.3 N for term III (Figure 20).

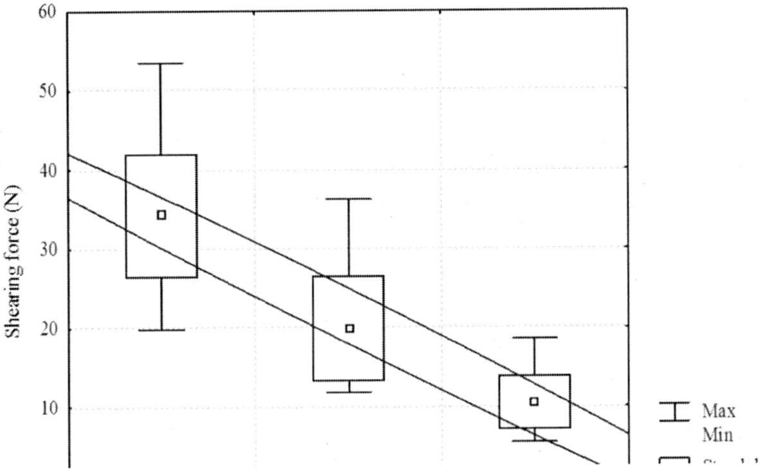

FIGURE 20. Cutting energy for different harvest terms: 1 – term I, 2 – term II, 3 – term III

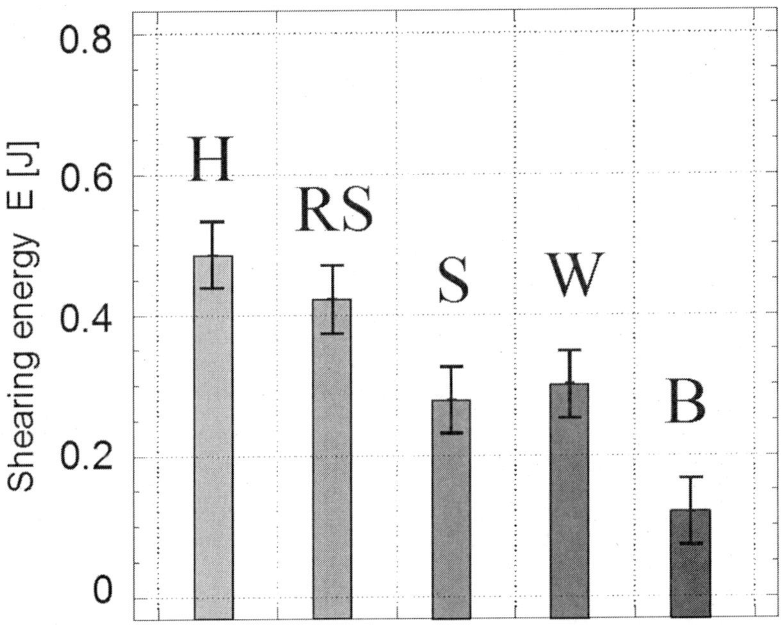

FIGURE 21. Energy of kernel cutting off from cob core. H– after harvest, W – wetted cobs, S – cobs from regular storage, RS – cobs from refrigerated storage and B – after blanching

The kernels are yellow-gold in the late milk ripeness. The period of corncobs suitability for consumption is relatively short – cobs get over-ripe rapidly, especially at high temperature. The process of over-ripening connected with transformation of sugars into starch, is delayed in with leaves and in cool conditions (Szymanek et al. 2005).

In order to estimate the effect of storage conditions on the possibility of extending the period of corn suitability for processing, tests were performed, consisting in storing corncobs for 7 days under the following conditions:

- conventional storage - temperature approx. 20⁰C (barn, umbrella roof),
- refrigerated storage - cold storage room, temperature approx. 2⁰C,
- soaking in water - temperature approx. 20⁰C.

Immediately after harvest the mechanical properties of cobs at late milk ripeness phase were compared, as well as those of cobs after blanching. Corncob blanching was made in a 4-minute water bath with a temperature of 80-90ºC. After the blanching, the cobs were cooled to a temperature of about 20ºC and dried. As follows from Figure 21, kernels from cobs subjected to the process of blanching need lower energy expenditure for the kernel cutting process (0.27J), which provided an inducement for a broader study of the process of cutting of blanched kernels.

REFERSNCES

1. J. A Abbott, and R Harker, 2003Texture, 1 19 http://usna.usda.gov /hb66 /021texture.pdf)
2. G. V Barbosa-Cánovas, P Jullano, and M Peleg, 2004Rev.2006), Engineering Properties of Foods, in Food Engineering. In: Barbosa-Cánovas, G.V., (ed.) Encyclopedia of Life Support Systems (EOLSS), Eolss Publishers, Oxford,UK, (http://www.eolss.net)
3. W Bilanski, B Szot, R. L Kushwaha, A Stepniewski, Comparition of strength features of rape siliques and seeds for varieties cultivated in various countries. International Agrophysics, 1994 1994 8 177 184
4. P Chen, and Z Sun, 1991A review of non-destructive methods for quality evaluation and sorting of agricultural products. Journal of Agricultural Engineering Research, 49: 85--98.
5. P Chen, 1996Quality evaluation technology of agricultural products. Proc. of ICAME'96, Seul, Korea, Vol. I, 171 190
6. E Davison, A. G Meiering, F. J Middendorf, 1979A theoretical stress model of rapeseed. Canadian Agricultural Engineering, 21 1
7. Dobrzański, jr. B., 1993Geometrical dependences of seed shell of compressed pea. Int. Agrophysics, 7 4 259 264

8. Dobrzański, jr. B., 1996The influence of moisture content on the shape deformation of soybean. AgEng'96, Madrid, Paper 96F-082 1 7
9. Dobrzański, jr. B., 2000The analysis of seed cracking at grain legume processing. CD-ROM of AgEng'2000 papers, Paper: 00-PH-050 1 10
10. Dobrzański, jr. B., 2002The cracking mechanisms of grain legume. rozdział- str.: 167 194in "Physical Methods in Agriculture", J. Blahovec, M. Kutilek (eds): Kluwer Academic Publishers.
11. B Dobrzański, jr., 2011Shrinkage and swelling phenomena in agricultural products. 730 733in Encyclopedia of Agrophysics. J. Gliński, J. Horabik, J. Lipiec (eds.) Springer
12. Dobrzański, jr. B., Szot B., 1997Strength of pea seed cover. Int. Agrophysics, 11 4 301 306
13. Dobrzański, jr. B., 1998Mechanizmy powstawania uszkodzeń nasion roślin strączkowych. Acta Agrophysicca, 13 1 96in polish)
14. Dobrzański, jr. B., Banak E., Grundas S., Sosnowski S., Pecen J., 2003Metoda rentgenograficzna w identyfikacji uszkodzeń wewnętrznych nasion fasoli szparagowej. Acta Agrophysica, 95, 2in polish)
15. Dobrzański, jr. B., Mazurek W., Rybczyński R., Geodecki M., Baranowski P., Walczak R., 2003A new method of the seed viability estimation. New Methods, Means and Technologies for Application of Agricultural Products. Agricultural Engineering LUA, 1(06), 105-110.
16. B Dobrzanski, A Stepniewski, 1991Modulus of elasticity of rapeseed shell. Proceedings of 8th. International Rapeseed Congress, Saskatoon, Canada, 1256-1260
17. B Dobrzanski, B Szot, A Stepniewski, Stress in the shell during large deformation of compressed rapeseed. Proceedings of International Conference on Agricultural Engineering, Beijing, China, 1992
18. D. G Dorell, M. W Adams, 1969Effect of some seed characteristic on mechanically included seedcoat damage in navy beans. Agron. J., 61 5 672 673
19. J Fornal, J Sadowska, A Ostaszyk, 1995Industrial drying of rapeseeds vs. oil extractivity. Proceedings of 9th. International Rapeseed Congress, Cambridge, England IV, 1406 1408
20. J Fornal, J Sadowska, R Jaroch, B Kaczynska, T Winnicki, 1994Effect of drying of rapeseeds on their mechanical properties and technological usability. International Agrophysics, 8
21. J Haman, B Dobrzanski, B Szot, A Stepniewski, 1994Strength of shell in compression test of rapeseed. International Agrophysics, 8 245 250
22. Ling ChY., Whitney L., 1975Analysis of spherical shell acted on by two concentric forces in opposite directions. Theory, determination and control of physical properties of food materials., D. Reidel Publ. Co., 189 196
23. M Liu, K Haghighi, R. L Stroshine, 1989Viscoelastic characterization of the soybean seedcoat. Trans. ASAE 32 3 946 952
24. Mac Donald, jr. M.B., 1985Physical seed quality of soybean. Seed Sci. Technol. 13 601 628
25. J. K Mensah, G. L Nelson, F. L Herum, T. G Richard, 1984Mechanical properties related to soybean seedcoat cracking during drying. Trans. ASAE, 27 2 550 560
26. R Misra, J. H Young, D. D Haman, 1981Finite element procedures for estimating shrinkage stresses during soybean drying. Trans. ASAE 24 3 751 753
27. N. N Mohsenin, Physical properties of plant and animal materials. Gordon and Breach Science Publishers, New York, London, Paris 1978
28. L Otten, R Brown, W. S Reid, 1984Drying of white beans-effect of temperature and relative humidity on seed coat damage. Canad. Agric. Eng., 26 101 104
29. D. G Overhults, H. E White, H. E Hamilton, I. J Ross, 1973Drying soybeans with heated air. Trans. ASAE 16 1 112 113
30. M. R Palusen, W. R Nave, L. E Gray, 1981Soybean seed quality as affected by impact damage. Trans. ASAE 24 6 1577 1582

31. A Stepniewski, B Dobrzanski, Higher temperature of rapeseeds as a factor affecting their mechanical resistance. Proceedings of 8-th. International Rapeseed Congress, Saskatoon, Kanada, 1991 1991 1246 1250
32. A Stepniewski, B Szot, Postharvest operations and quality of rapeseed. Proceedings of 9th. International Rapeseed Congress, Cambridge, England 1995I, 232 234
33. A Stepniewski, B Szot, Quality of rapeseed in postharvest handling. Preprints of workshop on Control Applications in Post-Harvest and Processing Technology, CAPPT'95, Ostend, Belgium 1995
34. A Stepniewski, B Szot, J Fornal, J Sadowska, Drying conditions and mechanical properties of rapeseed. J. Food Physics, 1994 1994 86 89
35. B Szot, A Stepniewski, Studies on mechanical properties of rape in the aspect of its quantity and quality losses. Zemedelska Technika, 1995 1995 41 133 136
36. B Szot, A Stepniewski, B Dobrzanski, R Rybczynski, The influence of higher temperature of rapeseeds on their mechanical resistance. Proceedings of 4th. International Conference Physical Properties of Agricultural Materials and Their Influence on Technological Processes, Rostock, Germany, 1989s. 815- 818
37. G. M White, T. C Bridges, O. J Loewer, I. J Ross, 1980Seed coat damage in thin-layer drying of soybeans. Trans. ASAE 23 1 224 227

CHAPTER 2

Seed Priming: New Comprehensive Approaches for an Old Empirical Technique

Stanley Lutts[1*], Paolo Benincasa[2], Lukasz Wojtyla[3], Szymon Kubala S[3], Roberta Pace[2], Katzarina Lechowska[3], Muriel Quinet[1] and Malgorzata Garnczarska[3]

[1] *Groupe de Recherche en Physiologie végétale (GRPV), Earth and Life Institute-Agronomy, Université catholique de Louvain, Louvain-la-Neuve, Belgium*

[2] *Department of Agricultural and Environmental Sciences, University of Perugia, Perugia, Italy*

[3] *Faculty of Biology, Department of Plant Physiology, Adam Mickiewicz University, Poznan, Poland*

ABSTRACT

Seed priming is a pre-sowing treatment which leads to a physiological state that enables seed to germinate more efficiently. The majority of seed treatments are based on seed imbibition allowing the seeds to go through the first reversible stage of germination but do not allow radical protrusion through the seed coat. Seeds keeping their desiccation tolerance are then dehydrated and can be stored until final sowing. During subsequent germination, primed seeds exhibit a faster and more synchronized germination and young seedlings are often more vigorous and resistant to abiotic stresses than seedlings obtained from unprimed seeds. Priming often involves soaking seed in predetermined amounts of water or limitation of the imbibition time. The imbibition rate could be somehow controlled by osmotic agents such as PEG and referred as osmopriming. Halopriming implies the use of specific salts while "hormopriming" relies on the use of plant growth regulators. Some physical treatments (UV, cold or heat,..) also provide germination improvement thus suggesting that priming effects are not necessarily related to seed imbibition. A better understanding of the metabolic events taking place during the priming treatment and the subsequent germination should help to use this simple and cheap technology in a more efficient way.

Keywords: germination, omics approaches, priming, seedling growth, stress resistance

1. INTRODUCTION

Efficient seed germination is important for agriculture. Successful establishment of early seedling indeed requires a rapid and uniform emergence and root growth. Germination of orthodox seeds commonly implies three distinct phases (**Figure 1**) consisting in (1) Phase I: seed hydration process related to passive imbibition of dry tissues associated with water movement first occurring in the apoplastic spaces; (2) Phase II: activation phase associated with the re-establishment of metabolic activities and repairing processes at the cell level; and (3) Phase III: initiation of growing processes associated to cell elongation and leading to radicle protrusion. Phases I and III both involve an increase in the water content while hydration remains stable during Phase II. It is commonly considered that before the end of Phase II, germination remains a reversible process: the seeds may be dried again and remain alive during storage and able to subsequently re-initiate germination under favorable conditions.

Water-based seed priming is defined as a pre-sowing treatment that partially hydrates seeds without allowing emergence [1]. Various treatments may indeed be applied during the reversible phase of germination (point 3). They widely differ according to the osmotic potential of the priming solution, the duration, the external temperature, and the presence of specific chemical compounds. The efficient treatments trigger metabolic processes activated during the phase II of germination, which are then temporally stopped before a loss of desiccation occurs (**Figure 1**) [2].

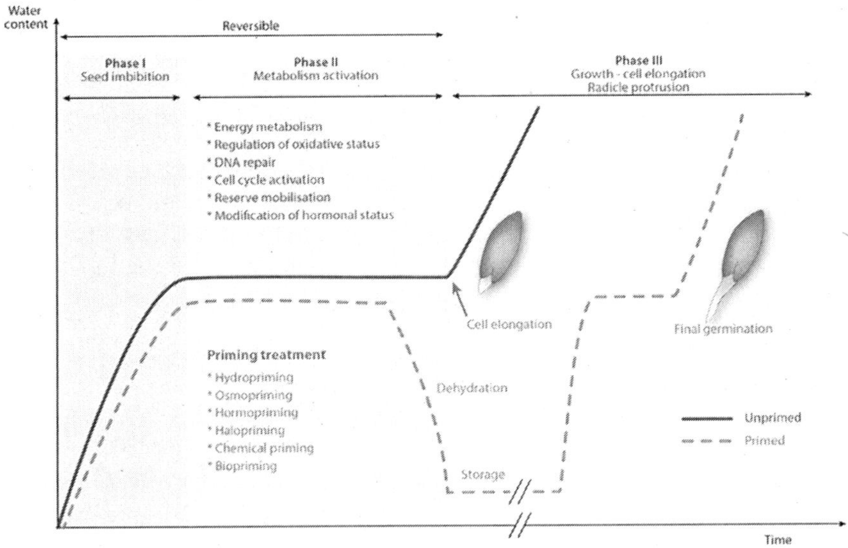

FIGURE 1. Seed hydration curves and germinating phases in unprimed and primed seeds.

The overall consequence of seed priming consists in an increased seed vigor defined as the whole set of properties conditioning seed lots performance in a wide range of environment [3]. Priming strategies may afford several economic and agronomic advantages to cultivated plants (point 4). Numerous data published in the literature indeed reported an improvement in the rate and uniformity of germination but also an obvious improvement in the behavior of the obtained seedlings in terms of plant growth and stress resistance.

Although priming is used since decades by farmers and seed companies to improve germination, it can also occur under natural plant conditions. This is more especially the case in serotinous plants growing in deserts and able to retain their seeds for a long time. These seeds indeed undergo several hydration-dehydration cycles enhancing subsequent germination after final seed dispersion caused by heavy rain [4]. From a general point of view, the priming process not only concerns seeds but also the whole plant system itself and may be defined as an induced state whereby a plant reacts more rapidly and more efficiently to a stress [5]. In this acception, plants exposed to a primary constraint are triggering a set of temporary metabolic adaptation leading to a stress memory and allowing them to adapt more efficiently to subsequent episodes of stress [6, 7].

Although the interest of seed priming has been demonstrated since a long time, the underlying physiological and biochemical basis of this fascinating process remain poorly understood. Holistic approaches related to omics tools now provide new opportunities to elucidate the molecular components of the priming phenomena. Similarly, nondestructive and noninvasive methods such as digital image technology may be used in a more precise way to study the kinetics of imbibition in relation to the modification of the seed ultrastructure. This chapter reviews the most recent progresses accomplished in the understanding of the seed priming-induced modifications.

2. A BRIEF HISTORY OF SEED PRIMING

Man established contact with seed physiology since the beginning of agriculture and quickly realized that many seeds do not germinate easily and uniformly. Ancient civilization was fascinated by the capacity of an apparently « dead seed » to resurrect and to produce a viable young and healthy seedling after germination. The Greek Theophrastus (ca. 372–287 BC) already focused on seed physiology and suggested that germination process may be temporarily interrupted [8]. Pre-hydration of legume seeds before sowing was performed by Roman farmers in order to increase the germination rate and synchronize germination as reported by the Roman naturalist Gaius Plinius Secundus. Several centuries later, these techniques were still used for a wide range of species according to the French agronomist Olivier de Serres (1539–1619) [8]. In 1664, Evelyn [9] mentioned that temperature prior sowing may have an impact on further germination while one century later, Ingenhousz [10] analyzed the impact of light on seedling emergence.

During the nineteenth century, numerous botanists started to describe morphological processes associated with seed germination [11, 12]. Sachs [13] experimented the impact of various compounds (including tyrosine and asparagine) before and during germination. The discovery of plant hormones in the 1920s underlined the crucial role of these compounds in seed desiccation tolerance, reserve mobilization, as well as cell division and cell elongation occurring during germination. The possibility to influence final germination as a consequence of pre-sowing treatment has led to a wide range of empirical methods for numerous cultivated plant species during the year 1970s [14].

3. PRIMING METHODS AND PRIMING AGENTS

Several methods of seed priming have been developed in order to invigorate seeds and alleviate the environmental stresses. A common feature of water-based priming techniques, which distinguishes them from other pre-sowing treatments, is partial seed pre-hydration and the activation of early germination events in seed. Priming efficiency is affected by many factors and strongly depends on treated plant species and chosen priming technique. Physical and chemical factors such as osmotica and water potential, priming agent, duration, temperature, presence or absence of light, aeration, and seed condition also influence priming success and determine germination rate and time, seedling vigor, and further plant development [15, 16].

3.1. Hydropriming

Hydropriming is the simplest method of seed priming, which relies on seed soaking in pure water and re-drying to original moisture content prior to sowing. No use of additional chemical substances as a priming agent makes this method a low-cost and environmentally friendly. The main disadvantage of hydropriming is uncontrolled water uptake by seeds. This is a consequence of free water availability to seeds during hydropriming, so that the rate of water uptake depends only on seed tissue affinity to water [17]. Moreover, this technique may result in unequal degree of seeds hydration thus leading to lack of simultaneous metabolic activation within seeds followed by unsynchronized emergence [18]. Considering these limiting factors, it is highly important to define accurate treatment duration, temperature, and water volume used in hydropriming to ensure desired level of seed hydration and to prevent radicle protrusion. Despite the aforementioned limitations, many reports indicated beneficial effect of hydropriming on seed germination and seedling growth under both optimal and stress conditions, in various crop plants such as chickpea, maize [19], wheat [20], Indian mustard [21], canola [22], sunflower [23], rice [24], mung bean [25], capsicum [26], and durum wheat [27].

One of the commercially used types of hydropriming is the system named "drum priming", patented in the early 1990s [28, 29]. In this technique, seeds are gently rotating in drum and gradually hydrated by addition of water in vapor form. Drum priming allows seed imbibition in a controlled manner and could be an attractive alternative to conventional hydropriming. Specially designed apparatus enables monitoring of the seed weight, precise regulation of time, and water amount during hydration process, what ultimately results in an appropriate and uniform moisture level of the seeds [30]. Drum priming with 24-epibrassinolide shows positive effect on germination time and seedling growth of bell pepper concomitant with improved superoxide dismutase (SOD), catalase (CAT), and peroxidase (POX) activities [31]. Another variant of hydropriming, so-called "on-farm priming", consist of seed soaking in water followed by surface drying and subsequent sowing. The duration of treatment obligatorily cannot be longer than "safe limit" (maximum time of priming without risk of seed or seedling damage by premature germination) [32]. The positive impact of this method on crop emergence and yield was confirmed by Harris et al. [33]. On-farm priming is especially useful for resource-poor farmers in marginal tropical environment [34].

3.2. Osmopriming

Osmopriming involves soaking seeds in osmotic solution with low water potential instead of pure water. Due to low water potential of osmotic solutions, water enters seed slowly which allows gradual seed imbibition and activation of early phases of germination but prevents radicle protrusion [35]. Usually water potential of priming agent varies from −1.0 down to −2.0 MPa [36]. However, values of water potential together with duration of the priming treatment should be always adjusted to species, cultivar, and sometimes seed lot. Different compounds are used in osmopriming procedure including polyethylene glycol (PEG), mannitol, sorbitol, glycerol, and inorganic salts such as NaCl, KCl, KNO_3, K_3PO_4, KH_2PO_4, $MgSO_4$, and $CaCl_2$ [37]. Priming with salt solutions is often referred as "halopriming". Most common chemical employed in osmopriming treatment is PEG, mainly owing to its specific characteristic. Large molecular size of PEG prevents its penetration into the seed thus avoiding induction of potential cytotoxic effect and reduction of osmotic potential within seed [35]. Nevertheless, PEG exhibits some undesirable features including high viscosity, which restrict diffusion of oxygen in the solution so in PEG priming aeration system is preferred [2]. Seed priming with PEG has been shown as an effective method to improve seed germination, seedling emergence, and stress tolerance of several crop plants under unfavorable conditions such as salt, water, chilling, and nano-ZnO stresses [1, 3, 36–38].

3.3. Solid Matrix Priming

Solid matrix priming (SMP, matriconditioning), in which water uptake by seeds is controlled, has been developed as an alternative method to osmopriming because of high cost of osmotic agents and technical problems with aeration [2]. During solid matrix priming, seeds are mixed and incubated with wet solid water carrier for a certain period. Afterward, seeds are separated from matrix, rinsed, and back-dried. The use of solid medium allows seeds to hydrate slowly and simulates natural imbibition process occurring in the soil [18]. To successfully accomplish SMP, materials utilized as matrices should possess specific physical and chemical features such as low matrix potential, minimal water solubility, high water holding capacity and surface area, no toxicity to seeds, and ability to adhere to seed surface. In fact, vermiculite, peat moss, charcoal, sand, clay, and some commercially offered substrate such as Celie or Micro Cell are exemplary solid carries applied in solid matrix priming [2, 35]. In order to obtain the best priming performance, time of treatment and optimal water content must be determined separately for each matrix [39].

Positive effects of SMP on crop seeds have been noted in many reports. Solid matrix priming enhanced field performance of carrot [40] as well as improved germination and seed vigor of soybean [41]. Study on onion showed that matriconditioning improved seed germination rate, seedling emergence, and growth under optimal and low temperature conditions [42]. Sand priming increased the activities of antioxidant enzymes such as catalase (CAT), peroxidase (POX), and soluble sugar content in waxy maize concomitant with improved rate of germination and seedling growth under high-salt stress conditions [43].

It is well established that integration of SMP with biological and chemical factors may greatly enhance seed performance [18]. Adoreoli and de Adnrade [44] indicated that inclusion of gibberellins/fungicide/*Bacillus subtilis* to matriconditioning leads to improved stand establishment and productivity of some vegetable crops under tropical conditions. Similarly, matriconditioning with GA_3enhanced the quality of hot pepper seeds [45]. More recently published data demonstrated that solid matrix priming with *Trichoderma viride* improved seedling emergence and yield of okra under low temperatures [46].

3.4. Hormopriming

During hormopriming, seeds imbibition occurs in the presence of plant growth regulators, which can have direct impact on seed metabolism. The following regulators are commonly used for hormopriming: abscisic acid, auxins, gibberellins, kinetin, ethylene, polyamines, and salicylic acid (SA). Gibberellic acid (GA_3) and PEG priming improved photosynthetic properties, antioxidant system, seedling emergence, and growth of white clover on heavy metal polluted soil [47]. Priming spring wheat seeds with GA_3 increased grain yield and salt tolerance by modulating hormone homeostasis together with alterations of ion uptake and accumulation between shoots and roots [48]. Enhanced salt tolerance, growth, and grain yield of wheat were also observed after kinetin-priming [49]. Among the different techniques of seed priming (hydro-, osmo-, and halopriming), spermidine pretreatment appeared to be the most effective method for induction drought tolerance in rice [50]. High efficiency of polyamines-priming on the improvement of rice tolerance to drought has been demonstrated also by Farooq et al. [51]. Critical role of phytohormones exogenously supplied into seeds for plant response to salinity stress was stated in wheat seeds primed with ascorbic acid and salicylic acid, as this pretreatment method increases the ability of wheat to grow successfully under salt stress, whereas hormonal priming with ABA was not effective in this case [52].

3.5. Biopriming

Biopriming involves seed imbibition together with bacterial inoculation of seed [53]. As other priming method, this treatment increases rate and uniformity of germination, but additionally protects seeds against the soil and seed-borne pathogens. Hydration of seeds infected with pathogens during priming can result in a stronger microbial growth and consequently impairment of plant health. However, applying antagonistic microorganisms during priming is an ecological approach to overcome this problem [54]. Moreover, some bacteria used as biocontrol agents are able to colonize rhizosphere and support plant in both direct and indirect way after germination stage [55]. It was found that biopriming is a much more effective approach to disease management than other techniques such as pelleting and film coating [56]. Nowadays, the use of biopriming with plant growth-promoting bacteria (PGPB) as an integral component of agricultural practice shows great promise [57, 58]. In pearl millet, biopriming with *Pseudomonas fluorescens* isolates enhanced plant growth and resistance against downy mildew disease [59]. Biopriming with rhizobacteria improved germination parameters of radish seeds under saline conditions [60].

3.6. Others

Chemical priming refers to seed treatment with different chemical solutions used as priming agents. This approach includes priming with wide range of both natural and synthetic compounds such as antioxidants (ascorbic acid, glutathione, tocopherol, melatonin, and proline), hydrogen peroxide, sodium nitroprusside, urea, thiourea, mannose, selenium, chitosan, fungicide etc. Positive impact of chemical priming with various priming agents in a wide range of environmental conditions was indicated by numerous studies [26, 61–64]. Seed priming with β-amino butyric acid increased drought and salt tolerance of green gram [65]. Application of ascorbic acid as a seed priming agent induced drought and salt resistance of wheat [66, 67]. Analysis conducted by Fercha et al. [67] revealed that priming with ascorbate counteracts the negative effects of salinity stress by changes in abundance of proteins involved in metabolism, protein destination, and storage.

Nutripriming is a technique in which seeds are soaked with solutions containing the limiting nutrient instead of pure water. The idea of this method is to obtain nutritional effect together with biochemical advantages of priming in order to improve seed quality, germination parameters, and seedling establishment [68]. Seed priming with Zn improved productivity of chickpea and wheat [69], germination and early seedling growth of rice [70], development and root growth of maize seedling exposed to low root zone temperatures [71], while K-priming brought favorable effect on growth and nutrient status of cotton seedling under saline conditions [72]. Some nutripriming techniques are commonly used by seed companies in the process of seed production and preparation for growers. One of this methods, broad spectrum nutrient seed priming (BSN), is based on imbibing seeds in mixture of minerals, such as zinc, copper, manganese, molybdenum, and phosphorus, which has been proved to fertilize the seed and provides the nutrients for early growth, which positively affects germination, seedling vigor, and root system development (http://seedprimer.com/).

4. SEED PRIMING AND AGRICULTURE

Pre-sowing priming induces a particular physiological status in seeds and has emerged as a promising strategy to improve plant behavior in the field. There is a strong interest for farmers and seed companies to find suitable cheap priming treatments but also to precisely identify the agronomical properties improved as a result of priming in cultivated species.

4.1. Hastening and Synchronization of Germination

Primed seeds often exhibit an increased germination rate and greater germination uniformity. An enhanced and uniform seedling emergence may contribute to regular crop establishment. Priming may enhance events taking place at the beginning of the germination, but the whole process is interrupted at a given state, which is the same for all concerned seeds. Priming may also induce structural and ultrastructural modifications that could facilitate subsequent water uptake and attenuate initial differences between the seeds in terms of imbibition, thus resulting in a more uniform germination [47].

A faster emergence may help to improve competitivity of cultivated plants against weed species as recently demonstrated by Jalali and Salehi [73] for sugar beets. In mung bean plants, a faster seedling establishment resulting from priming may contribute to a total increase in yield up to 45% [74].

Priming-induced increase in germination may be associated to a change in plant hormone biosynthesis and signaling. Priming has been reported to increase gibberellins (GA)/abscisic acid (ABA) ratio [75], and this may be a direct consequence of a priming impact in gene expression pattern [76]. A more uniform GA endogenous concentration in primed seeds may help to synchronize endosperm weakening, embryo cell elongation, and reserve mobilization [77]. Ethylene also directly influences germination speed and percentage. Increase in ethylene production during priming may promote endo-β-mannase activity facilitating endosperm weakening and post-priming germination [78]. Priming has been reported to initiate repair and reactivation of pre-existing mitochondria and to initiate the biogenesis of new ones [79]. It may thus afford a higher level of energy over a short time to sustain final germination [80].

4.2. Plant Growth

Plants issued from primed seeds often exhibit a faster growth than those issued from unprimed ones. Determine whether such growth stimulation is the consequence of a more rapid seedling establishment or result from a long-term specific physiological status induced by priming still remains an unresolved question. In numerous cases, the beneficial impact of priming on plant growth is more obvious under nonoptimal than under optimal conditions, leading to the global concept that a major advantage of priming consists in an increase in stress resistance (point 4.10). Thus, in direct relation to memory events, the main question is related to the remanence of priming-induced modifications. Imram et al. [71] showed that such modifications remain intact several weeks after germination in maize.

In rice, priming with 5-aminolevulinic acid improved shoot elongation [81] while priming with picomolar rutin augmented both root and shoot length in relation to an increase in photosynthetic pigments, phenolic and flavonoid contents [82]. In wheat, priming with sodium prusside stimulated plant growth as a consequence of improved capacity to scavenge free radicals by antioxidants [83], and a similar observation was reported for rice as a result of an increase in glutathione peroxidase (GPX) activity [24] and other antioxidant enzyme activities [84].

The beneficial impact of priming on plant growth may be due to an improved nutrient use efficiency allowing a higher relative growth rate [85] and to an improved regulation of the plant water status [86]. Jisha and

Puthur [65] confirmed that the priming effect of β-aminobutyric acid on seeds of *Vigna radiata* further get carried over the seedlings. A higher growth of seedlings issued from primed seeds may also be analyzed in relation to a direct impact of pretreatment on cell cycle regulation and cell elongation processes (point 7) [77, 78].

4.3. Mineral Nutrition

Modification of nutrient uses efficiency by young seedlings may be a consequence of priming-induced overexpression of genes encoding for specific transporters, although only few transporters appear specifically induced by priming itself [36]. An efficient strategy to improve mineral nutrition of young seedling is to use nutrient-based seed priming strategy. Phosphorous seed priming supported crop development at early stages and may compensate for P deficiency in the soil [87, 88]. Jamil et al. [89] demonstrated that improvement of mineral status of P-primed cereals reduced strigolactone exudation and thus sensitivity to the parasite weed *Striga hermonthica*. Muhammad et al. [85] recently performed experiments using Zn, Mn, B, and P priming. These authors demonstrated that nutrient seed priming allowed maize plants to maintain Zn and Mn supply for at least 3 weeks in highly calcareous soils characterized by a low nutrient availability. Similarly, Pame et al. [90] showed that P accumulation in rice may be increased by using P-primed seeds, which is of special interest in Asia where about one-third of the area of rainfed rice is situated on P-deficient soils. Such a higher absorption could not be explained only by nutrient accumulation in the seeds during the primed phase since it is still observed in plants several weeks after sowing. It may therefore be hypothesized that priming interferes with regulation of acquisition mechanisms and further research is crucially needed to identify the molecular mechanisms involved in these processes. Priming with boron improves seedling emergence in rice and, on a long-term basis, increases panicle fertility in relation to an improvement in stigma receptivity [91]. Seed priming may also contribute to improve N nutrition, mainly through an enhanced nitrate reductase activity in plants [40]. Priming with nonessential beneficial elements, such as Si, leads to an increase in Si content of cultivated plants and has a protective impact on plant development [86].

Beside the improvement of essential elements uptake, priming also helps to reduce accumulation of putatively toxic elements. Chromium (VI) accumulation is reduced in maize seedlings issued from salicylic acid primed seeds and cultivated in the presence of this toxic element [82]. Osmopriming with PEG and hormopriming with GA improved germination and early seedling growth of white clover maintained on a heavy metal-contaminated soil, but the impact on Cd accumulation by plants may differ according to the considered treatment since GA_3 increased Cd accumulation while PEG reduced it [47]. Liu et al. [92] demonstrated that PEG increases Ca^{2+} cytosolic concentration through hyperpolarization-activated calcium permeable channels, which could explain a lower Cd accumulation as a consequence of an improved selectivity toward calcium.

Numerous data are also available concerning the priming effect on the plant behavior exposed to salinity. It is frequently reported that priming-induced stress resistance may be a consequence of an improved discrimination for K^+ over Na^+ nutrition. Both osmo- and hydropriming were efficiently used to influence K^+ selectivity of seedlings, but the underlying molecular basis of this improvement still needs to be identified, especially in terms of regulation of monovalent cation transporters.

4.4. Yield-Related Parameters

A huge amount of studies is devoted to the impact of seed priming on the seed germination phase and early seedling growth. Most of those studies are conducted under controlled environmental conditions in plant growth chambers or greenhouses. Data reporting a real improvement under field conditions remain rare. Yield effect may be linked to a faster plant establishment allowing a longer growth period. Khan et al. [93] reported that plant issued from primed seed benefits from a longer period of assimilates accumulation in sugar beet. Conversely, in some cases, phenological evolution of cultivated plants may be modified by priming: in chickpea, plants issued from priming encountered an earlier seed maturity allowing them to escape disease or heat terminal stress in the season [94]. Yield increase may also result from a higher plant density observed as a consequence of priming-induced increase in germination percentage [95].

Since less than a decade, several data started to be available for priming-induced yield improvement in rice. Shah et al. [96] demonstrated that priming had a positive effect on the weight of 1000 grains in this species. Boron priming induced an obvious decrease in panicle sterility and consequently improved the number of grains per inflorescence [91]. Binang et al. [97] also demonstrated that priming had a significant effect on the number of tillers, number of fertile panicles, and consequently grain yield of new NERICA rice varieties. Promising yield improvement has also been reported for maize [85, 98], onion [99], okra [100], and sugar beet [73]. Beside its impact on quantitative parameters, priming may also improve the quality of harvested plants, as recently reported by Janecho et al. [101] for the vitamin content and nutritional value of legumes.

4.5. Stress Resistance

Most of the studies performed on the seedlings issued from primed seeds demonstrated a clear improvement of resistance to environmental constraints. **Table 1** is providing a nonexhaustive list of recent publications dealing with stress resistance improvement on cultivated plant species. Frequently, such improvement is obvious just after emergence at the seedling level, but progressively disappears at the adult stage. For example, some young plants issued from priming treatments displayed improvement of resistance to chilling [84], low temperature [75], salinity [43, 102], high temperature [80], drought [24, 65, 103], and UV exposure [82]. Some interesting studies also demonstrated that priming may afford resistance to biotic stresses such as *Fusarium oxysporum* in tomato [104], viral disease in *Brassica rapa* [105], and downy mildew in pearl millet [106]. Such a large set of data suggests that seed priming may elicit numerous pathways contributing to stress resistance. The molecular basis involved in this stress resistance remains intact during the dehydration phase following priming and may contribute to stress resistance during the final germination step. Moreover, some data suggest that a single priming treatment may induce resistance to various stresses.

The priming procedure itself implies frequently the use of stressing agent, as it is the case for PEG and salt. In some cases, priming may be performed at low temperature to reduce the kinetics of seed hydration. A slow hydration may be considered as a stressing process since the water content is too low to allow radicle elongation (see point 5). It may thus induce defense responses within embryos. This is especially the case for biochemical processes involved in protection against reactive oxygen species (point 8). Several components of the ROS-mediated signaling pathways are activated during the first hydration phase of the priming process. The ultimate stress resistance in the seedlings may then be linked to the persistence of the antioxidative defenses after final germination. Since management of oxidative stress is an important component of resistance to a wide range of stress, this observation may, at least partly, explain the cross-resistance phenomena.

The dehydration step that follows the partial hydration phase is also a major stressing phase. Numerous studies focus on late embryogenesis abundant proteins (LEA) normally involved in the acquisition of desiccation tolerance. Chen et al. [107] showed that the major dehydrins disappear during osmopriming while Maia et al. [108] conversely suggested that PEG may induce LEA synthesis. Water deprivation associated with the dehydration phase may also trigger accumulation of transcription factors, some of them being specifically involved in stress resistance [109, 110]. Molecular chaperones such as heat shock proteins (HSP) also explain a priming-induced improvement of stress resistance [78]. It may also be hypothesized that priming-induced modification of the seed hormonal status, mainly an increase in ABA, may somewhat influence the seed and young seedling response to environmental constraints in relation to a faster activation of ABA-responsive genes involved in stress acclimation [108, 109].

TABLE 1. Nonexhaustive list of recent studies devotes to priming-induced increase in the stress resistance of cultivated plant species.

Environmental constraint	Plant species	Priming treatment	Reference
Salinity	*Brassica juncea*	hydro/osmopriming	[21]
	Brassica napus	PEG	[120]
		Halopriming	[22]
	Helianthus annuus	KNO_3	[23]
	Triticum durum	ascorbic acid	[27]
	Medicago sativa	PEG	[37]
	Zea mays	Sand priming	[43]
	Triticum aestivum	ABA + SA	[52]
		NO (nitroprusside)	[63]
		Ascorbic acid	[67]
		Biopriming	[102]
		$KCl + CaCl_2$	[167]
	Raphanus sativus	Biopriming	[60]
	Gossypium arvense	Potassium	[72]
	Citrus sinensis	NO (nitroprusside)	[185]
Drought/water stress	*Oryza sativa*	KH_2PO_4	[24]
		Dry priming	[50]
		Polyamines	[51]
		Salicylic acid	[134]

	Zea mays	Urea; KNO3	[64]
	Vigna radiata	BABA	[65]
	Triticim aestivum	Ascorbic acid	[66]
		Silicon	[86]
	Brassica napus	PEG	[178]
		Hydropriming	[117, 118]
	Glycine max	Osmoconditioning	[187]
	Cicer arietinum	Osmo/hydropriming	[170]
Chilling and low temperatures	*Vigna radiata*	Hydropriming/proline	[25]
	Zea mays	*chitosan*	[62]
		Nutrient priming	[71]
	Glycine max	Osmopriming	[79]
	Oryza sativa	Salicylic acid	[84]
	Beta vulgaris	Osmopriming	[93]
	Spinacea oleracea	Osmopriming	[107]
	Nicotiana tabacum	Putrescine	[166]
High temperatures	*Lactuca sativa*	Hydropriming	[76]
	Daucus carota	PEG	[80]
Heavy metals	*Trifolium repens*	PEG	[47, 179]
	Poa pratensis	PEG, Gibberellins	[179]
Biotic stresses			
Phythium ultimum	*Zea mays*	Biopriming	[53]
Verticillium	*Brassica napus*	Biopriming	[56]
Downy mildew	*Pennisetim glaucum*	Biopriming	[59]
		BABA	[106]
Yellow mosaic virus	*Vigna radiata*	On farm priming	[75]
Fusarium oxysporum	*Solanum lycopersicum*	Methyl jasmonate	[104]
Double-stranded DNA virus	*Brassica rapa*	Biopriming	[105]

The beneficial impact of priming treatments relies on numerous properties as indicated in **Figure 2**.

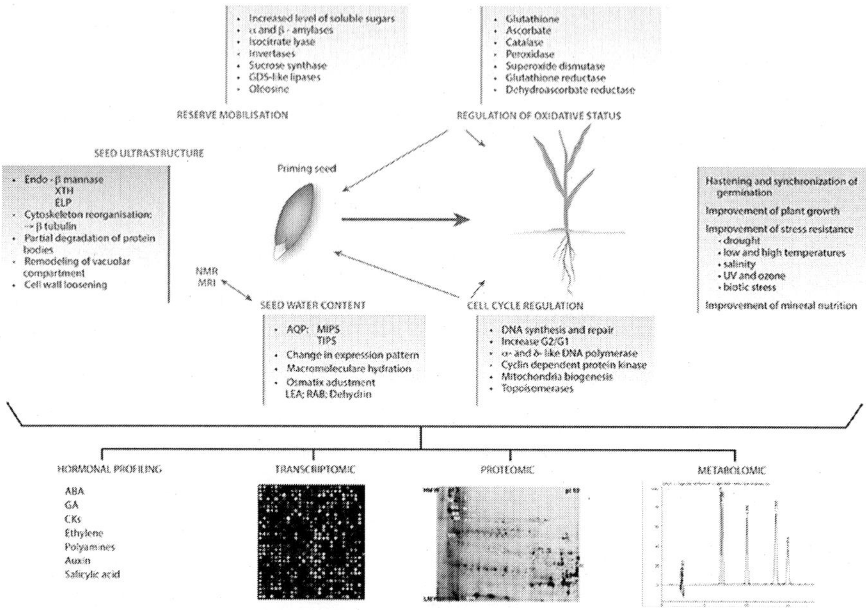

FIGURE 2. General overview of biochemical and physiological basis of priming effects. Priming modifies seed ultrastructure, reserve mobilization, regulation of oxidative status and cell cycle, and seed water content. The obtained seedlings may be improved for growth, mineral nutrition, and stress resistance. The components of priming effect may be revealed through an integrated convergent proteomic, transcriptomic, and metabolomics holistic approach.

5. PRIMING AND SEED WATER CONTENT

Seed germination is characterized by triphasic kinetic of water uptake with a rapid initial uptake (imbibition), followed by an apparent lag phase. A further increase in water uptake occurs only after germination is completed, as the embryonic axis elongates [111] (**Figures 1** and **2**). Early imbibition of seeds involves the fastest and most drastic changes in tissue hydration observed during germination. The water content of seeds and tissues within seeds depends of the composition of stored reserves. Seed imbibition and subsequent embryo growth depend on water exchanges and water potential gradients represent the motive force for water flow and finally tissue expansion. Water transport across cell membranes is essential for the initiation of metabolism. This intracellular water transport is mediated by aquaporins.

Aquaporins (AQPs) are transmembrane proteins, members of the major intrinsic protein (MIP) family that facilitate rapid and passive water transport across cell membranes and play a crucial role in plant water relations [112–114]. Plant aquaporins are remarkably diverse with several subfamilies of MIPs identified in dicots and monocots. Among them, the plasma membrane intrinsic proteins (PIPs) and the tonoplast intrinsic proteins (TIPs) subfamilies constitute the largest number of AQPs and correspond to AQPs that are abundantly expressed in the plasma and vacuolar membranes, respectively. Both PIP and TIP subfamilies are believed to play a key role in transcellular and intracellular plant water transport.

To gain insight into the role of water channel in germination, the expression profiles of *AQP* genes were studied in *Arabidopsis* [115], *Oryza sativa* [116], and *Brassica napus* [117] during seed imbibition and early embryo growth. These results have demonstrated the possible role of several AQPs in seed germination also in response to abiotic stresses. Moreover, Liu et al. [116] have shown the reduced seed germination rate via *OsPIP1;3* silencing and the promotion of seed germination via *OsPIP1;3* over-expression under drought conditions demonstrating that OsPIP1;3 is required for normal seed germination.

Seed priming involves imbibing seeds with restricted amounts of water to allow hydration sufficient to permit pre-germinative metabolic events to proceed while preventing radicle protrusion. This treatment can extend phase II of water uptake while preventing seeds from entering into phase III. The completion of radicle emergence is prevented by restricted amount of water provided to the seed (hydropriming, solid matrix priming) or decreased water potential (Ψw) of the imbibition medium by the use of osmotic solutes such as PEG or salts (osmopriming) [111]. In a study on *Brassica napus* osmopriming, Kubala et al. [36] revealed that at the beginning of the soaking period and at the end of drying phase, the seed water content was as low as 5%. The soaking treatment allowed seed imbibitions up to 50%, which should be enough to re-initiate metabolism.

As AQPs regulate water movement, it can be supposed that these proteins play an important role both in priming treatment (seed soaking) and post-priming germination under both favorable and unfavorable conditions. A role for aquaporin-controlled water transport across cell membranes in primed seeds of *Brassica napus* during germination was demonstrated by Gao et al. [118]. Seed priming with PEG or ABA resulted in an enhanced germination, particularly under salt and osmotic stresses at low temperature. Priming treatment induced expression of *BnPIP1* but had no effect on transcript level of Bnγ-*TIP2*. However, transcripts of both Bn*PIP1* and Bnγ-*TIP2* genes during germination were present earlier in primed seeds than nonprimed ones. Gao et al. [118] speculated that Bn*PIP1* was involved in water transport required for the activation of enzymatic metabolism of storage nutrients in the early stages of rapeseed germination, while Bnγ-*TIP2* expression was correlated with cell growth during radicle emergence. Changes in the expression pattern of *SoPIP1;1, SoPIP1;2, SoPIP2;1*, and *SoδTIP* during *Spinacia oleracea* seeds osmopriming and post-priming germination under optimal conditions, chilling and drought have been reported by Chen et al. [119]. The authors have stated that all these genes were up-regulated within 2–4 d of priming (phase II-imbibition). Therefore, the high expression of those *AQPs* might contribute to water transport across plasma and vacuolar membranes to facilitate water supply to expanding tissues and to increase germination potential of primed seeds. The down-regulation of all *AQPs* genes expression was observed under chilling and drought. However, the expression of some *AQPs* genes was elevated in primed seeds that also exhibited greater chilling and drought tolerance [119]. Kubala et al. [36] revealed up-regulation of two genes encoding tonoplast AQPs (*TIP4.1* and *TIP1.2*) in *Brassica napus* seeds in relation to osmopriming. In this study, expression of *TIP1.2* increased approximately 20 fold during post-priming germination as compared to unprimed seeds. In addition, the same authors have also stated facilitated water uptake and higher stress tolerance of germinating primed *Brassica napus* seeds [26, 120]. The above-mentioned results have demonstrated that water transport and sufficient water supply for embryo during post-priming germination regulated by AQPs may be one of the crucial components modulated by pre-sowing seed priming that influences germination rate and stress resistance.

As PIPs, but not TIP, are generally found at the plasma membrane, PIPs are thought to play a key role in seed water uptake. Nevertheless, both microarray [121] and macroarray experiments [115] with the complete set of genes encoding major intrinsic proteins revealed that out of 13 PIPs encoded by the *Arabidopsis* genome, transcripts for only three isoforms (PIP1;2, PIP1;4, and PIP1;5) were detectable in seeds. Mapping of TIPs in germinating *Arabidopsis* seeds has revealed that isoforms TIP3;1 and TIP3;2 detected in embryos, appear to localize to both the plasma membrane and tonoplast [122]. Vander Willigen et al. [115] have observed that during germination, very high level of TIP3 protein was coincident with decreased level of PIP1;2 and PIP2;1 polypeptides until phase III of water uptake. As stated by Vander Willigen et al. [115], it is intriguing how such low concentrations of PIP protein during the early phases of germination can achieve basic transcellular water transport in the seed. Gattolin et al. [122] have speculated that TIP3 may be the only AQP involved in seed water intake, and that the presence of TIP3 at the plasma membrane may compensate for the absence (or low concentration) of PIPs. In the light of these results, the enhanced germination potential of primed *Brassica napus* seeds could be partially explained by the up-regulation of TIPs during priming and post-priming germination [36]. However, the involvement of apoplastic water movement and simple diffusion of water across membranes during seed imbibition cannot be ruled out. The treatment of *Arabidopsis* seeds with mercury, a general blocker of aquaporins, reduced the speed of seed germination but did not affect its developmental sequence or basic aspects of seed water relations. Vander Willigen et al. [115] suggested that aquaporins functions are not involved in early seed imbibition but would rather be associated with water uptake accompanying expansion and growth of the embryo.

The transmembrane water transport via the regulation of AQP quantity and activity endows seeds with a remarkable capacity to modulate water absorption, transport, and compartmentation within tissues. Nuclear magnetic resonance (NMR) spectroscopy has provided insights into changes in the physical states of seed water during germination [123–126]. In particular, magnetic resonance imaging (MRI) has revealed a precise spatial distribution of water within tissues of germinating seeds and different patterns between species [125, 127, 128] highlighting the tight control of water transport. Water status of primed seeds was characterized by Nagarajan et al. [129] in study on tomato halo-and osmopriming. Nagarajan et al. [129] pointed out that better performance of primed seeds may be attributed to the modifications of seed water-binding properties and reorganization of seed water during imbibition, so as to increase the macromolecular hydration water required for various metabolic activities related to the germination process. In the future, it will be crucial to see how the spatial pattern of aquaporin expression can fit the hydration pattern revealed by MRI during both priming and post-priming germination, therefore enabling a comprehensive understanding of water transport in seeds.

Several studies have reported that water uptake is improved by priming treatment as primed seeds exhibited a faster imbibition in comparison with nonprimed ones, although pre-treated seeds were dried after priming to reach the same water content as nonprimed ones [36, 47, 107]. Although MRI studies revealed that water penetrates seeds through the hilum and micropyle [125, 130], Galhaut et al. [47] did not observe any particular modification of these structures after *Trifolium repens* priming, despite a faster seed hydration. However, scanning electron microscopy analysis showed that primed seeds of *Trifolium repens* exhibited seed coat tears and circular depressions that can favor seed imbibition. Moreover, X-ray photographs revealed tissue detachment in dry primed seeds that formed free space between the cotyledons and radicle, making water flow easier, thus contributing to tissue hydration [47]. Similarly, formation of free space surrounding the embryo in dry primed seeds of tomato was noticed by Liu et al. [131]. In brief, these observations suggest that structural modifications might contribute to rapid seed germination by improving water uptake.

The maintenance of favorable water status is critical for survival of germinating seeds under environmental stresses leading to tissue dehydration. The accumulation of nontoxic, compatible solutes within seed tissues, that is, osmotic adjustment is a major trait associated with maintenance of high cell turgor pressure potential in response to stress conditions. Priming treatment itself may generate a moderate abiotic stress during soaking (e.g. osmotic stress, salt, and drought created by the priming agents) [36, 132]. The accumulation of osmotically active solutes such as amino acids (e.g. proline) ammonium compounds (e.g. glycine betaine), sugars (e.g. glucose, fructose, sucrose) during priming was noticed in several species and was shown to improve seed germination under subsequent water stress [3, 21, 133, 134].

Seeds can also experience dehydration in the course of priming treatment, that is, during drying after soaking. Late embryogenesis abundant proteins (LEAs) can stabilize cell structure and macromolecules upon cell dehydration by preventing inactivation and aggregation of proteins and the loss of membranes integrity. This could be realized due to the ability of most LEA proteins to either coat intracellular macromolecules with a coherent layer of water or to interact with the surface of proteins and thus acting as water replacement [135]. As LEA proteins accumulate at a high level in response to cell/tissue dehydration, they may contribute to acquisition of tolerance to drought and related stresses such as osmotic, salt, and cold stress. In support of this, several studies revealed changes in the pattern of expression/accumulation of LEA transcript/protein in seeds caused by priming treatment and suggested their association with improved stress tolerance of primed seeds [36, 107, 136–138]. For example, the transcripts of two genes: *Em6*, encoding LEA group 1 protein and *RAB18*, encoding responsive to ABA 18 protein, belonging to LEA group 2, declined during osmopriming (soaking in PEG solution), reaccumulated after slow drying and again degraded during *Brassica oleracea* seeds germination [138]. The up-regulation of *RAB18* and *Em6* expression during slow seed drying suggests that they play a role in drought tolerance. Chen et al. [107] reported transient accumulation of four dehydrins-like proteins (32, 30, 26, 19-kD) in seeds of *Spinacia oleracea* during early stages of osmopriming followed by progressive degradation to a lower level in primed dry seeds compared to unprimed ones. A similar trend was confirmed to cold acclimation protein CAP85. In contrast to protein concentration, relative expression of *CAP85* was greater in primed dry seeds than in unprimed ones. Recently, Kubala et al. [36] revealed accumulation of LEA transcripts (*LEA4-1, LEA4-5*) and LEA3 proteins during soaking in PEG solution. The authors proposed that soaking in PEG with a low osmotic potential should not be considered only as a rehydration phase: water uptake may be sufficient to reinitiate a physiological activity from a previous quiescent stage but water content of 50% remained low enough to represent a water stress situation, especially when it is maintained during several days [36].

6. PRIMING AND SEED ULTRASTRUCTURE

In general, the ability of seeds to germinate seems to be critically determined by a change in the balance between growth potential of the embryo and the mechanical resistance of the surrounding tissues. In many species, the endosperm tissue enclosing the embryo restrains the germination process acting as a physical barrier, which restricts radicle emergence.

Weakening of tissues surrounding elongating radicle by cell separation due, for instance, to the activity of cell wall hydrolases may occur as a consequence of priming (**Figure 2**). It was determined that osmopriming induced hydrolysis of the endosperm tissue of *Cucumis melo* seeds [139] and increased the endo-β-mannanase activity in the endosperm cap and decreased its mechanical restraint on the elongating tomato embryo [140]. A strong correlation was observed between lowering of the mechanical restraint and the activity of endo-β-mannanase [141].

Penetration of the structures surrounding the embryo is a consequence of radicle cells elongation. Up-regulation of the gene encoding xyloglucan endotransglucosylase/hydrolase (XTH) in response to osmopriming and accumulation of transcript for extensin-like protein (ELP) during post-priming germination was observed in rapeseed [36]. As XTHs have the ability to cleave xyloglucans and rejoin the cut ends with new partners, they are engaged in cell wall loosening during growth and in the restructuring of the cell walls after extension.

Cytoskeleton reorganization is also necessary to achieve large rates of cells elongation that precedes radicle protrusion. The component of microtubules (β-tubulin) accumulated in tomato seeds during germination and priming and the expression preceded visible germination [142]. Higher level of β-tubulin protein accumulation was shown in rapeseed during PEG soaking, drying, and post-priming germination. The up-regulation of genes encoding γ- and β-tubulins was also noticed during post-priming germination [36].

Ultrastructural observations performed during the 6-d period of solid matrix priming (SMP) of carrot (*Daucus carrota*) seeds indicated the breakdown of storage materials, specific to the catabolic phase of germination *sensu stricto*, both in the axis and in the micropylar endosperm covering the radicle tip [143]. It was found that complete degradation of storage protein and lipid bodies and subsequent starch accumulation occurred in the radicles of carrot seeds after 8-d SMP. In the endosperm, the catabolic changes were limited to the micropylar area, where extensive breakdown of storage cell walls, partial degradation of protein bodies, and no storage lipid hydrolysis were observed [144].

During seed germination, storage proteins, which provide a source of reduced nitrogen, and inorganic minerals need to be mobilized to support seedling growth. In addition, a lytic aqueous vacuolar compartment building up the turgescence necessary for cell expansion and to promote radicle protrusion and embryo elongation has to be formed (**Figure 2**). Bolte et al. [145] investigated the features and the dynamics of the vacuoles during the early stages of *Arabidopsis* seed germination and indicated the successive occurrence of two different lytic compartments in the protein storage vacuoles (PSV). The first one corresponds to globoids specialized in mineral storage and the second one is at the origin of the central lytic vacuole in these cells [145]. Different mechanisms for the transformation of PSV into lytic vacuole in the root tip cell of germinating tobacco (*Nicotiana tabacum*) seeds were proposed by Zheng and Staehelin [146]. Ultrastructural studies demonstrated that the radicle cells of tobacco contain only one type of vacuole at particular time of development. Upon rehydration, the radicle cells only contain PSVs, but during subsequent root development, the PSVs are systematically transformed into lytic vacuoles via cell type specific pathways.

At present, we do not have a complete view of ultrastructural changes occurring during seed priming. One would expect that similar vacuole remodeling might occur during priming, especially in embryonic axis. The maintenance of metabolism over, for instance, several days of seed priming requires mobilization of embryo reserves. It is speculated that accumulation of endogenous osmotica, cell vacuolization, together with cell wall loosening initiated during priming, might determine the embryonic axis extension and radicle protrusion during post-priming germination. Deeper and more detailed studies should be continued in order to completely clarify this phenomenon.

7. SEED PRIMING AND CELL CYCLE REGULATION

Some of the hypotheses proposing explanation for priming-induced improvement are based on its effect on DNA in relation to activation of DNA repair mechanisms, synchronization of the cell cycle in G_2 and preparation to cell division (**Figure 2**). During seed maturation, most of the embryo cells are stopped at G_1 or G_0 phase of cell cycle and only some species have a small proportion of cells in the G_2 phase [111]. During seed imbibition, meristematic activity is limited; however, some preparation to cell division occurs. In embryos of dry tomato seeds, most cells have 2C DNA level and are in G1 phase of nuclear division [147]. The authors observed that DNA synthesis preceded germination as during imbibition in water, 4C signal was found mostly in the embryonic root tip, which suggests that cell enters S phase. They also primed seeds for 14 d in PEG-6000, which enhanced the rate and uniformity of germination. The 4C DNA signal of root tip cells increased during priming starting from 3 d incubation in PEG and was constant after re-drying the seeds to the initial moisture content. This observation suggests that priming increased the ratio of cells in G_2 phase to G_1 phase and indicates that the beneficial effects of priming on seedling performance are associated with the replicative DNA synthesis prior to germination [147]. This is accompanied by increase in α- and δ-like DNA polymerase activities in primed seeds and during germination.

The initiation of cell cycle and proceeding the cell to S phase may depend also on a G_1 checkpoint control. Most, if not all, cell cycle proteins responsible for cell cycle control appear to be already present in dry mature seeds, although some of them should be synthesized *de novo*. However, not only protein synthesis but also their modification may play a regulatory function for cell cycle control [148]. Cell division starts just after radicle protrusion, thus, seed priming, which prolongs Phase II of seed germination and is finished just before Phase III, does not affect cell division in itself [16]. Seed priming extends Phase II, when DNA repair mechanisms and expression of genes encoding proteins needed in cell cycle control and commencement are activated and overreach the level observed in unprimed seeds. Pre-activation of cell cycle by priming could be through regulation of the activity of cell cycle proteins such as cyclin-dependent protein kinases and proliferating-cell nuclear antigens [16]. It was found that osmopriming of *Brassica napus* seeds induced expression of cell division control protein 48 homolog C, cyclin P4;1, cyclin like protein and topoisomerase II in dry seeds, as well as proliferating-cell nuclear antigen 2 and cyclin dependent kinase 3;2 during imbibition [36]. Accumulation of proliferating-cell nuclear antigen during maize seed imbibition was associated with transition of the cells from G_1 to G_2 [149]. Moreover, microtubules, apart from cytoskeleton formation, cytoplasmic streaming, organellar movement, and cell wall formation, function in mitotic spindle formation during mitosis. Microtubules in dry seeds are depolymerized and form discrete granular bodies, which become organized into the cytoskeleton during imbibition [111, 142]. Higher expression of genes encoding microtubule-associated protein 65-1 and 70-2 as well as tubulin subunits γ-1, β-1, β-3 and microtubule motor activity proteins belonging to kinesin family was also observed during PEG soaking and in dry osmoprimed *Brassica napus* seeds [36]. Enhanced expression of tubulin genes was associated with accumulation of β-tubulin protein during osmopriming and subsequent germination [36]. Also in pre-hydrated seeds of *Arabidopsis thaliana* and *Solanum lycopersicum* accumulation of tubulins (mainly β-tubulin) was stated during germination as compared to unprimed seeds [136, 142].

During Phase II of seed germination, when water uptake is severely limited, major metabolic processes are activated [111]. One of the most important events undergoing during Phase II is DNA repair, which precedes cell cycle activation [142, 150]. The process of DNA replication is preceded by repair of DNA damage caused mainly by reactive oxygen species, which are accumulated during seed storage and aging [151]. DNA repair covers first period of DNA synthesis, while the second period of DNA synthesis (replication) is observed before cell division. DNA synthesis in Phase II of germination and also during seed priming corresponds rather to DNA repair, mainly in organelle such as plastids and mitochondria [152]. Increased number of mitochondria in leek embryo cell of osmoprimed seeds was observed by Ashraf and Bray [153]. Mitochondria biogenesis before mitochondria division involved the transition of promitochondria to mature mitochondria. This process is accompanied by the expression of genes of nucleotide biosynthesis, transport, and organelle RNA- and DNA-related functions [154, 155]. Pre-sowing seed osmopriming induced higher expression of genes corresponding to mitochondria biogenesis such as translocases of the inner membrane (TIM) complex TIM10 and TIM23-1, mitochondrial ribosomal protein and translational elongation factor EF2, which is targeted into mitochondria [36].

There are still some gaps in comprehensive understanding of pre-sowing seed priming impact on DNA repair and cell cycle regulation. Activation of different DNA repair mechanism has been observed during seed imbibition preceding germination and they are believed to be essential for successful reactivation of cell cycle [111]. They include α and β tyrosyl-DNA phosphodiesterase 1, α and β DNA topoisomerase I [156], 8-oxoguanine DNA glycosylase/lyase and formamidopyrimidine-DNA glycosylase [157], transcription elongation factor II-S [158], DNA ligase VI and IV [159]. Varier et al. [16] have suggested that in primed seeds DNA damage is repaired before replication, primarily through DNA synthesis. However, in a study on *Cicer arietinum* primed seeds, the role of DNA repair genes in enhancing the physiological quality of seeds was postulated [160]. The authors tested the expression level of genes encoding proteins with already proved function on DNA repair mechanisms in relation to priming methods and seed size. Moreover, enhanced accumulation of transcripts was found in dry and imbibed osmoprimed *Brassica napus* seeds [36] for genes involved in DNA repair according to function description in databases, such as DUTP-pyrophosphatase-like 1, endonuclease V family protein, ribonucleoside-diphosphate reductase subunit M2 (TSO), casein kinase II, replicon protein A2, DNA glycosylase DEMETER (DME), BARD1, RECQ helicase L4B, and MUTS homolog 2. Thus, activation of DNA repair mechanisms in seeds occurs prior to their germination and contributes to enhanced germination rate and better quality of seeds undergoing pre-sowing seed priming.

8. MANAGEMENT OF OXIDATIVE STATUS

Management of oxidative status is also considered as an important part of primed seeds physiology [18, 36, 161, 162] (**Figure 2**). Beginning with the seed development through maturation and germination, the seed moisture content as well as seed metabolic activity is subjected to dramatic changes. The biochemical and cellular events triggered by water uptake and subsequent loss are accompanied by a generation of reactive oxygen species (ROS) [151, 163]. During seed imbibition and early stages of germination, ROS production occurs mainly through respiratory activities of mitochondria, activities of β-oxidation pathways and enzymes such as NADPH oxidases, extracellular peroxidases, and oxalate oxidases [163]. ROS accumulation and associated oxidative damage can be regarded as a source of stress that may affect the successful completion of germination. As ROS, particularly H_2O_2 can act as signaling molecules, seeds must be endowed with a ROS removing system that tightly regulates their concentration. Scavenging of ROS is carried out by antioxidant system, a multifunctional machine, which includes enzymes (i.e. catalase (CAT), superoxide dismutase (SOD), and ascorbate peroxidase (APX), monodehydroascorbate reductase (DHAR), and glutathione reductase (GR)) as well as nonenzymatic compounds (i.e. ascorbic acid (AsA) or reduced glutathione (GSH) [1, 151]. Metabolism of ROS, mainly H_2O_2, which is believed to play a central role in oxidative status signaling, is strictly associated to other reactive species and signaling molecules such as nitric oxide and hydrogen sulfide, which contribute and regulate the transition from dormant to germination phase [151, 163, 164].

Kubala et al. [36] have considered the priming as a process, which consists of two main phases: controlled hydration of the seeds and drying back to the initial moisture content. Seed pre-hydration followed by re-drying during priming treatment, similarly to seed maturation and germination, exerts changes in moisture content, which leads to ROS production and activation of the antioxidant system. The activation of APX and the accumulation of AsA and GSH during osmopriming of spinach seeds (*Spinacia oleracea*) were accompanied by the repression of SOD and CAT activity [1]. These results indicated that activation of AsA-GSH cycle during osmopriming of spinach seeds can decrease the level of lipid peroxidation products in primed seeds. Moreover, the same authors have suggested that the renewal of antioxidant defense system, possibly required by seedling establishment, occurred during the late stages of germination as a result of up-regulation of CAT activity after initial reduction and overall antioxidant activation [1]. Kubala et al. [36] have indicated through an integrated transcriptomic and proteomic approach that the priming-induced germination can be linked with the activation of antioxidant system. The authors showed that during osmopriming of *Brassica napus* seeds *CAT2* and *PER21* encoding peroxidase 21 were up-regulated and GR protein was accumulated. Moreover, the same authors have observed up-regulation of *PER13* gene and accumulation of peroxidase 12, DHAR and peroxiredoxin proteins during post-priming germination. Furthermore, Kubala et al. [162] have postulated a correlation between activation of antioxidant metabolism in osmoprimed *Brassica napus* seeds and increased tolerance to salt stress during germination. The enhanced activity of APX, SOD, and CAT corresponded with increased expression rate of *APX, SOD*, and *CAT* genes. Similar result has been obtained by Nouman et al. [165], who showed that priming of *Moringa oleifera* seeds with *Moringa* leaf extract (MLE) improves growth under saline condition mainly by activation of antioxidant system (SOD, CAT, and POD). Priming of mung bean seeds (*Vigna radiata* L. Wilczek) with β-amino butyric acid (BABA) enhanced the activities of SOD and POX leading to improved tolerance to NaCl and PEG 6000 stresses [65]. Enhanced chilling stress tolerance in two tobacco varieties (MSk326 and HHDJY) was due to increased activity of antioxidant enzymes (SOD, POD, CAT, and APX) as a result of priming seeds with putrescine [165]. Results obtained by Islam et al. [167] showed that in haloprimed wheat (*Triticum aestivum*) seeds, increased activity of CAT, POD, and APX enhanced tolerance to salinity stress. Osmopriming with PEG has improved sorghum (*Sorghum bicolor*) seed germination and seedling establishment under adverse soil moisture conditions and has been correlated with antioxidant system activation (APX, CAT, POD, and SOD) [3]. Rice (*Oryza sativa*) seeds primed with polyethylene glycol (PEG) showed increased activity of APX in parallel with decreased activity of SOD, POD, and CAT under ZnO nanoparticles stress [38]. The same authors have also observed down-regulation of genes encoding the antioxidant enzymes (*APXa, APXb, CATa, CATb, CATc, SOD2*, and *SOD3*) in PEG primed seeds under nano-ZnO stress. They have concluded that priming with PEG significantly alleviates the toxic effects of nano-ZnO through improved cell structures of leaf and roots.

Seed aging during storage is associated with ROS production. Appearance of oxidative stress results in a decrease of seed quality. Kibinza et al. [161] showed that priming plays an important role in seed recovery from aging through CAT activation. Their results revealed accumulation of hydrogen peroxide (H_2O_2) and reduction of CAT at the gene expression level and protein content during sunflower (*Helianthus annuus* L) seed aging. Interestingly, the adverse results of aging were recovered by seed osmopriming, which led to induction of CAT synthesis by activating gene expression and translation of the enzyme.

Summing up, the management of oxidative status in primed seeds plays a very important role as a machinery, which leads to protection against oxidative stress, recovery from aging, and regulation of ROS production/accumulation. Alleviation in ROS level exerts a signal, which could be perceived, transduced, and crosstalk with other signaling pathways, thus executing physiological response by activation or repression of molecular processes.

9. RESERVE MOBILIZATION

Recent transcriptome and proteome study of cabbage (*Brassica oleracea*) and *Arabidopsis* seeds as well as integrated transcriptomic and proteomic approach in study of rapeseeds (*Brassica napus*) osmopriming revealed that germination and priming altered similar processes [36, 136, 138]. Indeed, it is proposed that germination-related processes such as respiration, energy metabolism, and early reserve mobilization can also occur during priming [16] (**Figure 2**). Faster and uniform rice (*Oryza sativa*) seed germination due to priming was related to improved activity of α-amylase, resulting in increased level of soluble sugars in primed kernels [168]. Sung and Chang [169] have shown that priming of maize (*Zea mays*) seeds leads to increased activity of enzymes for carbohydrates (α and β amylases) and lipids (isocitrate lyase, ICL) mobilization. Priming of chickpea (*Cicer arietinum* L. Cv PBG-1) seeds with mannitol led to increased activity of amylase, invertases (acid and alkaline), sucrose synthase (SS), and sucrose phosphate synthase (SPS) in shoots of primed seedlings. The higher amylase activity in shoots suggests a rapid hydrolysis of transitory starch formed in the shoots of primed seedlings leading to more availability of glucose for seedling growth. [170]. Higher content of soluble protein, aldolase, and ICL activity has been observed in haloprimed pepper (*Capsicum annum* L) seeds than in control seeds [171]. Moreover, the α-glucosidase accumulation and increased level of globulin degradation products were observed during the priming process of sugar beet (*Beta vulgaris* L.) seed [172]. Similar observation on primed sugar beet seed has been done by Capron et al. [173] who showed increased solubilization of 11S-globulin-β-subunit in response to hydro- and osmopriming. Gallardo et al. [136] have also observed higher polypeptides content in both hydro- and osmoprimed *Arabidopsis* seeds. They were identified as products of 12S-cruciferin-β-subunit degradation.

Kubala et al. [36] have shown that during *Brassica napus* seeds osmopriming and post-priming germination accumulation of transcript and proteins for seed storage proteins occurred. The authors have observed up-accumulation of cruciferin CRU1. The group of six genes encoding GDSL-like lipases, playing a role in triacylglycerols (major storage lipids in rape seeds) catabolism, were strongly up-regulated during post-priming germination while the other three genes for GDSL-like lipases as well as extracellular lipase 6 were up-regulated in osmoprimed seeds [36]. The activation of lipid catabolism-related genes was correlated with the activation of genes involved in lipid transport such as genes encoding bifunctional inhibitor/lipid-transfer protein/seed storage 2S albumin superfamily protein [36]. Priming can also reduce a level of oleosins-proteins, which surround oil bodies. In osmoprimed *Brassica napus* seeds as well as during post-priming germination, down accumulation of oleosin S4-3 protein and down-regulation of *OLEOSIN2* gene were observed, respectively [36].

Activation of respiration and rapid ATP production is primary metabolic events occurring during priming [18], and higher respiratory activity is required to cover energy pool for speed up germination. The increased ATP level/energy charge after priming was observed in tomato (*Solanum lycopersicum*), eggplant (*Solanum melongena*), araucaria (*Araucaria columnaris*), spinach, oat (*Avena sativa*), and cabbage (*Brassica oleracea*) [174, 175]. Primed seed needs a large amount of fuel, which supplies energy required for higher reserve mobilization rate. All together lead to improved energy turnover and increased metabolism rate of primed seed and contribute to better germination and stress tolerance.

10. HOLISTIC "OMIC" APPROACHES OF SEED PRIMING

The identification of biomarkers of seed priming effect is a relevant goal for plant physiologists. The sequence of events associated with water-based priming involves a limited hydration step during the soaking period followed by a more or less rapid dehydration phase (**Figure 1**). It may be considered that each phase has its own profile of activation/deactivation.

Transcription is not necessarily coordinately associated with translation, resulting in some cases in a limited correspondence between mRNA levels and protein abundance. Some proteins synthesized during incubation may also be degraded during dehydration, although the rate of degradation might somewhat be influenced by the rate of drying. Finally, protein synthesis may also occur as a result of the translation of long-lived mRNA previously synthesized during seed maturation [17, 37, 103, 110].

The advantage of these complementary holistic approaches is that they provide a large set of data allowing physiologist to obtain a global view of all parameters of the metabolism associated with priming. The major disadvantage, however, is that these techniques are rather expensive. As a consequence, samples are frequently analyzed after a given duration of treatment and the time-dependent evolution of parameters are rarely considered. It is still difficult to reconstitute a kinetic approach, but published studies rather provide a large set of information at a given moment while the priming procedure is a dynamic process.

10.1. Transcriptomics

Transcriptomics is the study of the transcriptome—the complete set of RNA transcripts that are produced by the genome under specific circumstances—using high-throughput methods. Several techniques are available for transcriptomic approaches, including microarray, c-DNA amplified fragment length polymorphism (cDNA-AFLP), expressed sequence tag sequencing, serial analysis of gene expression, RNAseq, and massive parallel signature sequencing [176]. According to Buitink et al. [109], more than 1300 genes may be differentially regulated during priming with PEG at -1.7 MPa in *Medicago truncatula*. Genes whose expression are regulated during the priming process are commonly categorized according to the function of the corresponding protein (metabolism and regulation of metabolism, cell cycle regulation, DNA processing, transcription regulation, cellular transport and communications, stress responses, etc.). However, a consistent part of the identified genes is still not annotated.

In *Brassica napus*, numerous priming-regulated genes are involved in gluconeogenesis, which is essential for triacylglycerol breakdown into small molecules. Another important category of genes involved in water-based priming of rapeseed encodes for transcription factors, and this is especially the case after hormopriming with ABA [138]. In the same species, it was demonstrated that germination of primed seeds involves a specific set of genes comparatively to germination of unprimed ones [36]. In young seedlings obtained from primed seeds, the expression of stress-related genes is often more rapid than in plants from unprimed ones, as recently exemplified for cold tolerance [26]. Not only genes related to protein synthesis but also genes involved in protein degradation may be induced by priming. Some transcription factors may be down-regulated during soaking and up-regulated during drying [36]. Genes involved in the synthesis of osmoprotectant, such as proline, may also be regulated at various steps of priming [120].

In *Brassica oleracea*, transcripts that are abundant in dry seeds rapidly decline during osmopriming and germination expression programs are initiated during osmopriming [138]. For these authors, genes expressed during slow drying following the soaking period are correlated with the stress resistance properties of primed-material. This may especially concern cell-cycle-regulated genes, enzymes involved in C metabolism, and components of the translation machinery. Acquisition of the desiccation tolerance after the soaking period is a crucial trait for priming. Transcriptomic data demonstrated that the molecular determinism associated with such trait may be, at least partly, similar to those involved during a « normal » seed dehydration process during maturation. In *Medicago truncatula*, several regulatory genes expressed during drought stress are also up-regulated during seed maturation. During priming, these genes are involved in post-soaking dehydration while genes conditioning cell cycle, cell wall biogenesis, and energy metabolism are repressed [109]. Transcripts accumulated after PEG-priming treatment comprise those encoding for LEA and seed storage proteins, or genes controlling seed dormancy while genes involved in photosynthesis and cell wall modifications are commonly repressed [108].

Beside involvement of transcription factors, it could be of primary importance in the future to analyze the priming-induced epigenetic changes. Indeed, DNA methylation may be directly involved in « stress memory » and some changes, such as cytosine methylation or overexpression of histone deacetylase was reported to occur during germination [78].

10.2. Proteomics

Proteome is the entire set of proteins present at a given moment in a given biological sample and proteomic is the large-scale study allowing identification and quantification of these proteins. Proteomics has now been practiced in different plant systems through the separation of proteins by two-dimensional gel electrophoresis followed by peptide mass fingerprinting. Tandem mass spectrometry can get sequence information from individual isolated peptides. Quantitative proteomics represent an important extension enabling the comparison of changes in protein levels across different samples [177]. Gel-free shotgun proteomic is an alternative for identification and quantification of protein in large-scale studies. This method was recently used to reveal potential biomarkers of priming-induced salt tolerance in durum wheat with a concomitant use of protein fractionation and hydrogel nanoparticles enrichment technique [27]. According to these authors, hydropriming was accompanied by a significant change in 72 proteins. Most of them are involved in proteolysis, protein synthesis, metabolism regulation, and disease or defense response. Priming with ascorbic acid changed the pattern of proteome signature and most of identified proteins are involved in metabolism regulation, antioxidant protection, repair processes, and methionine-related metabolism.

Cycloheximide, a potent inhibitor of protein synthesis, may be added to priming or to germination solution to analyze the precise requirement of protein synthesis during these steps. It appears that protein synthesis is not necessarily required for normal germination process, which could be related to the fact that in some cases, germination relies only on translation of pre-stored long-lived mRNA produced during seed maturation [14, 178]. Cycloheximide may drastically affect osmopriming-induced improvement of Cd resistance in *Poa pratensis* and *Trifolium repens* in relation to a decrease in α and β-amylase [179]. Protein concentration in primed seeds is the result of two contrasting pathways: the first implies specific protein synthesis while the second involves the breakdown of storage proteins through protease activation. It was demonstrated that priming may act on protease activities and also influence the expression of protease encoding-gene [180].

Some strategies are also used to separately identify proteins in embryos and in embryo-surrounding tissues. In monocots, a special attention is often paid to aleurone layers involved in hydrolytic enzyme synthesis [181]. Aleurone layer is easily separated in some plant material such as wheat or barley but is more difficult with other monocot such as turfgrasses with very small seeds. In wheat, it was demonstrated that the set of proteins regulated by priming is quite different in embryos and in surrounding tissues and the number of proteins affected by priming was by far higher in the embryos than in the other parts of the seed. Proteins regulated in embryos involve those directly linked to metabolism regulation (especially methionine biosynthesis, glutamate/glutamine metabolism, amino acids synthesis, etc.), energy supply, cell growth and maintenance of cell structure while proteins up-regulated in surrounding tissues are mainly involved in reserves mobilization or in the management of the oxidative stress [67]. In barley, however, ascorbate peroxidase was observed only in the embryo while several other redox-related proteins differed in spatio-temporal patterns at the onset of radicle elongation [181].

Priming is considered as an invigoration treatment and several proteomic approaches were performed to unravel the biomarker of seed vigor. Catusse et al. [182] found that 18 proteins are accumulated during hydropriming of sugarbeet seeds and that the same proteins appear directly reduced during aging. Seed vigor appears directly related to lipid and starch mobilization, protein synthesis and methionine cycle. In poplar, more than 81 proteins showed a significant change in abundance when comparing the proteomes among seed with different vigor [183]. According to these data, the decrease in seed vigor is an energy-dependent process, which requires protein synthesis and degradation as well as cellular defense and rescue. Salicylic acid (SA) was proposed as an invigorating elicitor promoting seed germination under saline conditions: SA re-induced the late maturation program during early stages of germination, induced the synthesis of antioxidant enzymes, and improved the quality of protein translation [184]. Similarly, the "prime-ome" approach performed by Tanou et al. [185] confirmed the importance of redox proteomic and processes such as N-nitrosilation, tyrosine nitration, and mitogen-activated protein kinase MPK3 signaling in priming effects.

Osmotic priming was reported to trigger desiccation tolerance in *Medicago truncatula* [110]. Proteomic analysis demonstrated that such trait is directly linked to the synthesis of late-embryogenesis abundant proteins from different groups. Secondary structure of some proteins was compared in the hydrated and dry state after fast or slow drying using Fourier transform infrared spectroscopy, which confirms that these proteins adopted α helices and β-sheets conformation during drying process.

Some proteomic approaches are conducted on plant species whose genomes are not sequenced. A recent 2DE-MS/MS-based proteomic study was conducted on pearl millet seeds primed by β-aminobutyric acid and showed an over-representation of proteins belonging to glucose metabolism, and a majority of induced proteins are directly related to energy [106]. Seedlings issued from those seeds are more resistant to downy mildew (*Sclerospora graminicola*). It is interesting to note that several of the elicited proteins are present in the extracellular space and in organelles (mainly mitochondrion and chloroplast).

10.3. Metabolomics

Metabolome refers to the complete set of small-molecule metabolites present within a plant tissue or organ at a moment. Metabolomic should thus be considered as the quantitative measurements of the whole set of compounds involved in the metabolism of a given biological sample. Plant metabolomics has become an essential part of functional genomics. It often appears difficult to analyze the entire range of metabolites by a single analytical method, and several tools are thus commonly combined for this purpose: high performance liquid chromatography (HPLC), gas chromatography (GC), electrospray ionization coupled with mass spectrometry, capillary electrophoresis, atmospheric pressure chemical ionization (APCI), and secondary ion mass spectrometry (SIMS). Metabolomic approach, however, still remains difficult to perform on seed material considering the high proportion of molecule issued from reserve mobilization. An excess of soluble sugar in graminaea and an excess of small lipids issued from oil digestion in oleaginous plant material may greatly hamper isolation of other compounds involved in hastening the germination of primed seeds. Similarly, some important compounds might be present only in specific tissues or cell compartments and react with sugars during extraction processes; this is especially the case of gibberellins which play a crucial role in germination but are believed to conjugate with sugar or phenol compounds, which greatly compromise isolation of and identification procedures of numerous metabolites [186].

Priming process may drastically modify the synthesis and accumulation of endogenous antioxidant such as glutathione, ascorbic acid, and even α-tocopherol. A modification in reducing sugars concentration, such as glucose resulting from partial starch digestion during priming, may influence protein glycation, a nonenzymatic reaction between reducing sugars and amino groups in protein [150]. The amount of proline was reported to be modified in osmoprimed seeds [120]. Some other data reported modifications in aspartate, leucine, threonate, glutamate, fumarate, or pinitol content in primed seeds from different species [1, 2, 35, 104, 152]. Methionine may also play a key role in priming process, mainly as a precursor of both ethylene and polyamine synthesis. Polyamines are small aliphatic molecules influencing all aspects of plant metabolism and development, including the germination processes. Ethylene is also involved in various aspects of seed germination, and priming was reported to modify the kinetics of ethylene synthesis from its precursor 1-aminocyclopropane-1-carboxylic acid (ACC) [186]. Other phytohormones are expected to assume important functions in priming effect (ABA, gibberellins, cytokinins, auxin, etc.) [50, 52], but the quantitative impact of priming procedure on these compounds is still not well established.

11. MAIN LIMITATIONS OF PRIMING TECHNIQUES

Seed priming has emerged as an effective approach for increasing seed vigor. The optimal treatment differs between species, cultivar, and seed lots. Such variability is a major limitation of the priming method since numerous trials are required to identify the most appropriate strategy for each situation. There is no "general rule" concerning modalities of seed priming and there is no clear trend of priming response according to the taxonomic position of the species [188]. This, undoubtedly, limits the commercial implementation of priming treatments.

Some priming treatments may imply a risk of medium contamination by fungi and bacteria, which may heavily impair subsequent seed germination [189]. This may especially be the case for PEG and sometimes requires the simultaneous use of pesticide, although the impact of these compounds on the priming efficiency remains unknown. After priming treatment, seeds are dried back to their initial moisture content, but this dehydration phase is usually performed rapidly and occurs faster than classical dehydration of maturing seeds. It has been hypothesized that this brutal desiccation procedure alters the beneficial effect of priming [2].

The major drawback of priming is that it reduces the longevity of primed seeds as compared with the nonprimed seeds [190–192]. Storability of primed seed material is consequently reduced, and this results in higher costs for material maintenance for farmers and seed companies. The loss of viability, however, appears quite variable depending on the species, cultivars, and seed lots. In extreme cases, priming-induced advantages may even disappear after only 14 d of storage and the obtained seedling may then perform worse than those issued from unprimed seeds [193]. One of the limitations of these studies, however, is that experiments are commonly performed on artificially aged seeds using short-term exposure to high temperature in a moist environment, which is not necessarily fulling relevant from a real aging process. Some studies, using classical long-term strategies, reported that longevity is not necessarily affected or may even be increased by priming treatments [194]. Hussain et al. [195] recently demonstrated that post-priming temperature plays a key role in maintenance of seed longevity, which is indeed rapidly compromised in rice seeds maintained at room temperature while it remains intact when the material is stored at 4°C. According to these authors, the deleterious effect observed at 25°C storage was related to hampered starch metabolism in primed seeds. It was also suggested that maintenance of the seed longevity at 4°C may be due to a high viscosity that strongly reduces molecular mobility in the cytoplasm and thus limits the impact of deteriorative process even in seeds exhibiting low water content [196]. Oxygen during storage may trigger metabolic processes in primed seeds, which did not re-establish a true quiescent stage after dehydration when stored at room temperature while oxygen had no influence at low temperatures [195].

In some cases, repeat priming treatment after storage may partly remove the damaging effect on seed viability [193] while, in other cases, such a loss is permanent and not reversible [194]. Whatever, the fact that an additional treatment may be required to restore full germination potential represents an additional cost and source of variability.

12. CONCLUSIONS

Seed priming is an old empirical strategy used since centuries by farmers, and since decades by seed companies, to improve germination processes in cultivated plant species. The underlying mechanisms involved in this positive impact of pre-sowing treatments remained obscure for a long time. The present review aimed to summarize recent information provided by various tools allowing the identification of molecular cues conditioning priming efficiency.

Data obtained from molecular approaches applied to some well-known plant species (rice, rapeseed, tomato, etc.) are now available. The role of genes associated to metabolic and cell cycle events now starts to be deciphered, mainly those encoding for translational components such as ribosomal subunits, translational initiation factor, enzyme involved in carbon metabolism, histones, and transcription factors. A putative epigenetic basis of priming effects should also be considered. Some authors identified enzyme activity changes in relation to priming while others also reported numerous changes in the seed storage protein. Priming has also been reported to increase proteins related to the cell cycle activities such as α- and δ-DNA polymerase. Protecting proteins like dehydrins or HSP is expected to assume protective functions during the dehydration step. Similarly, proteins involved in water transport, cell wall modification, cytoskeletal organization, and cell divisions, may be to some extent regulated during priming. Gene expression and enzyme activities involved in osmocompatible solute synthesis may be of primary importance to regulate tissue protection during the dehydration step and water fluxes during the final germination phase. Measurements of water uptake by primed seeds suggest a reduction in the lag time of imbibition. Water uptake and its subsequent cell-to-cell movement during germination might be controlled by aquaporins and expression of the corresponding genes constitutes a specific target of the priming treatment.

Among the different hypothesis proposed to explain the biochemical basis for germination improvement, DNA replication and cell cycle advancement during priming treatment as well as synchronization of the cell cycle at the G2 phase are supported by some experimental evidences. DNA synthesis is involved during priming treatment itself but also during post-germinative events. Changes are also observed with the modification of the membrane structure and reorganization of mitochondrial integrity. Activation of antioxidative properties by priming treatment may also explain the improved behavior of plant material, especially when final germination and/or growth occur under stress condition.

The priming-induced decrease of the storage capacity is a major limitation for the application of the priming technique by seed companies. Partial vacuum storage may be useful for extending the longevity of primed seeds. Improved longevity may be related to enhanced anti-oxidative activity that minimizes the accumulation of total peroxide during long-term storage. Another challenge for private seeds company is to identify appropriate treatments able to restore vigor of old dry seed lots in order to increase their mean percentage of germination to values compatible with commercial purposes.

If priming may undoubtedly be considered as a valuable strategy to improve stand establishment, its impact on final yield and crop production has not be always confirmed. Most studies devoted to "omics" approaches of seed priming are performed on young seedling cultivated under fully controlled environmental conditions. The link between seedling behavior in plant growth chambers and the adult plant performance in field conditions is far from being clear. As a consequence, there is an urgent need to focus on transcriptomic, proteomic, and metabolomics of adult plants issued from primed seeds, especially at the reproductive stage, in order to assess the long-term impact of priming on cultivated plants throughout the plant cycle.

In cultivated plant species, a given priming treatment also has contrasting effects on various cultivars. It may be hypothesized that the ability to respond to priming treatment might be genetically controlled but, to the best of our knowledge, no data are available concerning this important aspect. Thus, further progresses are needed not only to identify the set of genes that are regulated by priming, but also the set of genes that putatively regulate priming response and efficiency themselves.

REFERENCES

1. Chen K, Arora R. Dynamics of the antioxidant system during seed osmopriming, post-priming germination, and seedling establishment in spinach (*Spinacia oleracea*). Plant Science. 2011;180:212-220.
2. Paparella S, Araújo SS, Rossi G, Wijayasinghe M, Carbonera D, Balestrazzi A. Seed priming: state of the art and new prospectives. Plant Cell Reports. 2015;34:1281-1293.
3. Zhang F, Yu J, Johnston CR, Wang Y, Zhu K, Lu F, Zhang Z, Zou J. Seed priming with polyethylene glycol induces physiological changes in sorghum (*Sorghum bicolor* L. Moench) seedlings under suboptimal soil moisture environments. PLoS One. 2015;10: e0140620.
4. Santini BA, Martorell C. Does retained-seed priming drive the evolution of serotiny in drylands? An assessment using the cactus *Mammilaria hernandezii*. American Journal of Botany. 2013;100:365-373.

5. Balmer A, Pastor, Gamir J, Flors V, Mauch-Mani B. The « prime-ome »: towards a holistic approach to priming. Trends in Plant Science. 2015;20:443-452.
6. Tanou G, Fotopoulos V, Molassiotis A. Priming against environmental challenges and proteomics in plants: update and agricultural perspectives. Frontiers in Plant Science. 2012;3:216.
7. Gamir J, Sánchez-Bel P, Flors V. Molecular and physiological stages of priming: how plants prepare for environmental challenges. Plant Cell Reports. 2014;33:1935-1949.
8. Evenari M. Seed physiology: its history from Antiquity to the beginning of the 20th century. Botanical Review. 1984;50:119-142
9. Evelyn JS. A Discourse of Forest Trees and the Propagation of Timber. John Martin (Royal Society). London; 1664.
10. Ingenhousz J. Experiments Upon Vegetables. 1779. P. Elmsly and H. Payne, London.
11. Amici GB. Note on pollen action mode on stigmata. Annales Sciences Naturelles Botanique. 1830;21:329-332
12. Sachs J. About treatments modifying germination strength in seeds. Botanical Zeitung. 1859;17:177-188.
13. Sachs J. Vorlesungen Lectures on Plant Physiology. Berlin; 1887.
14. Khan AA, Peck NH, Samimy C. Seed osmoconditioning: physiological and biochemical changes. Israel Journal of Botany. 1981;29:133-144.
15. Hussain M, Farooq M, Basra SMA, Ahmad N. Influence of seed priming techniques on the seedling establishment, yield and quality of hybrid sunflower. International Journal of Agriculture & Biology. 2006;8:14-18.
16. Varier A, Vari AK, Dadlani M. The subcellular basis of seed priming. Current Science. 2010;99: 450-456.
17. Taylor AG, Allen PS, Bennett MA, Bradford JK, Burris JS, Mishra MK. Seed enhancements. Seed Science Research. 1998;8:245-256.
18. McDonald MB. Seed priming. In: Black M, Bewley JD, editors. Seed Technology and its Biological Basis. Sheffield, Sheffield Academic Press; 2000. p 287-325.
19. Rahman MM, Ahammad KU, Alam MM. Effect of soaking condition and temperature on imbibition rate of maize and chickpea seeds. Research Journal of Seed Science. 2011;4:117-124.
20. Basra SMA, Zia MN, Mahmood T, Afzal I, Khaliq A. Comparison of different invigoration techniques in wheat *(Triticum aestivum* L.) seeds. Pakistan Journal of Arid Agriculture. 2002;5:11-16.
21. Srivastava AK, Lokhande VH, Patade VYwx, Suprasanna P, Sjahril R, D'Souza SF. Comparative evaluation of hydro-, chemo-, and hormonal priming methods for imparting salt and PEG stress tolerance in Indian mustard (*Brassica juncea* L.). Acta Physiologiae Plantarum. 2010;32:1135-44. 10.1007/s11738-010-0505-y
22. Omidi H, Khazaei F, Hamzi Alvanagh S, Heidari-Sharifabad H. Improvement of seed germination traits in canola (*Brassica napus* L.) as affected by saline and drought stresses. Plant Ecophysiology. 2009;3:151-158.
23. Kaya MD, Okçu G, Atak M, Çıkılı Y, Kolsarıcı Ö. Seed treatments to overcome salt and drought stress during germination in sunflower (*Helianthus annuus* L.). European Journal of Agronomy. 2006;24: 291-295.
24. Goswami A, Banerjee R, Raha S. Drought resistance in rice seedlings conferred by seed priming: role of the anti-oxidant defense mechanisms. Protoplasma. 2013;250:1115-29.
25. Posmyk MM, Janas KM. Effects of seed hydropriming in presence of exogenous proline on chilling injury limitation in *Vigna radiata* L. seedlings. Acta Physiologiae Plantarum. 2007;29:509-517.
26. Patade VY, Khatri D, Manoj K, Kumari M, Ahmed Z. Cold tolerance in thiourea primed capsicum seedlings is associated with transcript regulation of stress responsive genes. Molecular Biology Reports. 2012;39:10603-10613.
27. Fercha A, Capriotti AL, Caruso G, Cavaliere C, Gherroucha H, Samperi R, Stampachiacchierea S, Laganaa A. Gel-free proteomics reveal potential biomarkers

of priming-induced salt tolerance in durum wheat. Journal of Proteomics. 2013;91:486-499.

28. Rowse HR. Methods of priming seeds. 1992. United States Patent No. 5,119,589.
29. Rowse HR. Methods of priming seeds. 1991. UK Patent No. 2192781.
30. Warren JE, Bennett MA. Seed hydration using the drum priming system. HortScience. 1997;32:1220-1221.
31. Da Silva CB, Marcos-Filho J, Jourdan P, Bennett MA. Performance of bell pepper seeds in response to drum priming with addition of 24-Epibrassinolide. HortScience. 2015;50:873-878.
32. Harris D, Joshi A, Khan PA, Gothkar P, Sodhi, PS. On-farm seed priming in semi-arid agriculture: development and evaluation in maize, rice and chickpea in India using participatory methods. Experimental Agriculture. 1999;35:15-29.
33. Harris D, Raghuwanshi BS, Gangwar JS, Singh SC, Joshi KD, Rashid I, Hollington PI. Participatory evaluation by farmers for on-farm seed priming in wheat in India, Nepal and Pakistan. Experimental Agriculture. 2001;37:403-415.
34. Harris D, Breese WA, Kumar Rao JVDK. The improvement of crop yield in marginal environments using 'on-farm' seed priming: nodulation, nitrogen fixation and disease resistance. Australian Journal of Agricultural Research. 2005;56:1211-1218.
35. Di Girolamo G, Barbanti L. Treatment conditions and biochemical processes influencing seed priming effectiveness. Italian Journal of Agronomy. 2012;7:8-18.
36. Kubala S, Garnczarska M, Wojtyla Ł, Clippe A, Kosmala A, Żmieńko A, Lutts S, Quinet M. Deciphering priming-induced improvement of rapeseed (*Brassica napus* L.) germination through an integrated transcriptomic and proteomic approach. Plant Science. 2015;231:94-113.
37. Yacoubi R, Job C, Belghazi M, Chaibi W, Job D. Proteomic analysis of the enhancement of seed vigour in osmoprimed alfalfa seeds germinated under salinity stress. Seed Science Research. 2013;23:99-110.
38. Salah SM, Yajing G, Dongdong C, Jie L, Aamir N, Qijuan H, Weimin H, Mingyu N, Jin H. Seed priming with polyethylene glycol regulating the physiological and molecular mechanism in rice (*Oryza sativa* L.) under nano-ZnO stress. Scientific Reports. 2015;5:14278.
39. Mereddy R, Wu L, Hallgren SW, Conway KE. Solid matrix priming improves vigor of okra seeds. Proceedings of the Oklahoma Academy of Science. 2000;80:1-5.
40. Singh PK., Pandita VK, Tomar BS, Seth R. Standardization of priming treatments for enhancement of seed germination and field emergence in carrot. Indian Journal of Horticulture. 2015;72:306- 309.
41. Mercado MFO, Fernandez PG. Solid matrix priming of soybean seeds. Philippine Journal of Crop Science. 2002;2:27-35.
42. Kępczyńska E, Piękna-Grochała J, Kępczyński J. Effects of matriconditioning on onion seed germination, seedlings emergence and associated physical and metabolic events. Plant Growth Regulation. 2003;41:269-278.
43. Zhang CF, Hu J, Lou J, Zhang Y, Hu WM. Sand priming in relation to physiological changes in seed germination and seedling growth of waxy maize under high-salt stress. Seed Science and Technology. 2007;35:733-738.
44. Andreoli C, Andrade RV. Integrating matriconditioning with chemical and biological seed treatments to improve vegetable crop stand establishment and yield under tropical conditions. Seed Technology. 2002;24:89-99.
45. Ilyas S, Sutariati GA, Suwarno FC, Sudarsono S. Matriconditioning improved quality and protein level of medium vigor hot pepper seed. Seed Technology. 2002;24:67-77.
46. Pandita VK, Anand A, Nagarajan S, Seth R, Sinha SN. Solid matrix priming improves seed emergence and crop performance in okra. Seed Science and Technology. 2010;38:665-674.
47. Galhaut L, Lespinay A, Walker DJ, Bernal MP, Correal E, Lutts S. Seed priming of *Trifolium repens* L. improved germination and early seedling growth on heavy metal-contaminated soil. Water Air Soil Pollution. 2014;225:1-15.

48. Iqbal M, Ashraf, M. Gibberellic acid mediated induction of salt tolerance in wheat plants: growth, ionic partitioning, photosynthesis, yield and hormonal homeostasis. Environmental and Experimental Botany. 2013;86:76-85.
49. Iqbal M, Ashraf M, Jamil A. Seed enhancement with cytokinins: changes in growth and grain yield in salt stressed wheat plants. Plant Growth Regulation. 2006;50:29-39.
50. Zheng M, Tao Y, Hussain S, Jiang Q, Peng S, Huang J, Cui K, Nie L. Seed priming in dry direct-seeded rice: consequences for emergence, seedling growth and associated metabolic events under drought stress. Plant Growth Regulation. 2016;78:167-178.
51. Farooq M, Wahid A, Lee DJ. Exogenously applied polyamines increase drought tolerance of rice by improving leaf water status, photosynthesis and membrane properties. Acta Physiologia Plantarum. 2009;31:937-945.
52. Afzal I, Basara SMA, Farooq M, Nawaz A. Alleviation of salinity stress in spring wheat by hormonal priming with ABA, salicylic acid and ascorbic acid. International Journal of Agriculture and Biology. 2006;8:23-28.
53. Callan NW, Marthre DE, Miller JB. Bio-priming seed treatment for biological control of *Pythium ultimum* preemergence damping-off in sh-2 sweet corn. Plant Disease. 1990;74: 368-372.
54. Reddy PP. Bio-priming of seeds. In: Reddy PP, editor. Recent Advances in Crop Protection. India, Springer; 2013. p 83-90. DOI: 10.1007/978-81-322-0723-8_6
55. Callan NW, Mathre DE, Miller J.B, Vavrina CS. Biological seed treatments: factors involved in efficacy. Horticultural Science.1997;32:179-183.
56. Müller H, Berg G. Impact of formulation procedures on the effect of the biocontrol agent *Serratia plymuthica* HRO-C48 on Verticillium wilt in oilseed rape. BioControl. 2008;53: 305-316. DOI 10.1007/s10526-007-9111-3
57. Gulick BR. Plant Growth-Promoting Bacteria: Mechanisms and Applications. Hindawi Publishing Corporation, Scientifica. 2012;1-15.
58. Timmusk S, Abd El-Daim IA, Copolovici L, Tanilas T, Kännaste A, Behers L, Nevo E, Seisenbaeva G, Stenström E, Niinemets Ü. Drought-tolerance of wheat improved by rhizosphere bacteria from harsh environments: enhanced biomass production and reduced emissions of stress volatiles. PLoS One. 2014;9(5):e96086.
59. Raj NS, Shetty NP, Shetty HS. Seed bio-priming with *Pseudomonas fluorescens* isolates enhances growth of pearl millet plants and induces resistance against downy mildew. International Journal of Pest Management. 2004;50:41-48.
60. Kaymak HC, Guvenc I, Yarali F, Donmez MF. The effects of bio-priming with PGPR on germination of radish (*Raphanus sativus* L.) seeds under saline conditions. Turkish Journal of Agriculture and Forestry. 2009;33:173-179.
61. Khaliq A, Aslam F, Matloob A, Hussain S, Geng M, Wahid A, Rehman H. Seed priming with selenium: consequences for emergence, seedling growth, and biochemical attributes of rice. Biological Trace Element Research. 2015;166:236-244.
62. Guan YJ, Hu J, Wang XJ, Shao CX. Seed priming with chitosan improves maize germination and seedling growth in relation to physiological changes under low temperature stress. Journal of Zhejiang University Science B. 2009;10:427-443.
63. Duan P, Ding F, Wang F, Wang BS. Priming of seeds with nitric oxide donor sodium nitroprusside (SNP) alleviates the inhibition on wheat seed germination by salt stress. Journal of Plant Physiology and Molecular Biology. 2007;33:244-250.
64. Anosheh HP, Sadeghi H, Emam Y. Chemical priming with urea and KNO3 enhances maize hybrids (*Zea mays* L.) seed viability under abiotic stress. Journal of Crop Science and Biotechnology. 2011;14:289-295.
65. Jisha KC, Puthur JT. Seed priming with BABA (β-amino butyric acid): a cost-effective method of abiotic stress tolerance in *Vigna radiata* (L.) Wilczek. Protoplasma. 2015.
66. Farooq M. Irfan M, Aziz Y, Ahmad I, Cheema SA. Seed priming with ascorbic acid improves drought resistance of wheat. Journal of Agronomy and Crop Science. 2013;199:12-22.
67. Fercha A, Capriotti AL, Caruso G, Cavaliere C, Samperi R, Stampachiacchiere S, Laganà A. Comparative analysis of metabolic proteome variation in ascorbate-

primed and unprimed wheat seeds during germination under salt stress. Journal of Proteomics. 2014;108:238-25.

68. Farooq M, Wahid A, Siddique KHM. Micronutrients application through seed treatments – a review. Journal of Soil Science and Plant Nutrition. 2012;12:125-142.
69. Arif M, Waqas M, Nawab K, Shahid M. Effect of seed priming in Zn solutions on chickpea and wheat. African Crop Science Conference Proceedings. 2007;8:237-240.
70. Abbas SQ, Hassan MU, Hussain B, Rasool T, Ali Q. Optimization of zinc seed priming treatments for improving the germination and early seedling growth of *Oryza sativa*. Advancements in Life Sciences. 2014;2:31-37.
71. Imran M, Mahmood A, Romheld V, Neumann G. Nutrient seed priming improves seedling development of maize exposed to low root zone temperatures during early growth. European Journal of Agronomy. 2013;49:141-148.
72. Shaheen HL, Iqbala M, Azeema M, Shahbazb M, Shehzadia M. K-priming positively modulates growth and nutrient status of salt-stressed cotton (*Gossypium hirsutum*) seedlings. Archives of Agronomy and Soil Science. 2015.
73. Jalali AH, Selehi F. Sugar beet yield as affected by seed priming and weed control. Archives of Agronomy and Soil Science. 2013;59:281-288.
74. Rashid A, Harris D, Hollington P, Ali S. On-farm seed priming reduces yield losses of mungbean (*Vigna radiata*) associated with mungbean yellow mosaic virus in the North West Frontier Province of Pakistan. Crop Protection. 2004;23:1119-1124.
75. El-Araby MM, Moustafa SMA, Ismail AI, Hegazi AZA. Hormones and phenol levels during germination and osmopriming of tomato seeds, and associated variation in protein patterns and anatomical seed features. Acta Agronomica Hungarica. 2006;54:441-458.
76. SchwemberAR, Bradford KJ. A genetic locus and gene expression pattern associated with the priming effect on lettuce seed germination at elevated temperature. Plant Molecular Biology. 2010;73:105-118.
77. Sung Y, Cantliffe DJ, Nagata RT, Nascimiento WM. Structural changes in lettuce seed during germination at high temperature altered by genotype, seed maturation temperature and seed priming. Journal of the American Society of Horticultural Science. 2008;133:300-311.
78. Chen K, Arora R. Priming memory invokes seed stress-tolerance. Environmental and Experimental Botany. 2013;94:33-45.
79. Sun H, Lin L, Wang X, Wu S, Wang X. Ascorbate-glutathione cycle of mitochondria in osmoprimed soybean cotyledons in response to imbibitional chilling injury. Journal of Plant Physiology. 2011;168:226-232.
80. Nascimento WM, Huber DJ, Cantliffe DJ. Carrot seed germination and respiration at high temperature in response to seed maturity and priming. Seed Science and Technology. 2013;41:164-169.
81. Kanto U, Jutamanee K, Osotsapar Y, Chai-Arree W, Jattupornpong S. Promotive effect of priming with 5-aminolevulinic acid on seed germination capacity, seedling growth and antioxidant enzyme activity in rice subjected to accelerated ageing treatment. Plant Production Science. 2015;18:443-454.
82. Singh A, Gupta R, Pandey R. Rice seed priming with picomolar rutin enhances rhiospheric *Bacillus subtilis* CIM colonization and plant growth. PLoS One. 2016;11:e0146013.
83. Hameed A, Farooq T, Hameed A, Ibrahim M, Sheikh MA, Basra SMA. Wheat seed germination, antioxidant enzymes and biochemical enhancements by sodium nitroprusside priming. Agrochimica. 2015;59:93-107.
84. Pouramir-Dashtmian F, Khajel-Hosseini M, Esfahani M. Improving chilling tolerance of rice seedling by seed priming with salicylic acid. Archives of Agronomy and Soil Science. 2014;60:1291-1302.
85. Muhammad I, Kolla M, Volker R, Günter N. Impact of nutrient seed priming on germination, seedling development, nutritional status and grain yield of maize. Journal of Plant Nutrition. 2015;38:1803-1821.

86. Ahmed M, Qadeer U, Ahmed ZI, Hassan FU. Improvement of wheat (*Triticum aestivum*) drought tolerance by seed priming with silicon. Archives of Agronomy and Soil Science. 2016;62:299-315.
87. Taylor A, Seel WE. Do *Striga hermonthica*-induced changes in soil matric potential cause the reduction in stomatal conductance and growth of infected maize? New Phytologist. 1998;138:67-73.
88. Zelonka L, Veneranda S, Māra V. Effect and aftereffect of barley seed coating with hoshors on germination, photosynthetic pigments and grain yield. Acta Universitatis Ltviensis Biology. 2005;691:111-119.
89. Jamil M, Charnikhova T, Verstappen F, Ali Z, Wainwright H, Bouwmeester H. Effect of phosphate-based seed priming in strigolactone production in *Striga hermonthica* infection in cereals. Weed Research. 2014;54:207-313.
90. Pame AR, Kreye C, Johnson D, Heuer S, Becker M. Effects of genotype, seed P concentration and seed priming on seedling vigor of rice. Experimental Agriculture. 2015;51:370-381.
91. Atique-ur-Rehman M, Farooq M, Cheema ZA, Wahid A. Seed priming with boron improves growth and yield of fine grain aromatic rice. Plant Growth Regulation. 2012;68:189-201.
92. Liu Z, Ma Z, Guo X, Shao H, Cui Q, Song W. Changes of cytosolic Ca2+ fluorescent intensity and plasma membrane calcium channels of maize root tip cells under osmotic stress. Plant Physiology and Biochemistry. 2010;48:860-865.
93. Khan AA, Peck NH, Taylor AG, Samimy C. Osmoconditioning of beet seed to improve emergence and yield in cold soils. Agronomy Journal. 1983;75:788-794.
94. Musa AM, Harris D, Johansen C, Kumar J. Short duration chickpea to replace fallow after aman rice: the role of on-farm seed priming in the High Barind Tract of Bangladesh. Experimental Agriculture. 1981;37:509-521.
95. Murungu FS, Chiduza C, Nyamugafa P, Clarck LJ, Whalley WR, Finch-Savage WE. Effect of on-farm seed priming on consecutive daily sowing occasions on the emergence and growth of maize in semi-arid Zimbabwe. Field Crop Research. 2004;89:49-57.
96. Shah Z, Haq IU, Rehman A, Khan A, Afzal M. Soil amendments and seed priming influence nutrients uptake, soil properties, yield and yield-components of wheat (*Triticum aestivum* L.) in alkali soils. Soil Science and Plant Nutrition. 2013;59:262-270.
97. Binang WB, Shiyam JQ, Ntia JD. Effect of seed priming method on agronomic performances and cost effectiveness of rainfed, dry-seeded NERICA rice. Research Journal of Seed Science. 2012;5:136-143.
98. Priya P, Patil VC, Aravind Kumar BN. Effect of seed priming practices on growth, yield and economics of maize (*Zea mays* L.) based cropping systems under rainfed conditions in Northern Karnataka, India. Plant Archives 2013;1:207-213.
99. Patil BD, Manjare MR. Effect of seed priming on germination and bulb yield in onion (*Allium cepa* L.). Ecology, Environment and Conservation. 2013;19:23-246.
100. Sharma AD, Rathore SVS, Srinivasan K, Tyago RK. Comparison of various seed priming methods for seed germination, seedling vigour and fruit yield in okra (*Abelmoschus esculentus* L. Moench). Sciantia Horticulturae. 2014;165:75-81.
101. Janecko A, Dziurka M, Ostrowska A, Biesaga-Koscielniak J, Koscielniak J. Improving vitamin content and nutritional value of legume yield through water and hormonal seed priming. Legume Research. 2015;38:185-193.
102. Fallahi J, Rezvani-Moghaddam P, Nassiri Mahallati M, Behdani MA, Shajari MA, Amiri MB. 2013. Influence of seed nitrogen content and biofertilizer priming on wheat germination in salinity stress conditions. Archives of Agronomy and Soil Scence. 2013;59:791-891.
103. Jisha KC. Vijayakumari K, Puthur JT. Seed priming for abiotic stress tolerance: an overview. Acta Physiologia Plantarum. 2013;35:1381-1396.
104. Krol P, Igielski R, Pollmann S, Kepczynska E. Priming of seeds mith methyljasmonate induced resistance to hemibiotroph *Fusarium oxysporum* f. sp. *Lycopersici* in tomato

via 12-oxo-phytodienoic acid, salicylic acid, and flavonol accumulation. Journal of Plant Physiology. 2015;179:122-132.

105. Kalischuk ML, Johnson D, Kawchuk LM. Priming with a double-stranded DNA virus alters *Brassica rapa* seed architecture and facilitates a defense response. Gene. 2015;557:130-137.

106. Anup CP, Prasad M, Nataraj S, Mayuri NG, Manbali J, Hussain A, Kukkundoor RK. Proteomic analysis of elicitation of downy mildew disease resistance in pearl millet by seed priming with β-aminobutyric acid and *Pseudomonas fluorescens*. Journal of Proteomics. 2015;120:58-74.

107. Chen K, Fessehaie A, Arora R. Dehydrin metabolism is altered during seed osmopriming and subsequent germination under chilling and desiccation in *Spinacia oleracea* L. cv. Bloomsdale: possible role in stress tolerance. Plant Science. 2012;183:27-36.

108. Maia J, Dekkers BJW, Provart NJ, Ligterink W, Hilhorst HWM. The re-establishement of desiccation tolerance in germinated *Arabidopsis thaliana* seeds and its associated transcriptome. PLoS One. 2011;6:e29123.

109. Buitink J, Jegger JJ, Guisle I., Wu BL, Wuillème Lamirault G, Bars AL, Meur NL, Becker A, Küster H, Leprince O. Transcriptome profiling uncovers metabolic and regulatory processes occurring during the transition from desiccation sensitive to desiccation-tolerant stages in *Medicago truncatula* seeds. Plant Journal. 2006;47:735-750.

110. Boudet J, Buitnink J, Hoekstra FA, Rogniaux H, Larré C, Satour P, Leprince O. Comparative analysis of the heat stable proteome of radicles of *Medicago truncatula* seeds during germination identifies late embryogenesis abundant proteins associated with desiccation tolerance. Plant Physiology. 2006;140:1418-1436.

111. Bewley JD, Bradford KJ, Hilhorst HWM, Nonogaki H. Seeds. Physiology of Development, Germination and Dormancy. 3rd ed. New York, Heidelberg, Dordrecht, London: Springer; 2013. 392 p.

112. Maurel C. Plant aquaporins: novel functions and regulation properties. FEBS Letters. 2007;581:2227-22236.

113. Gomes D, Agasse, A., Thiebaud P, Delrot S, Geros H, Chaumont F. Aquaporins are multifunctional water and solute transporters highly divergent in living organisms. Biochimica Biophysica Acta. 2009;1788:1213-1228.

114. Alleva K, Chara O, Amodeo G. Aquaporins: another piece in the osmotic puzzle. FEBS Letters. 2012;586:2991-2999.

115. Vander Willigen VC, Postaire O, Tournaire-Roux C, Boursiac Y, Maurel C. Expression and inhibition of aquaporins in germinating *Arabidopsis* seeds. Plant and Cell Physiology. 2006;47:1241-1250.

116. Liu HY, Yu X, Cui DY, Sun MH, Sun WN, Tang ZC, Su WA. The role of water channel proteins and nitric oxide signaling in rice seed germination. Cell Research. 2007;17:638-649.

117. Ge FW, Tao P, Zhang Y, Wang JB. Characterization of *AQP* gene expressions in *Brassica napus* during seed germination and in response to abiotic stresses. Biologia Plantarum. 2014;58:274-282.

118. Gao YP, Young L, Bonham-Smith P, Gusta LV. Characterization and expression of plasma and tonoplast membrane aquaporins in primed seed of *Brassica napus* during germination under stress conditions. Plant Molecular Biology. 1999;40:635-644.

119. Chen K, Fessehaie A, Arora R. Aquaporin expression during seed osmopriming and post-priming germination in spinach. Biologia Plantarum. 2013;57:193-198.

120. Kubala S, Wojtyla Ł, Quinet M, Lechowska K, Lutts S, Garnczarska M. Enhanced expression of the proline synthesis gene *P5CSA* in relation to seed osmopriming improvement of *Brassica napus* germination under salinity stress. Journal of Plant Physiology. 2015;183:1-12.

121. Schmid, M, Davison TS, Henz SR, Pape UJ, Demar M, Vingron M, Schölkopf B, Weigel D, Lohmann JU. A gene expression map of *Arabidopsis thaliana* development. Nature Genetics. 2005;37:501-506.

122. Gattolin S, Sorieul M and Frigerio L. Mapping of tonoplast intrinsic proteins in maturing and germinating *Arabidopsis* seeds reveals dual localization of embryonic TIPs to the tonoplast and plasma membrane. Molecular Plant. 2011;4:180-189. DOI: 10.1093/mp/ssq051

123. Krishnan P, Joshi DK, Nagarajan S, Moharir AV. Characterisation of germinating and non-viable soybean seeds by nuclear magnetic resonance (NMR) spectroscopy. Seed Science Research. 2004;14:355-362.

124. Manz B, Müller K, Kucera B, Volke F, Leubner-Metzger G. Water uptake and distribution in germinating tobacco seeds investigated in vivo by nuclear magnetic resonance imaging. Plant Physiology. 2005;138:1538-1551.

125. Garnczarska M, Tomasz Zalewski T, Kempka M. Water uptake and distribution in germinating lupine seeds studied by magnetic resonance imaging and NMR spectroscopy. Journal of Experimental Botany. 2007;58:3961-3969.

126. Falourd X, Natali F, Peters J, Foucat L. Molecular mobility in *Medicago truncatula* seed during early stage of germination: neutron scattering and NMR investigations. Chemical Physics. 2013;428:181-185.

127. Terskikh VV, Feurtado JA, Ren C, Abrams SR, Kermode AR. Water uptake and oil distribution during imbibition of seeds of western white pine (Pinus *monticola* Dougl. ex D. Don) monitored *in vivo* using magnetic resonance imaging. Planta. 2005;221:17-27.

128. Wojtyla Ł, Garnczarska M, Zalewski T, Bednarski W, Ratajczak L, Jurga S. A comparative study of water distribution, free radical production and activation of antioxidative metabolism in germinating pea seeds. Journal of Plant Physiology. 2006;163:1207-1220.

129. Nagarajan S, Pandita VK, Joshi DK, Sinha JP, Modi BS. Characterization of water status in primed seeds of tomato (*Lycopersicon esculentum* Mill.) by sorption properties and NMR relaxation times. Seed Science Research. 2005;15:99-111.

130. Pietrzak LN, Frégeau-Reid J, Chatson B, Blackwell B. Observation on water distribution in soybean seed during hydration processes using nuclear magnetic resonance imaging. Canadian Journal of Plant Science. 2002;82:513-519.

131. Liu Y, Van der Burg WJ, Aartse JW, Van Zwol RA, Jalink H, Bino RJ. X-ray studies on changes in embryo and endosperm morphology during priming and imbibition of tomato seeds. Seed Science Research. 1993;3:171-178.

132. Ashraf M, Foolad MR. Pre-sowing seed treatment: a shotgun approach to improve germination growth and crop yield under saline and non-saline conditions. Advances in Agronomy. 2005;88:223-71.

133. Farooq M, Basra SMA, Wahid A. Priming of field-sown rice seed enhances germination, seedling establishment, allometry and yield. Plant Growth Regulation. 2006;49:285-294.

134. Li X, Zhang L. SA and PEG-induced priming for water stress tolerance in rice seedling. In: Zhu E, Sambath S, editors. Information Technology and Agricultural Engineering. Berlin, Heidelberg: Springer; 2012. 134. p 881-887.

135. Battaglia M, Covarrubias AA. Late embryogenesis abundant (LEA) proteins in legumes. Frontiers in Plant Science. 2013;4:190.

136. Gallardo K, Job C, Groot SPC, Puype M, Demol H, Vandekerckhove J, Job D. Proteomic analysis of Arabidopsis seed germination and priming. Plant Physiology. 2001;126:835-848.

137. Li F, Wu X, Tsang E, Cutler AJ. Transcriptional profiling of imbibed *Brassica napus* seed. Genomics. 2005;86:718-30.

138. Soeda Y, Konings MCJM, Vorst O, van Houwelingen AMM, Stoopen GM, Maliepaard CA, Kodde J, Bino RJ, Groot SPC, van der Geest AHM. Gene expression programs during *Brassica oleracea* seed maturation, osmopriming, and germination are

indicators of progression of the germination process and the stress tolerance level. Plant Physiology. 2005;137:354-368.

139. Oluoch MO, Welbaum GE. Viability and vigor of osmotically primed muskmelon seeds after nine years of storage. Journal of the American Society for Horticultural Science. 1996;121:408-413.

140. Toorop PE, Van AC, Hilhorst HWM. 1998. Endosperm cap weakening and endo-β-mannanase activity during priming of tomato (*Lycopersicon esculentum* cv. Moneymaker) seeds are initiated upon crossing a threshold water potential. Seed Science Research. 1998;8:483-491.

141. Nawaz J, M, Abdul Jabbar A, Nadeem GA, Sajid M, Subtain MUI, Shabbir I. Seed priming a technique. International Journal of Agriculture and Crop Sciences. 2013;6:1373-1381

142. de Castro RD, Zheng X, Bergervoet JHW, de Vos CHR, Bino RJ. β-tubulin accumulation and DNA replication in imbibing tomato seeds. Plant Physiology.1995;109:499-504.

143. Dawidowicz-Grzegorzewska A. Ultrastructure of solid matrix-primed endospermic and nonendospermic seeds. In: Ellis RH, Black M, Murdoch AJ, Hong TD, editors. Basic and Applied Aspects of Seed Biology. Dordrecht: Kluwer Academic Publishers; 1997. p. 479-487.

144. Dawidowicz-Grzegorzewska A. Ultrastructure of carrot seeds during matriconditioning with Micro-Cel E. Annals of Botany. 1997;79:535-545.

145. Bolte S, Lanquar V, Soler MN, Beebo A, Satiat-Jeunemaître B, Bouhidel K, Thomine S. Distinct lytic vacuolar compartments are embedded inside the protein storage vacuole of dry and germinating *Arabidopsis thaliana* seeds. Plant and Cell Physiology. 2011;52:1142-1152.

146. Zheng H, Staehelin A. Protein storage vacuoles are transformed into lytic vacuoles in root meristematic cells of germinating seedlings by multiple, cell type-specific mechanisms. Plant Physiology. 2011;155:2023-2035. DOI: 10.1104/pp.110.170159

147. Bino RJ, de Vries JN, Kraak L, van Pijlen JG. Flow cytometric determination of nuclear replication stages in tomato seeds during priming and germination. Annual Botany. 1992;69:231-236.

148. Vázquez-Ramosa JM. de la Paz Sáncheza M. The cell cycle and seed germination. Seed Science Research. 2003;13:113-130.

149. Sanchez MDL, Gurusinghe SH, Bradford KJ, Vazquez-Ramos JM. Differential response of PCNA and Cdk-A proteins and associated kinase activities to benzyladenine and abscisic acid during maize seed germination. Journal of Experimental Botany. 2005;56:515-523.

150. Ventura L, Donà M, Macovei A, Carbonera D, Buttafava A, Mondoni A, Rossi G, Balestrazzi A. Understanding the molecular pathways associated with seed vigor. Plant Physiology and Biochemistry. 2012;60:196-206.

151. Bailly C, El-Maarouf-Bouteau H, Corbineau F. From intracellular signaling networks to cell death: the dual role of reactive oxygen species in seed physiology. Comptes Rendus Biologies. 2008;331:806-814. DOI: 10.1016/j.crvi.2008.07.022

152. Osborne DJ. Biochemical control-systems operating in the early hours of germination. Canadian Journal of Botany. 1983;61:3568-3577.

153. Ashraf M, Bray CM. DNA synthesis in osmoprimed leek (*Allium porrum* L.) seeds and evidence for repair and replication. Seed Science Research. 1993;3:15-23.

154. Law SR, Narsai R, Taylor NL, Delannoy E, Carrie C, Giraud E, Millar AH, Small I, Whelan J. Nucleotide and RNA metabolism prime translational initiation in the earliest events of mitochondrial biogenesis during *Arabidopsis* germination. Plant Physiology. 2012;158:1610-1627.

155. Law SR, Narsai R, Whelan J. Mitochondrial biogenesis in plants during seed germination. Mitochondrion. 2014;19:214-221.

156. Macovei A, Balestrazzi A, Confalonieri M, Carbonera D. The Tdp1 (tyrosyl-DNA phosphodiesterase) gene family in barrel medic (*Medicago truncatula* Gaertn.): bioinformatic investigation and expression profiles in response to copper- and PEG-mediated stress. Planta. 2010;232:393-407.

157. Macovei A, Balestrazzi A, Confalonieri M, Faé M, Carbonera D. New insights on the barrel medic MtOGG1 and MtFPG functions in relation to oxidative stress response in planta and during seed imbibition. Plant Physiology and Biochemistry. 2011;49:1040-1050.

158. Macovei A, Balestrazzi A, Confalonieri M, Buttafava A, Carbonera D. The TFIIS and TFIIS like genes from *Medicago truncatula* are involved in oxidative stress response. Gene. 2011;470:20-30.

159. Waterworth WM, Masnavi G, Bhardwaj RM, Jiang Q, Bray CM, West CE. A plant DNA ligase is an important determinant of seed longevity. Plant Journal. 2010;63:848-886. DOI: 10.1111/j.1365-313X.2010.04285.x

160. Sharma SN, Maheshwari A. Expression patterns of DNA repair genes associated with priming small and large chickpea (*Cicer arietinum*) seeds. Seed Science and Technology. 2015;43:250-261.

161. Kibinza S, Bazin J, Bailly C, Farrant JM, Corbineau F, El-Maarouf-Bouteau H. Catalase is a key enzyme in seed recovery from ageing during priming. Plant Science. 2011;181:309-315.

162. Kubala S, Wojtyla Ł, Garnczarska M. Seed priming improves salt stress tolerance during germination by modulation of antioxidative capacity. In: Proceedings of the 11th International POG Conference Reactive Oxygen and Nitrogen Species in Plants, Warszawa, 17–19 June 2013, BioTechnologia. 94:223.

163. Wojtyla Ł, Lechowska K, Kubala S, Garnczarska M. Different modes of hydrogen peroxide action during seed germination. Frontiers Plant Science. 2016;7:66.

164. Krasuska U, Gniazdowska A. Nitric oxide and hydrogen cyanide as regulating factors of enzymatic antioxidant system in germinating apple embryos. Acta Physiologiae Plantarum. 2012;34:683-692.

165. Nouman W, Basra SMA, Yasmeen A, Gull T, Hussain SB, Zubair M, Gul R. Seed priming improves the emergence potential, growth and antioxidant system of *Moringa oleifera* under saline conditions. Plant Growth Regulation. 2014;73:267-278.

166. Xu S, Hu J, Li Y, Ma W, Zheng Y, Zhu S. Chilling tolerance in *Nicotiana tabacum* induced by seed priming with putrescine. Plant Growth Regulation. 2011;63:279-290.

167. Islam F, Yasmeen T, Ali S, Ali B, Farooq MA, Gill RA. Priming-induced antioxidative responses in two wheat cultivars under saline stress. Acta Physiologiae Plantarum. 2015;37:153.

168. Farooq M, Basra SMA, Khalid M, Tabassum R, Mahmood T. Nutrient homeostasis, metabolism of reserves, and seedling vigor as affected by seed priming in coarse rice. Canadian Journal of Botany. 2006;84:1196-1202.

169. Sung FJM, Chang YH. Biochemical activities associated with priming of sweetcorn seed to improve vigor. Seed Science and Technology. 1993;21:97-105.

170. Kaur S, Gupta AK, Kaur N. Effect of osmo- and hydropriming of chickpea seeds on seedling growth and carbohydrate metabolism under water deficit stress. Plant Growth Regulation. 2002;37:17-22.

171. Smith PT, Cobb BG. Physiological and enzymatic characteristic of primed, re-dried, and germinated pepper seeds (*Capsicum annum* L). Seed Science and Technology. 1992;20:503-513.

172. Job C, Kersulec A, Ravasio L, Chareyre S, Pepin R, Job D. The solubilization of the basic subunit of sugarbeet seed 11-S globulin during priming and early germination. Seed Science Research. 1997;7:225-243.

173. Capron I, Corbineau F, Dacher F, Job C, Côme D, Job D. Sugarbeet seed priming: effects of priming on germination, solubilization of 11-S globulin and accumulation of LEA proteins. Seed Science Research. 2000;10:243-254.

174. Mazor L, Perl M, Negbi M. Changes in some ATP-dependent activities in seeds during treatment with polyethylene glycol and during the redrying process. Journal of Experimental Botany. 1984;35:1119-1127.

175. Corbineau F, Özbingol, N, Vineland D, Côme D. Improvement of tomato seed germination by osmopriming as related to energy metabolism. In: Black M, Bradford KJ, Vasquez-Ramos J, editors. Seed Biology Advances and Applications: Proceedings of the Sixth International Worshop on Seeds. Merida, Mexico, CABI, Cambridge; 2000. p. 467-474.
176. Haider S, Pal R. Integrated analysis of transcriptomic and proteomic data. Current Genomics. 2013;14:91-110.
177. Wong JW, Cagney G. An overview of label-free quantitation methods in proteomics by mass spectrometry. Methods Molecular Biology. 2010;604:273-283.
178. Pace R, Benincasa P, Ghanem ME, Quinet M, Lutts S. Germination of uintreated and primed seeds in rapeseed (*Brassica napus* var *oleifera* Del.) under salinity and low matric potential. Experimental Agriculture. 2012;48:238-251.
179. de Lespinay A, Lequeux H, Lambillotte B, Lutts S. Protein synthesis is differentially required for germination in *Poa pratensis* and *Trifolium repens* in the absence or in the presence of cadmium. Plant Growth Regulation. 2010;61:205-214.
180. de Lespinay A. Study of seed priming mechanisms of three plant species used in revegetation of industrial sites. PhD Thesis, Université catholique de Louvain, Belgium. 2009. 253 p.
181. Bonsager BC, Finnie C, Roepstorff P, Svensson B. Spatio-temporal changes in germination and radicle elongation of barley seeds tracked by proteome analysis of dissected embryo, aleurone layer, and endosperm tissues. Proteomics. 2007;7:4528-4540.
182. Catusse J, Meinhard J, Job C, Strub JM, Fischer U, Pestova E, Westohoff P, Van Dorselaer A, Job D. Proteomics reveals potential boiomarkers of seed vigor in sugarbeet. Proteomics. 2011;11:1569-1580.
183. Zhang H, Wang WQ, Liu SJ, Moller JM, Song SQ. Proteome analysis of poplar seed vigor. PLoS One. 2015;10:e0132509.
184. Rajjou L, Belghazi M, Huguet R, Robin C, Moreau A, Job C, Job D. Proteomic investigation of the effect of salicylic acid an Arabidopsis seed germination and establishment of early defense mechanisms. Plant Physiology. 2006;141:910-923.
185. Tanou G, Job C, Rajjou L, Arc E, Belghazi M, Diamentidis G, Molasiotis A, Job D. Proteomics reveals the overlapping roles of hydrogen peroxide and nitric oxide in the acclimation of citrus plants to salinity. Plant Journal. 2009;60:795-804.
186. Wu X, Li N, Li H, Tang H. An optimized method for NMR-based plant seed metabolomics analysis with maximized polar metabolite extraction efficiency, signal-to-noise ratio, and chemical shift consistency. Analyst. 2014;139:1769-1778.
187. Posmyk MM, Corbineau F, Vinel D, Bailly C, Côme D. Osmoconditioning reduces physiological and biochemical damage induced by chilling in soybean seeds. Physiologia Plantarum. 2001;111:473-482.
188. Tallowin JRB, Rook AJ, Brookman SKE. The effect of osmoting resowing treatment on laboratory germination in a range of wild flower species. Annals of Applied Biology. 1994;124:363-370.
189. Wright B, Rowse H, Whipps JM. Microbial population dynamics on seeds during drum and steeping priming. Plant and Soil. 2003;255:631-640.
190. Chiu KY, Chen CL, Sung JM. Effects of priming temperature on storability of primed sh-2 sweet corn seeds. Crop Science. 2002;42:1996-2003.
191. Hill HJ, Cunningham JD, Bradford KJ, Taylor AG. Primed lettuce seeds exhibit increased sensitivity to moisture content during controlled deterioration. HortScience. 2007;42:1436-1439.
192. Schwember AR, Bradford KJ. Drying rate following priming affect temperature sensitivity of germination and longevity of lettuce seeds. HortScience. 2005;40:778-781.
193. Hacisalihoglu G, Taylor AG, Paine DH, Hildebrand MB, Khan AA. Embryo elongation and germination rate as sensitive indicators of lettuce seed quality: priming and ageing studies. HortScience. 1999;34:1240-1243.

194. Butler LH, Hay FR, Ellis RH, Smith RD, Murray TB. Priming and redrying improve the survival of mature seeds of *Digitalis purpurea* during storage. Annals of Botany. 2009;103;1261-1270.

195. Hussain S, Zheng M, Khan F, Khaliq A, Fahad S, Peng S, Huang J, Cui K, Nie L. Benefits of rice seed priming are offset permanently by prolonged storage and the storage conditions. Scientific Reports. 2015;5:8101.

196. Buitink J, Hemminga MA, Hoekstra FA. Is there a role for oligosaccharides in seed longevity? An assessment of intercellular glass stability. Plant Physiology. 2000;122:1217-1224.

CHAPTER 3

Seed Size as Key Factor in Germination and Seedling Development of Copaifera langsdorffii (Fabaceae)

Matheus Lopes Souza[1,2], Marcílio Fagundes[1*]

[1]*Laboratório de Biologia da Conservação, Universidade Estadual de Montes Claros, Montes Claros, Brasil*

[2]*Laboratório de Fisiologia Vegetal, Universidade Federal de Minas Gerais, Belo Horizonte, Brasil*

ABSTRACT

Seed size is a plastic trait of the plants that directly affect seed germination and seedling recruitment. This study aimed to investigate the relationship between seed size, seed germinability and seedling performance of Copaifera langsdorffii by testing four hypotheses: 1) larger seeds have higher germination percentage; 2) smaller seeds require less time to germinate and for initial development of the seedlings; 3) larger seeds produce more vigorous seedlings and 4) seed size negatively affects seedling root/shoot ratio. In 2011, we selected 30 individuals of C. langsdorffii from which 300 seeds were randomly collected in the plant canopy. All these seeds were weighted and placed in germination tray using vermiculite as substrate. Seed germinability and initial development of seedlings were monitored daily until cotyledons fell. Small seeds have higher germination percentage and germinate faster when compared to large seeds. Nonetheless, seedlings originated from larger seeds have longer development times, resulting in more vigorous seedlings. In addition, seedlings originating from small seeds allocate proportionally greater amount of resources to roots when compared to larger seeds. The fact that small seeds have higher germination percentage and faster germination favors the colonization of transient habitats. However, larger seeds produce more vigorous seedlings, favoring the seedling establishment in more stable habitats. Thus, we argue that high variability in seed size of C. langsdorffii favors its widespread geographic distribution.

Keywords: Community Organization, Plant Distribution, Seed Biometry, Seed Reserves, Seedling Vigor

1. INTRODUCTION

Seed germination and seedling establishment are critical stages of the plant life cycles, influencing both distribution and abundance of plant species in communities [1] [2] . In fact, geographic distribution and abundance of some plant species may be associated with variations in seed germinability and seedling recruitment between different habitats due to environmental limitations, such as altitude, topography, soil quality and climate [3] -[5] .

Several ecological and evolutionary factors can affect the process of seed germination and further plant establishment [6] . Seed size has been considered an important evolutionary trait that affects the reproductive outcome of many plant species [7] . Seed size directly influences the germination time [8] , germination percentage [9] and seedling vigor [10] , which can indirectly determine plant distribution and abundance across different habitats [11] . However, seed size produced by plants varies between and within plant species, sometimes by several orders of magnitude [12] -[14] .

Smaller seeds generally germinate faster providing greater competitive advantage especially in early successional stages [15] . Nevertheless, larger seeds, although germinating slowly, often have higher percentage of germination than small seeds, being favored in predictable habitats [16] -[18] . Moreover, asynchronous and heterogeneous seed germination observed in several plants species can be associated with intra-specific variation in seed size and allow the species to colonize different habitats and expand their geographic distribution limit [5] [6] [19] .

The weight and size of the seeds are directly related to the amount of nutritional reserves that will be allocated for the initial seedling growth [20] . Large seeds tend to produce more vigorous seedlings when compared to small seeds [10] . Greater amount of stored reserves allow a higher probability of seedling establishment at sites with lower resource availability [17] . However, plants respond to their environmental in such way as to optimize their resource use. Thus, according the resource optimization hypothesis, plants allocate relatively less resource to their root system when nutrient availability increases [21] [22] . In this scenario, it is expected that seedlings originated from larger seeds would present smaller root: shoot ratio, as large seeds have more nutritional reserves.

Variation in seed size is an important area of plant ecology because seed size can directly affect the processes of germination and seedling recruitment, influencing the plant performance under different environmental [23] [24] . In Brazil, Copaifera langsdorffii (Fabaceae) presents wide geographic distribution, occurring at different biomes (Brazilian Savannah, Dry Forest, Amazon Rainforest and Atlantic Forest) [25] [26] . Hence, the objective of this study was to evaluate the relationship between seed size, germination and seedling performance of C. langsdorffii by testing three hypothesis: 1) smaller seeds require less time to germinate and for initial development of the seedlings; 2) larger seeds have higher germination percentage and produce more vigorous seedlings and 3) seedling root/shoot ratio is negatively affected by seed size.

2. MATERIAL AND METHODS

2.1. Studied Plant

Copaifera langsdorffii (Fabaceae) is a heliophytic, deciduous tropical tree, growing up to 10 to 15 m height. The species have a wide distribution, occurring from northern Argentina to southern Bolivia, throughout the Brazilian Cerrado [25] . C. langsdorffii presents supra-annual fruiting, alternating years of high and low or no fruit production. The flowering occurs from November to January and fruits are mature by July to September, coinciding with the time of greatest leaves' deciduousness [26] . Upon opening, each fruit exposes a single ellipsoid seed, which is black and shiny and is partially covered by a yellow-orange aril. The seeds of C. langsdorffii have orthodox behavior and slow germination, which extends up to 70 days after sown [27] .

2.2. Seed Collection

Seeds used in this study were collected from 30 individuals of Copaifera langsdorffii located in an area of Cerrado stricto sensu (16°40'26"S, 43°48'44"W) in Montes Claros municipality, north of Minas Gerais State, Brazil. The region have semi-arid climate with well-defined dry and wet seasons. The average annual temperature is about 23°C and average rainfall is about 1000 mm/year. The study area presents dystrophic soils and developed herbaceous-subshrub strata often affected by fire [4] [28] .

During August 2011, 30 reproductive individuals of C. langsdorffii were selected at study area. Those trees were from five to seven meters high, with a well-formed crown and in a good phytosanitary state (e.g. without presence of lianas or parasitic plants). A total of 100 fruits were haphazardly collected throughout the canopy from each selected tree and taken to Laboratory of Conservation Biology of State University of Montes Claros (UNIMONTES). All collected fruits were manually benefited, eliminating malformed seeds and those with visual signals of attacks by predators or pathogens. Thus, a set of 300 seeds (ten seeds per plant) were selected and weighed with an analytical scale. Finally those seeds were categorized in two size classes: smaller seeds (seeds < 0.394 g, n = 150) or larger seeds (seeds > 0.397 g, n = 150).

2.3. Germination Experiment

After seeds were weighed, all 300 seeds were individually sown in germination tray formed by 300 cells (2 cm length, 2 cm width and 3 cm height) and vermiculite was used as substrate for seed germination. All seeds were subjected to disinfection in a 1% sodium hypochlorite solution for two minutes before sown. The germination experiment was conducted in a germination chamber with controlled photoperiod and temperature (12 h/light at 30°C e 12 h/dark at 25°C). The humidity of the substrate was maintained daily by adding three ml of distilled water in each germination cell. The seeds were monitored daily to determine the percentage and time for germination. Seed was considered germinated when they presented primary root protrusion [18] .

At the same time, a soaking test was conducted under same germination conditions using others 80 seeds (40 seed for each size class). At this case, dry mass of all seeds was determined and so, all those 80 seeds were immersed in distilled water and weighed after 2, 4, 8, 18, 24, 30 and 48 hours of water absorption. Relative increase in fresh weight (Wr) of seeds was calculated as $Wr = [(Wf - Wi)/Wi] \times 100$, where Wi is the initial seed weight and Wf the weight after each time of water absorption [29] . Thus, imbibitions curves for smaller and larger seeds were based on increase seed mass at different times of seed immersion in distilled water.

2.4. Seedling Development

The seedlings originated from seed germination experiment were used to evaluate the effects of seed size on time for seedling development and seedling vigor. The time for seedling development was defined as the time from seed germination (root protrusion) until seedling cotyledon fell. After cotyledon falling, seedlings were removed from substrate to determinate their length and dry mass of shoot and root system. The root and shoot lengths were determined with a digital caliper. In order to get dry biomass of root and shoot all seedlings were individually placed in paper bags, identified and transferred to an oven with air circulation at 60°C for 48 hours. After this time the dry weight of shoot and root system was determined on an analytical scale. Finally, the ratio of root: shoot biomass (RSR) of all seedlings was calculated as $RSR = (W_{root}/W_{shoot})$, where W_{root} is dry mass of the root and W_{shoot} is dry mass of shoots.

2.5. Data Analysis

Survival analysis was performed in order to test the effect of seed size on percentage and time for germination. Thus, germination percentage within each size class was used as response variables while the germination time was the explanatory variable. Survival analysis evaluates the likelihood of germination at a certain time avoiding the temporal pseudo-replication inherent to the data. The effect of seed size on seed water absorption was tested by constructing Generalized Linear Mixed Models (GLMM), since these data also showed temporal pseudo-replication. Thus, increased seed mass (Wr) at different times (2, 4, 8, 18, 24, 30 and 48 hours of water absorption) was used as the response variable and seed size classes as explanatory variable.

The effect of seed size on seedling development time, seedling vigor and resource allocation between root: shoot (RSR) was tested by constructing GLMs. Thus, to evaluate the effect of seed size on time of development of the seedlings, time required for the cotyledon to fall was the response variable and seed weight was the explanatory variable. To evaluate the effect of seed size on seedling vigor, length and dry weight of shoot and root system were the response variables and seed weight was the explanatory variable. To evaluate the differences in resource allocation, the root: shoot ratio (RSR) was used as the response variable and seed weight as explanatory variable. The hypotheses proposed in this study were tested on R software [30] and all of the models were constructed using an appropriate error distribution for each response variable [31] .

3. RESULTS

Seed germination percentage of Copaifera langsdorffii reached 72%. Seed germination started in the eighth day after sawn and continued until the 55th day. The percentage of seed germination varied between seed size classes (χ^2 = 8.9, P < 0.01) and small seeds have higher germination percentage (80%) compared with large seeds (64.4%). Time required for seed germination also varied between seed size classes (χ^2 = 11.02, P < 0.001) with small seeds germinating faster than large seeds. The average time taken to 50% of small seeds germinate was approximately 29 days while the time for the same proportion of larger seeds germinate was about 42 days (Figure 1). The soaking test also showed that water absorption varied between seed size class ($F_{x,y}$ = 22.56, P = 0.003). In fact, small seeds were approximately two times more permeable to water than large seeds (Figure 2).

Seed weight positively affected time for cotyledon fall and ratio of root: shoot biomass of the seedlings (Table 1). Thus, seedlings originated from larger seeds retained their cotyledons for longer time compared with seedlings originated from smaller seeds (Figure 3(a)) and small seeds allocate proportionally more resources to root system compared to large seeds (Figure 3(b)). Furthermore, seed size positively affected dry weight of root and shoot, and shoot length of C. langsdorffii seedlings (Table 1, Figures 4(a)-(d)).

4. DISCUSSION

Seed germination percentage of Copaifera langsdorffii was affected by seed size. However, the result was different than expected because the percentage of germination was 15.6% higher in small seeds. In addition, small seeds require less time for germination than large seeds, corroborating the general patterns (see [8]). At this point it is important to emphasize that increase in time for germination can cause seed viability loss due to deterioration and microorganisms attack [32] -[34] . Thus, this fact can be used to explain the variation in percentage of seed germination observed in this study.

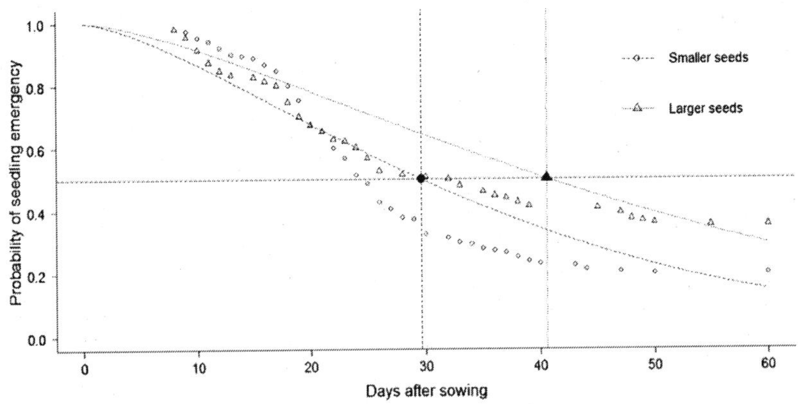

Figure 1. Germinability of Copaifera langsdorffii seeds between different size classes. Vertical lines indicate the time required for germination of 50% of smaller and larger seeds.

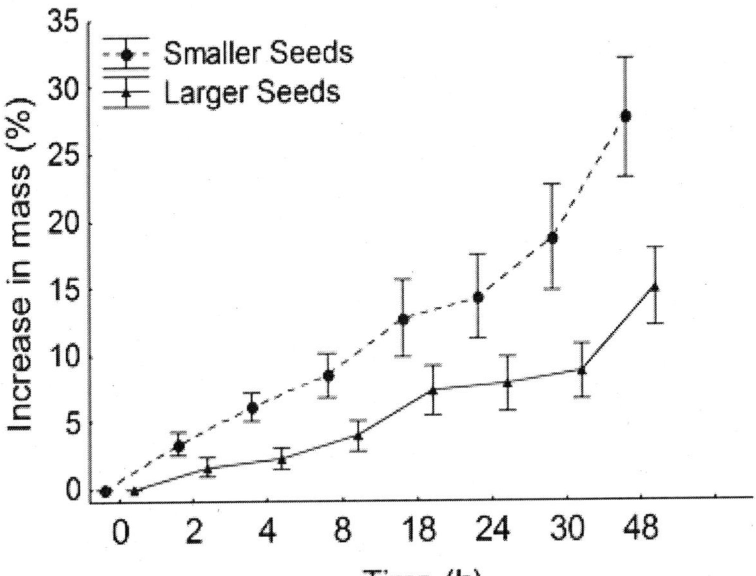

Figure 2. Imbibition curves for smaller and larger seeds of Copaifera langsdorffii.

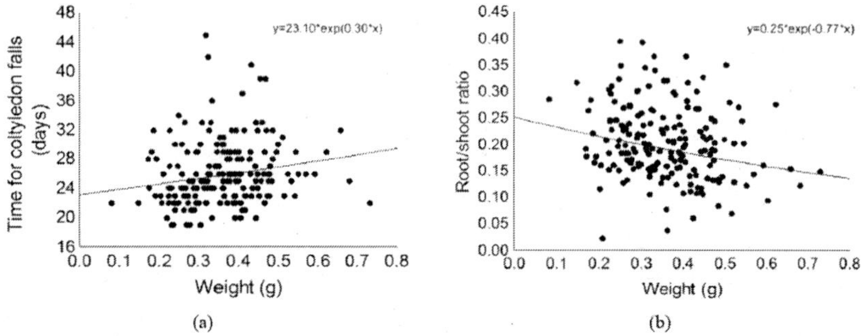

Figure 3. Relationship between seed weight and time for cotyledons fall (a) and ratio of root: shoot biomass (b) of Copaifera langsdorffii seedling.

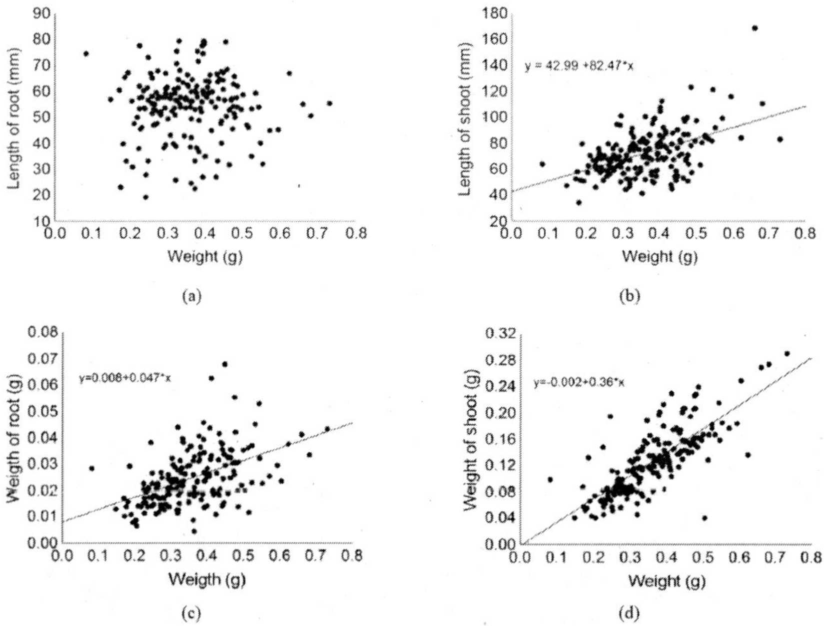

Figure 4. Relationship between seed weight and root length (a), shoot length (b), root dry weigh (c) and shoot dry weigh (d) of Copaifera langsdorffii seedling.

Small seeds of C. langsdorffii are more water permeable and germinate faster than large seeds. In general, seed size is directly related with seed coat thickness and inversely related with water absorption [35] . The increase in seed size also implies decrease in the surface: volume ratio, resulting in lower relative ability to absorb water and initiate the process of germination [36] . Thereby, smaller seeds have thinner coats and higher relative surface. This seed trait ensures greater permeability in small seeds and, consequently, less time for germination [37] . The fact that smaller seeds have thinner coat and high surface: volume ratio can justify effectively imbibition and variation in germination time between seed size classes of C. langsdorffii.

Table 1. Deviance analysis of the complete models evaluating the effect of seed weight on initial development and resource allocation of Copaifera langsdorffii seedlings.

Explanatory variable	Response variable	Deviance	Residual Df.	Residual deviance	X^2	P
Seed weight	Time for cotyledons fall	5.0181	180	116.32	5.0181	<0.01
	Length of shoot (mm)	13659	180	43381	56.676	<0.001
	Length of root (mm)	5.6645	180	26246	0.0386	0.8444
	Weight of shoot (g)	0.0048	180	0.015	58.52	<0.001
	Weight of root (g)	0.2773	180	0.153	325.79	<0.001
	root: shoot ratio	0.0184	180	0.768	0.0184	<0.05

After germination, C. langsdorffii seedlings originated from larger seeds retained their cotyledons for longer time. Large seeds have higher amount of reserves in theirs cotyledons and need longer time to incorporate theses nutrients in seedling tissues [20] . Thus, seedlings originated from larger seeds have both more time for development and nutrients available for growth. Thus, the result observed in this study agrees with the general prediction that larger seeds produce seedlings with larger initial size as reported by others authors [4] [10] [23] [38] [39] . In this scenario it would be reasonable to think that larger seeds/seedlings have higher competitive ability relative to small seeds. However, several other forces act synergistically to shape seed size. For example, nutritionally stressed habitat acts in favor of larger seeds/seedlings because larger seedlings (originated from larger seeds) have more nutritional reserves [2] [19] [40] [41] . On the other hand, larger seeds are preferentially attacked by predators, favoring smaller seeds survivorship [42] .

Another important aspect of the relation between seed and seedling size is related with resource allocation for root or shoot. In general, seedlings originated from smaller seeds must allocate a greater proportional amount of resources to roots development, resulting in higher root: shoot ratio [22] as observed in this study. The high investment in root tissues promotes greater development of root systems that reach deeper levels of substrate with more water and nutrients [43] . Thus, this resource reallocation between root and shoot could increase the survivorship chance of seedlings with poor cotyledons reserves.

Within a given population, plants should develop seeds of similar size because there is often a single best range of seed size that optimizes the local fitness [44] . However, seed size within species is one of the more plastic components of plant life history [45] . It is plausible that the variation in seed size within species is associated with differential seed performance among habitats [46] . For example, smaller seeds are favored in transient habitats, because they present greater percentage of germination and germinate more quickly. In contrast, larger seeds have more reserves, producing vigorous seedlings that present greater competitive ability in predictable habitats. The C. langsdorffii seeds showed broad variation in biomass. This variation in seed size allows the species to colonize different habitats, helping to justify C. langsdorffii widespread geographic distribution.

ACKNOWLEDGEMENTS

Matheus Lopes Souza, Marcílio Fagundes (FAPEMIG APQ 01926.11) and Programa de Pós-Graduação Stricto Sensu em Ciências Biológicas (PPGCB) da UNIMONTES. The authors also acknowledge CAPES and FAPEMIG by research grants.

REFERENCES

1. Wulff, R.D. (1986) Physiological Performance Seed Size Variation in Desmodium paniculatum. Journal of Ecology, 74, 99-114.
2. Armstrong, D.P. and Westoby, M. (1993) Seedlings from Large Seeds Tolerated Defoliation Better: A Test Using Phylogeneticaly Independent Contrasts. Ecology, 74, 1092-1100.
3. Burmeier, S. and Jensen, K. (2008) Is the Endangered Apium repens (Jacq.) Lag. Rare Because of a Narrow Regeneration Niche? Plant Species Biology, 23, 111-118.
4. Fagundes, M., Camargos, M.G. and Costa, F.V. (2011) A Qualidade do solo afeta a Germinação das Sementes e o Desenvolvimento das Plântulas de Dimorphandra mollis Benth. (Leguminosae: Mimosoidae). Acta Botanica Brasílica, 25, 908-915.
5. Ranieri, B.D., Pezzini, F.F., Garcia, Q.S., Chautems, A. and França, M.G.C. (2012) Testing the Regeneration Niche Hypothesis with Gesneriaceae (tribe Sinningiae) in Brazil: Implications for the Conservation of Rare Species. Austral Ecology, 37, 125-133.
6. Marques, M.C.M. and Oliveira, P.E.A.M. (2005) Características Reprodutivas das Espécies Vegetais da Planície Litorânea. História Natural e Conservação da Ilha do Mel. Editora da Universidade Federal do Paraná, Paraná.

7. Cordazzo, C.V. (2002) Effect of Seed Mass on Germination and Growth in Three Dominant Species in Southern Brazilian Coastal Dunes. Brazilian Journal of Biology, 62, 427-435.
8. Murali, K.S. (1997) Patterns of Seed Size, Germination and Seed Viability of tropical Tree Species in Southern India. Biotropica, 29, 271-279.
9. Mälken, T., Jorritsma-Wienk, L.D., Hoek, P.H. and Kroon, W.H. (2005) Only Seed Size Matters for Germination in Different Populations of the Dimorphic Tragopogonp ratensiss subsp. pratensis (Asteraceae). American Journal of Botany, 92, 432-437.
10. Yanlong, H., Mantang, W., Shujun, W., Yanhui, Z., Tao, M. and Guozhen, D. (2007) Seed Size Effect on Seedling Growth under Different Light Conditions in the Clonal Herb Ligularia virgaurea in Qinghai-Tibet Plateau. Acta Ecologica Sinica, 27, 3091-3108.
11. Silveira, F.A.O., Negreiros, D., Araújo, L.M. and Fernandes, G.W. (2012) Does Seed Germination Contribute to Ecological Breadth and Geographic Range? A Test with Sympatric Diplusodon (Lythraceae) Species from Rupestrian Fields. Plant Species Biology, 27, 170-173.
12. Leishman, M.R., Westoby, M. and Jurado, E. (1995) Correlates of Seed Size Variation: A Comparison among Five Temperate Floras. Journal of Ecology, 83, 517-529.
13. Silvertown, J. and Bullock, J.M. (2003) Do Seedlings in Gaps Interact? A Field Test of Assumptions in ESS Seed Size Models. Oikos, 101, 499-504.
14. Moles, A.T. and Westoby, M. (2006) Seed Size and Plant Strategy across the Whole Life Cycle. Oikos, 113, 91-105.
15. Baskin, C.C. and Baskin, J.M. (1998) Seeds—Ecology, Biogeography, and Evolution of Dormancy and Germination. Academic Press, New York.
16. Harper, J.L. (1977) Population Biology of Plants. Academic Press, New York.
17. Geritz, S.A. (1995) Evolutionarily Stable Seed Polymorphism and Small-Scale Spatial Variation in Seedling Density. The American Naturalist, 146, 685-707.
18. Ferreira, A.G. and Borghetti, F. (2004) Germinação: Do Básico ao Aplicado. Editora Artmed, Porto Alegre.
19. Mendes-Rodrigues, C., Oliveira, P.E. and Ranal, M.A. (2011) Seed Germination and Seedling Growth of two Pseudobombax species (Malvaceae) with Contrasting Habitats from Brazilian Cerrado. Revista de Biologia Tropical, 59, 1915-1925.
20. Primack, R.B. (1987) Relatioship among Flower, Fruits, and Seeds. Annual Review of Ecology and Systematics, 18, 409-430.
21. Agren, G.I. and Franklin, O. (2003) Root: Shoot Ratios, Optimization and Nitrogen Productivity. Annals of Botany, 92, 795-800.
22. Yang, Z. and Midmore, D.J. (2005) Modeling Plant Resource Allocation and Growth Partitioning in Responses to Environmental Heterogeneity. Ecological Modelling, 181, 59-77.
23. Leishman, M.R. (2001) Does the Seed Size/Number Trade-Off Model Determine Plant Community Structure: An Assessment of the Model Mechanisms and Their Generality. Oikos, 93, 294-302.
24. Rego, R.S., Silva, A.J.C., Brondani, G.E., Grisi, F.A., Nogueira, A.C. and Kuniyoshi, Y.S. (2007) Caracterização Morfológica do Fruto, Semente e Germinação de Duranta vestita Cham. (Verbenaceae). Revista Brasileira de Biociências, 5, 474-476.
25. Carvalho, P.E.R. (2003) Espécies Arbóreas Brasileiras. Embrapa Florestas, Colombo, Brasília.
26. Pedroni, F., Sanchez, M, and Santos, A.M. (2002) Fenólogia da Copaíba (Copaifera langsdorffii Desf.—Leguminosae, Caesalpinioideae) em Uma Floresta Semidecídua no Sudeste do Brasil. Revista Brasileira de Botânica, 25, 183-194.
27. Fagundes, M., Maia M.L.B., Queiroz, A.C.M., Fernandes, G.W. and Costa F.V.(2013) Seed predation of Copaifera langsdorffii Desf. (Fabaceae: Caesalpinioideae) by Rhinochenus brevicollis Chevrolat (Coleoptera: Curculionidae) in a Cerrado Fragment. Ecología Austral, 23, 218-221.

28. Costa, F.V., Siqueira, F.N., Oliveira, J.S. and Fagundes, M. (2011) Relationship between Plant Development, Tannin Concentration and Insects Associated with Copaifera langsdorffii (Fabaceae). Arthropod-Plant Interactions, 5, 9-18.
29. Baskin, J.M. and Baskin, C.C. (2004) A Classification System for Seed Dormancy. Seed Science Research, 14, 1-16.
30. R Development Core Team (2008) R: A Language and Environment for Statistical Computing. R Foundation for Statistical Computing, Vienna. http://www.rproject.org
31. Crawley, M.J. (2007) The R Book. John Wiley and Sons, New York.
32. Walters, C. (1998) Understanding the Mechanisms and Kinetics of Seed Aging. Seed Science Research, 8, 223-244.
33. Debeaujon, I., Léon-Kloosterziel, K.M. and Koornneef, M. (2000) Influence of the Testa on Seed Dormancy, Germination, and Longevity in Arabidopsis. Plant Physiology, 122, 403-414.
34. Gharoobi, B. (2011) Effects of Seed Size on Seedlings Characteristics of Five Barley Cultivars. Iranian Journal of Plant Physiology, 1, 265-270.
35. Beninger, C.W., Hosfield, G.L. and Nair, M.G. (1998) Physical Characteristics of Dry Beans in Relation to Seed Coat Color Genotype. HortScience, 33, 328-329.
36. Fowler, A.J.P. and Bianchetti, A. (2000) Dormência em sementes florestais. Embrapa Florestas, Colombo, Paraná.
37. Dolan, R.W. (1984) The Effect of Seed Size and Maternal Source on Individual Size in a Population of Ludwigia leptocarpa (Onagraceae). American Journal of Botany, 71, 1302-1307.
38. Dan, E.L., Mello, V.D.C., Wetzel, C.T., Popinigis, F. and Zonta, E.P. (1987) Transferência de Matéria seca Como Método de Avaliação do Vigor de Sementes de Soja. Revista Brasileira de Sementes, 9, 45-55.
39. Westoby, M., Leishman, M., Lord, J., Porter, H. and Schoen, D.J. (1996) Comparative Ecology of Seed Size and Dispersal. Philosophical Transactions of the Royal Society B, 351, 1309-1318.
40. Gross, K.L. (1984) Effects of Seed Size and Growth Form on Seedling Establishment of Six Monocarpic Perennial Plants. Journal of Ecology, 72, 369-387.
41. Foster, S.A. (1986) On the Adaptive Value of Large Seeds for Tropical Moist Forest Trees: A Review and Synthesis. Botanical Review, 52, 260-299.
42. Souza, M.L. and Fagundes, M. (2014) Predispersal Seed Predation of Copaifera langsdorffii (Fabaceae): A Tropical Tree with Supra-Annual Mass Fruiting in Brazilian Cerrado. Arthropod-Plant Interactions.
43. Canadell, J. and Zedler, P.H. (1995) Underground Structures of Woody Plants in Mediterranean Ecosystems of Australia, California and Chile. Ecology and Biogeography of Mediterranean Ecosystems in Chile, California and Australia. Springer-Verlag, New York.
44. Rees, M. and Westoby, M. (1997) Game-Theoretical Evolution of Seed Mass in Multi-Species Ecological Models. Oikos, 78, 116-126.
45. Harper, J.L., Lovell, P.H. and Moore, K.G. (1970) The Shapes and Size of Seeds. Annual Review of Ecology and Systematics, 1, 327-356.
46. Bu, H.Y., Du, G.Z., Chen, X.L., Wang, Y., Xu, X.L. and Liu, K. (2009) The Evolutionary Significance of Seed Germinability in an Alpine Meadow on the Eastern Qinghai-Tibet Plateau. Arctic, Antarctic, and Alpine Research, 41, 97-102.

CHAPTER 4

Climate Change: Seed Production and Options for Adaptation

John G. Hampton [1,*], Anthony J. Conner [2], Birte Boelt [3], Thomas G. Chastain [4] and Phil Rolston [1,2]

[1]*Seed Research Centre, Lincoln University, Lincoln 7647, New Zealand*

[2]*AgResearch Ltd., Private Bag 4749, Christchurch 8140, New Zealand*

[3]*Department of Agroecology, Aarhus University, DK-4200 Slagelse, Denmark*

[4]*Department of Crop and Soil Science, Oregon State University, Corvallis, OR 97331-3002, USA*

ABSTRACT

Food security depends on seed security and the international seed industry must be able to continue to deliver the quantities of quality seed required for this purpose. Abiotic stress resulting from climate change, particularly elevated temperature and water stress, will reduce seed yield and quality. Options for the seed industry to adapt to climate change include moving sites for seed production, changing sowing date, and the development of cultivars with traits which allow them to adapt to climate change conditions. However, the ability of seed growers to make these changes is directly linked to the seed system. In the formal seed system operating in developed countries, implementation will be reasonably straight forward. In the informal system operating in developing countries, the current seed production challenges including supply failing to meet demand and poor seed quality will increase with changing climates.

Keywords: abiotic stress; adaptation; agriculture; climate change; seed production; seed quality; seed systems

1. INTRODUCTION

The seed industry is the cornerstone of global food security; food security depends on seed security [1]. The global seed market is currently around US$ 54 billion [2]. Domestically, the largest seed market is in the USA (US$ 14 billion), followed by China (US$ 11.9 billion), France (US$ 3.4 billion), Brazil (US$ 3.1 billion) and Canada and India (each US$2.5 billion). Forty countries have domestic seed markets of US$ 100 million or greater, but others rely heavily on imports to supply their seed needs [2].

Seed worth US$ 10.9 billion was imported by 125 countries in 2013 [2]. The international seed market has tripled over the last three decades driven mainly by the evolution of multinational seed companies, the increased availability of F1 hybrids, protection of intellectual property, an increasing use of counter-season production, and the development of genetically engineered crops [3,4]. In US dollar values, seed exports are dominated by three countries: France, the Netherlands, and the USA. Outside Europe and the Americas, only China, Japan, Australia, New Zealand, South Africa, Thailand, and Turkey make the list of the largest twenty-five seed exporting nations [2].

The past fifty years have seen seed yield increases for the world's major crops of 1%–3% per year [1], due largely to genetic gains obtained with plant breeding [5,6], but also to improved agronomic inputs and specialized management systems [1]. Can these seed yield increases continue given that climate change will have direct impacts on land, water, temperature, atmospheric CO_2, and weather patterns [7]? These changes must impact agricultural activities [8], although there is debate as to whether the impacts will be negative or positive. Some authors [9,10] concluded that the overall impact of climate change on agriculture is expected to be negative, threatening global food security, whereas others (e.g., [11]) considered that global crop production could increase by approximately 50% by 2050 without extra land. However, the latter authors did stress that a large proportion of these yield increases will need to be delivered via plant breeders through the release of new cultivars adapted to cope with a changing environment [6,12,13]. Seed is the basic and most vital input of agriculture [14], and the seed industry must be able to continue to deliver the quantities of quality seed required for this purpose. Climate change poses several challenges for the continued production of high quality seed [15]. In this review, we examine the effects of climate change on seed production and explore options for adaptation of seed production in response to climate change, with a focus on the effects of increased CO_2, increased temperature, and their interaction.

2. SEED PRODUCTION

2.1. Seed Yield: Arable Crop Species

The impacts of changing climate and climate variability on seed production of arable crops such as rice, soybean, cotton, and peanut has been recently reviewed [15]. Enrichment of atmospheric CO_2 is likely to increase seed yields of C3 arable crops (e.g., wheat) by between 11% and 30% but leave yields of C4 crops (e.g., maize) unchanged [11,16]. Yield increases are usually because of a greater number of seeds per unit area [17,18,19] as a result of increased resource use efficiency for radiation, water, and nitrogen [20]. Elevated CO_2, through decreasing stomatal conductance, increases water use efficiency [15]. In water stress conditions, this may allow plants to retain more seeds, but yields will always be lower compared to those in non-resource stress environments [15]. Sometimes there is also an increase in seed mass [21], although the effects of elevated CO_2 on seed mass are highly variable [22].

Elevated temperature accelerates various aspects of plant metabolism and changes the rate of organ development, respiration, senescence, and source-sink relationships [23]. Most arable crop species are vulnerable to heat stress both pre- and post-fertilization, and any exposure to temperatures above the optimum for the species can result in significant seed yield losses [24,25,26] with the extent of the reduction in individual seed yield components depending on the timing and duration of the temperature stress experienced. Temperature increases, which reduce seed number, accelerate seed growth rate, reduce the duration of seed filling, and sometimes reduce seed mass [15] can, therefore, negate any positive seed yield responses to increased CO_2.

2.2. Seed Yield: Temperate Pasture Species

In contrast to the many reports of the effects of elevated CO_2 and high temperature on seed yield in arable crop species, little is known about the response of the temperate grass and forage legume species which underpin much of the world's milk, meat, and wool industries [27]. In white clover (*Trifolium repens* L.), increasing CO_2 concentration has been reported to increase flower numbers [28,29,30] and seeds per flower but not seed mass or seed yield [30]. In perennial ryegrass (*Lolium perenne* L.), elevated CO_2 decreased the time to ear emergence in one cultivar but increased the time to ear emergence by up to 10 days in four cultivars [31]. It also increased seed head numbers in a pasture sward, but not seeds per spikelet, seed mass [29], and seed yield [30]. While there are no reports on other pasture species managed for seed production, there have been a number of reports of seed increases for rangeland *Bromus* spp. [32,33,34] following exposure to elevated CO_2.

Hikosaka et al. [35] suggested that in grasses and legumes, seed production is limited by nitrogen rather than carbon and that seed yields would be increased only when nitrogen became more available at elevated CO_2 levels. In their study, legumes exhibited a more pronounced increase in nitrogen acquisition and enhanced seed production with elevated CO_2than non-leguminous species. Enhanced CO_2 has been reported to enhance nodulation and nitrogen fixation [36,37,38], and thus white clover may have been expected to demonstrate a seed yield increase. However, if soil nitrogen was not limiting, this may have negated any nitrogen fixation response [39]. Grass species increase seed production with elevated CO_2 by decreasing seed nitrogen concentration as well as increasing acquisition of plant available nitrogen [35].

White clover is generally regarded as a long-day plant, which requires a period of vernalization (cool, short days) before initiating reproductive growth [40]. After vernalization, and at 14 h or 16 h day lengths, inflorescence initiation is favored by high (27.5 °C day/19 °C night) temperatures [41,42]. However, a temperature of 30 °C with either 14 h or 18 h day lengths decreased florets per inflorescence and ovules per floret compared with those which developed at 20 °C in the same day length [41]. More reproductive buds per stolon and flowers per plant occurred at 26 °C than at 18 °C, both in a 16 h day length [43], but there was no change in mean floret or ovule number.

Perennial ryegrass tiller production is optimal at 18–24 °C [44], and higher temperature, especially at night, can depress tiller production [45], thus potentially reducing seed yield. The initiation of reproductive growth requires a dual induction: vernalization (ca. 6 °C in short days (primary induction) and then exposure to long days (secondary induction [46]). The optimum temperature during secondary induction is between 12 °C and 15 °C, and high temperatures at this time can result in devernilization of previously vernalized plants [47]. They can also reduce spikelet numbers [48]. Increasing temperature (range 13–23 °C) between spikelet initiation and ear emergence reduced both spikelet and floret number, thus reducing seed yield [49]. Increasing temperature also reduces the time to anthesis and the time for seed development, reducing seed mass [50]. Air temperature just before and after anthesis accounted for 70% of the variability in perennial ryegrass seed number [51].

2.3. Seed Yield: Vegetable Species

Few studies on the effects of elevated CO_2 and temperature on vegetable seed production have been reported. Carrot (*Daucus carota* L.) vegetative growth is increased by elevated CO_2 [52,53,54] although the effect is temperature dependent. At 11–12 °C, elevated CO_2 had no effect on carrot growth, but at mean air temperatures of >25 °C, vegetative biomass was nearly doubled [52]. However, there have been no reports on the effects of elevated CO_2 on carrot reproductive growth. Elevated CO_2 more than doubled radish (*Raphanus sativus* L.) vegetative dry matter [55], but the response was temperature dependent, increasing as air temperature increased [52]. In wild radish (*R. raphanistrum*), elevated CO_2 increased flower and seed production [56], but another report found that in the same species, elevated CO_2 reduced flower and seed numbers [57].

Increasing mean air temperature from 20 °C to 26 °C reduced carrot vegetative biomass, main stem height, and days to first flowering. Umbels per plant were increased at the higher temperature but the number of flowers per umbel was reduced, as was the fertility of the male flowers [58]. When temperature was increased from 20 °C day/10 °C night to 30 °C day/20 °C night, a 13%–20% reduction in mean carrot seed mass was found [59], which was associated with a reduction in seed filling time rather than any difference in seed relative growth rate. Daytime temperatures of 20 to 30 °C caused devernalization in radish [60], and a similar response has been recorded in vegetable brassicas [61,62]. Increasing temperature from 17 °C to 26 °C significantly reduced radish pod number and seeds per pod [63], and a reduction in cabbage (*Brassica oleracea* L.) seed yield was associated with increasing temperature during pod setting [64]. Although *B. napus* L. (canola) is not a vegetable brassica, the negative effects of a 5 °C temperature increase on seed yield could not be compensated by elevated CO_2 [65]. It is likely that this response would also occur in vegetable brassicas.

3. PLANT REPRODUCTIVE PROCESSES

The sexual reproductive phase in plants may be altered by both elevated CO_2 and temperature [15]. A plant grown at elevated CO_2 may be able to invest more resources in flowers [66], but most plants can only tolerate narrow temperature changes which, if exceeded, can reduce seed set and therefore seed yield [67,68].

3.1. Flowering Time

Elevated CO_2 can advance, delay, or impose no change on flowering time depending on genotypes and growing conditions [69]. Heat stress can also trigger either early or delayed flowering depending on the species and other environmental conditions [15]. Accelerated flowering may result in reproductive development before the plant has accumulated sufficient biomass resources required for the provisioning of developing seeds [70]. Heat stress can also cause asynchrony between male and female reproductive development [71], leading to problems for hybrid seed production which require that pollen from the male parent line be available when stigmas on the female parent line are receptive to pollen. The effect of elevated temperature in the presence of elevated CO_2 on flowering time is not yet known.

3.2. Floral Organs

Elevated CO_2 has been reported to both increase and decrease flower production [72,73] but to increase the time flowers remain open [66]. Variable responses for nectar production per flower and for nectar sugar composition and concentration have also been reported [66,72]. In one study, elevated CO_2 increased the concentration of glucose and fructose and reduced the ratio of sucrose to hexoses [73]. This may change the attractiveness of the nectar to pollinators (large bees such as *Bombus* spp. prefer sucrose-dominated nectars [74]) and the duration of pollinator visits received by flowers [73]. Heat stress can reduce the number, decrease the size, and cause deformity of floral organs [75] and accelerate stigma and ovule development, reducing the duration for which they are receptive to pollen and pollen tubes [76]. However, elevated temperature can reduce the negative effect of elevated CO_2 on the number of flowers produced [73].

3.3. Pollen

Exposure to elevated CO_2 has been reported to either increase or not change pollen production [77,78] but reduce pollen viability and pollen germination [78]. This decrease in pollen viability may have resulted from a nutrient limitation that developed as CO_2 became abundant [79]. Both nitrogen and phosphorous can become limiting under elevated CO_2 [80], and a reduction in their availability can reduce pollen quality [79].

The quantity and morphology of pollen, anther dehiscence, and pollen wall architecture, the chemical composition and metabolism of pollen, and pollen viability, germination, and pollen tube growth can all be negatively affected by heat [70,76]. Given the great amounts of pollen produced in most plant species, the production decreases recorded may not greatly impact reproductive output [76]. The effect of heat stress will be more important for seed production because of the negative effects on pollen viability, pollen germination, and pollen tube growth rate [77] leading to reduced seed set [68]. Increasing temperature decreased pollen production irrespective of CO_2 [78]. However, there was a negative interaction between temperature and CO_2 on pollen viability, with the decrease in pollen viability in response to increasing temperature being greater at elevated CO_2 than at ambient CO_2 levels [81].

3.4. Pollinators

Seed production from 87 of the world's leading food and forage crops depends on insect pollination [82], and climate change may be a threat to pollination services [83]. Any changes in nectar production and/or composition could have detrimental effects on the interactions of flowers and their pollinators. For example, a negative effect of elevated CO_2 on bumble bee (*Bombus terrestris* L.) survival has been reported and attributed to a change of sucrose to hexoses ratio (see Section 3.2). However, this effect on longevity was mitigated by increasing temperature [73].

In plant species where elevated CO_2 increases floral display (increased flower numbers and longevity), attractiveness to pollinators may be increased, and the increased nectar resources should benefit the populations of bees (and other insects) feeding on the flowers [66]. However, CO_2 enhancement can affect bees directly; within honey bee (*Apis mellifera* L.) hives, the highest flight traffic out of the hive is during the period of lowest CO_2 concentration [84]; CO_2 anaesthetized bees changed from hive to field activities at an earlier age after the treatment and their pollen collection almost ceased [85]. These observations suggest that CO_2 enhancement may affect bee foraging behavior [86] and possibly the motivation to forage [87] with negative effects on pollination.

It is likely that pollinators will change their activity patterns as temperature increases because, at least for bees, the thermal properties of their environment determine the extent of their activity [88]. Bees absorb heat at high temperatures and the time taken for thermoregulation comes at the cost of foraging, with a negative consequence for pollination [89] because of reduced efficiency of pollen removal and deposition. However, the current knowledge of the potential consequences of climate change, in particular increasing temperatures on plant/pollinator interactions, is limited. The effects of rising temperatures are more likely to be deleterious to pollinators in the tropics than to those at higher latitudes because the former are living in an environment already close to their optimal temperature while the latter have broader thermal tolerance and are currently living in cooler climates than their physiological optima [90]. A warming environment may actually enhance the performance of insects living at higher latitudes [89].

4. GENETIC PURITY OF CULTIVARS

A critically important impact of climate change on seed production involves the potential loss of genetic quality in open pollinated and F1 hybrid cultivars. Elevated CO_2 and temperature can induce changes in the sexuality in crops plants, including a breakdown in self-incompatibility or male sterility mechanisms. These breeding systems are used in seed production to promote outcrossing in order to maintain a high level of heterozygosity in many open pollinated crops or to guarantee F1 hybrid seed production [91]. Breakdown of these genetic mechanisms results in self-pollination. As a consequence, the genetic integrity of the cultivar is lost and any seed produced will no longer be true-to-type.

Many forage crops, such as clovers and perennial ryegrass, rely on self-incompatibility to maintain outcrossing and heterozygosity within open pollinated populations. A breakdown of self-incompatibility could lead to rapid inbreeding in only a few generations of seed production [91]. This would greatly reduce the frequency of heterozygosity at many loci, resulting in inbreeding depression which will significantly reduce performance of the resulting crop. For seed production in many vegetable crops, some form of male sterility is essential for F1 hybrid seed production. If the male sterile parent (the female parent) from which seed is harvested exhibits a slight reversion toward pollen fertility, then even a low frequency of seed from self-pollination greatly reduces the value of the F1 hybrid seed [91]. In many cases, the loss of male sterility will make the harvested seed worthless due to the loss of heterosis, uniformity, and performance of the subsequent crop.

The breakdown of self-incompatibility or male sterility as a consequence of heat stress (Table 1) results in a major source of pollen contamination for hybrid seed production. A potential solution involves the deployment of a single dominant herbicide resistance gene in the homozygous state in the intended pollen parent. All true hybrid seed harvested from the female parent is then heterozygous for herbicide resistance and any harvested seed arising from self-pollination of the female parent can be eliminated by sensitivity to the herbicide [92].

Table 1. Examples of effects of elevated temperature on the breeding systems in selected crop species.

Species	Response to Elevated Temperature	Reference
Carrot	Reduced male sterility at 26 °C	[58]
Brussels sprouts	Breakdown of male sterility above 17 °C	[93]
Radish	Breakdown of self-incompatibility at 26 °C	[63]
Alsike clover	Breakdown of self-incompatibility at 32 °C	[94]
Red clover	Breakdown of self-incompatibility at 40 °C	[95]
White clover	Breakdown of self-incompatibility at 35 °C	[96]
Perennial ryegrass	Breakdown of self-incompatibility at 34 °C	[97]

5. SEED QUALITY

The effects of elevated CO_2 and temperature on seed quality have been recently reviewed [22]. The following is therefore a summary of the known responses.

5.1. Seed Mass

Seed mass responses to elevated CO_2 are highly variable, both among and within species [35]. Increases are greater in legumes than non-legumes, and seed mass increases may result in a decreased seed nitrogen concentration in non-legumes [22]. Increasing temperature may decrease seed mass because of an accelerated seed growth rate and reduced seed filling duration [98,99], but lower seed mass does not necessarily reduce seed germination or vigour [22].

5.2. Germination

Seed germination in response to elevated CO_2 is also highly variable, with responses varying among species [100]. The reported change in carbon/nitrogen ratio [101] may reduce seed protein content, reducing seed viability by limiting the seed's ability to supply the amino acids required for protein synthesis during embryo growth [22]. Conversely, elevated CO_2 increases ethylene production [100] which can promote germination in many species. High temperature stress before seeds reach physiological maturity can reduce germination by inhibiting the ability of the plant to supply the assimilates necessary to synthesize the storage compounds required for the germination process [22], and/or the seeds suffer physiological damage to the extent that the ability to germinate is lost [102].

5.3. Seed Vigour

Nothing is known concerning the effects of elevated CO_2 on seed vigour, but seed vigour can be significantly reduced by high temperature stress both before and after physiological maturity [22]. High temperatures induce or increase the physiological deterioration of seeds [102], and limited evidence suggests that only short periods of high temperature stress at critical stages of seed development are required to reduce seed vigour [22].

6. POST SEED HARVEST

Internationally, present seed losses as a result of poor drying and storage are conservatively estimated to be between 10% and 40% depending on location, drying method, storage conditions, and species [103]. Information on seed drying, seed storage and maintaining seed viability is presented in [104,105].

Increasing atmospheric CO_2 is unlikely to have any negative effects on seed viability in storage. Seeds produce CO_2 as a byproduct of their respiration, and the accumulation of CO_2 from this source within sealed containers has been used to control seed storage insect pests [106], as has flushing of sealed containers with CO_2 [103].

Because seeds are good insulators they can absorb radiant heat from the sun, so that seed temperature can often be 10–12 °C higher than ambient air temperature [104]. Elevated air temperature and clear, cloudless days immediately prior to and/or during seed harvest can therefore lead to heating damage, even in seed lots harvested at a seed moisture content considered 'safe' for the species (e.g., 140–150 mg/g for wheat seed). For example, at an air temperature of 32 °C, perennial ryegrass at harvest had a seed temperature of 41 °C [104]. If this 'hot' seed is not cooled quickly before being processed or placed into storage, storage fungi will proliferate, resulting in further heating of the seed lot and the rapid loss of seed quality [104].

In countries where sun drying of seed is practiced, elevated ambient temperatures will increase the risk of rapid seed quality deterioration, again by increasing seed temperatures to levels (60–70 °C), which reduce germination. However, this damage can be mitigated by shading the seed from direct radiation from the sun during the drying process.

The effects of temperature on uncontrolled seed storage have been long known. For seed at moisture contents between 50 mg/g and 140 mg/g, seed storage life is generally doubled by every 5 °C decrease in storage temperature [107]. Increased ambient temperatures would, therefore, tend to reduce seed storage life, although this may be tempered because of the relationship between temperature and relative humidity (RH). Seeds are hygroscopic, meaning that air brought into contact with the seed exchanges water with the seed until the seed moisture content and water content of the air reach a balance: the equilibrium moisture content. In temperate environments, the water holding capacity of air is temperature dependent, so that RH decreases as temperature increases. Higher air temperatures may therefore assist in preventing seed moisture content in uncontrolled storage increasing to levels not 'safe' for storage [104].

7. OPTIONS FOR ADAPTATION

7.1. Moving Sites for Seed Production

Climate change modelling predicts gains in potential agricultural land globally [108], particularly in North America, Northern Europe, and the Russian Federation because of longer sowing windows and more favorable growing conditions as temperature increases. For example, it has been suggested that climate warming in Europe would expand the area of all cereals, and both cooler season (pea, canola) and warmer season (soybean, sunflower) seed crops northwards, while production would decrease in the Mediterranean region [20]. Significant losses in agricultural land are predicted for northern and Southern Africa due to a worsening of growing conditions from increased heat and moisture stress [108].

It may therefore be necessary to move seed production to the limits of adaptation, either in latitude or in elevation in order to reduce the chances of environmental stress negatively affecting seed yield and quality [22]. The impact of climate change on cabbage seed production in north-western India between 1981 and 2004 has been reported [64]. During this time, seed yield of the same cultivar decreased by around 40%, and this was associated with increasing minimum and maximum temperature during pod setting; when a mean maximum temperature exceeded 25 °C, a sharp decline in seed yield was observed. These authors considered that cabbage seed production would need to be shifted northward to minimize the effect of increasing temperature.

For the seed industries of North America, China, India, the countries of northern Europe, the southern South American countries, New Zealand, and Australia, moving seed production north or south will be possible, but will depend on water availability and the extent of competition for land use from other agricultural production [109]. However, the seed industries in countries at lower latitudes, particularly in Africa, may not have this option, forcing them to import even more of their seed requirements.

7.2. Changing Sowing Date

Changing sowing date to avoid the detrimental effects of high temperatures at flowering and during seed set has been recommended as an adaptation strategy for climate change [15]. A shift in sowing dates (e.g., earlier sowing of spring crops to avoid hot and dry periods during summer) is one of the expected adaptation responses to climate change in the cooler zones of Europe [110], while for wheat in northwest India, sowing date would need to be advanced by six days per degree rise in temperature [111]. One impact of increased temperatures has been the earlier onset of spring, and as a result of warmer spring temperatures in the northern Great Plains of the USA, earlier maize [112] and wheat [113] sowings have been possible, leading to significant yield increases by avoiding temperature stress during seed filling.

A day/night temperature of 38 °C/21 °C reduced pollen viability and seed set in sorghum [114]. These authors used years of climate data from one site in Queensland, Australia, and showed that the number of days when the maximum temperature was greater than 38 °C over the time of pollen production was greatest from mid-December to mid-January. There were many occasions (approximately 50%) with at least one such day, but many less (approximately 10%) with more than four days exceeding 38 °C during pollen production. Their conclusion was that for this site, sowings in mid-November were most at risk of experiencing high temperature during pollen formation, and that the risks would be lower for early November and early December sowings. They also made the point that as high temperature occurrences can occur over a wide time frame during summer, other sowing times would not be totally immune.

While changing sowing date to avoid/reduce the risk of temperature stress effects on seed yield and seed quality may be an effective management option, the effects of other factors will also need to be considered. These include soil temperature effects on germination and emergence [115], day length responses, particularly premature flowering in short days [116], the possibility of frosts during flowering [117], water availability [118], and increased weed, pest, and disease pressure [16,20].

7.3. Plant Breeding

Plant breeding has been a success story in increasing yields over the last fifty years [11], and will continue to be vitally important in developing cultivars adapted to climate-change conditions [6,15]. The desired traits include drought and temperature stress resistance, resistance to pests and diseases, and the ability to cope with salinity and water logging [6]. This can be achieved using a combination of conventional, molecular and transgenic approaches [119]. Conventional breeding methodologies, particularly phenotyping and double haploid technology, will play important roles in accelerating the development of climate-adapted germplasm [120]. Molecular technologies, including the use of marker-trait associations and genome-wide selection, are providing opportunities to develop germplasm with tolerance to multiple abiotic and biotic stresses [121]. Transgenic approaches are also being investigated [122]. An additional important target trait for seed production in highly heterozygous open pollinated and F1 hybrid cultivars will be more stable alleles conferring self-incompatibility and male sterility under elevated temperatures.

The adverse effects of heat stress can be mitigated by developing crop plants with improved thermotolerance [123]. Genetic diversity for heat tolerance has already been demonstrated for crops such as wheat [124], soybean [125], peanut ([126,127], Table 2), bean [128], cowpea [129], and cotton [130]. Breeders could also consider incorporating photoperiod insensitivity [131] so that seed filling can begin before the onset of the highest temperatures [132]. Both landraces and, when available, wild relatives contain substantial genetic variation, some of which will allow breeding for high temperature resistance [110].

Table 2. Total dry matter and pod and seed dry weight of four peanut genotypes grown at air temperatures of 28/22 °C (mean 25 °C) or 38/22 °C (mean 30 °C) from 21 to 90 days after planting (adapted from [127]).

	Susceptible [1]				Tolerant [2]				
	Genotype 1		Genotype 2		Genotype 3		Genotype 4		
	28/22 °C	38/22 °C	28/22 °C	38/22 °C	28/22 °C	38/22 °C	28/22 °C	38/22 °C	SED (28 df)
Total dry matter (g/plant)	148.9	114.3	179.6	115.9	158.6	129.7	185.4	138.0	4.56 ***
Pod dry matter (g/plant)	78.0	45.4	59.4	12.8	73.5	53.6	40.2	28.9	3.40 ***
Seed dry weight (g/plant)	63.8	26.6	45.0	10.0	61.7	42.8	36.9	28.7	2.51 ***

[1] genotypes susceptible to high temperature; [2] genotypes tolerant of high temperature; *** significant at $p < 0.001$.

Potential genetic strategies to improve plant heat-stress tolerance include traditional and contemporary molecular breeding protocols and transgenic approaches [123,133,134]. Heat stress tolerance is a polygenic character, and this makes it difficult to identify heat tolerance related traits by classical genetics. Bita and Gerats [134] reviewed strategies for crop improvement for heat stress tolerance and concluded there was a need for molecular biologists, plant physiologists, and plant breeders to work collaboratively to enable the identification of tolerance genes and elucidation of functional relationships between genotype and observed phenotypes. This would provide a system-wide phenome to genome analysis [123] allowing accurate trait mapping, introgression of superior alleles, or cloning of major quantitative trait loci (QLTs). Genome-wide analyses have started to provide some candidate genes involved in thermotolerance [135,136,137,138]. Understanding the genetic basis of heat tolerance is essential to allow breeders to facilitate the transfer of genes for heat tolerance to commercial cultivars [123,134].

A similar call has been made for further research to assess crop yield responsiveness to elevated CO_2 [12]. Several studies [139,140] suggest that traditional plant breeding has tended to select against CO_2 responsiveness. Ainsworth et al. [13] identified the need to better understand the mechanisms of yield responses to CO_2, to assess the genetic diversity available for improved responsiveness, and to establish efficient schemes to select for adaptation to elevated CO_2. These authors have proposed a new generation of free-air concentration enrichment experiments focused on adapting crops to elevated CO_2using the tools of molecular genetics.

Drought tolerance is also a highly desirable trait, and several studies have shown that cereal crops such as rice, maize and wheat and their sexually compatible wild relatives have substantial untapped genetic potential for breeding, including traits for withstanding drought [141,142]. Other studies have identified genetic markers and physiological traits associated with drought tolerance [141]; barley cultivars, for example, have been reported to have a 19-fold difference in the expression of genes associated with drought tolerance [143]. Drought tolerant maize cultivars developed through conventional breeding have been released in the USA [144]. The ability of genetic engineering to increase the ability of crops to withstand drought is unclear [141].

Climate-adapted germplasm will often have to incorporate packages of traits [119]. In Southern Africa this will be tolerance to drought stress at elevated temperatures [119], while in many parts of Asia, tolerance to both drought and water logging is becoming important [145].

8. SEED SYSTEMS

The ability for seed growers to make the adaptations required resulting from climate change is directly linked to the seed system within a country. In the formal system which operates in developed countries, where commercial seed companies supply most seed required, legislation and supporting activities are fully developed, and annual replacement of cultivars is high, growers will have access to seed of newly released cultivars with the desired traits whose genetic purity is maintained through seed certification. Seed crops will continue to be grown under contract for a seed company, and seed quality standards will continue to be met. Growers in developed countries are able to pay for the required climate change adaptations measures [15], can afford to insure for crop losses due to extreme weather events such as flooding, hail, or high winds, and have the expertise and/or access to expert advice to allow them to implement crop management changes.

The situation is vastly different in the developing world. As Singh et al. [15] put it, "for poor countries there is no climate safety net". In most developing countries, where there are few commercial seed companies, legislation and supporting activities are not developed and annual seed replacement is very low, informal sources still provide a large proportion of seeds to be sown, either saved from a farmer's own harvest or purchased from a local market (e.g., [146,147]). The use of improved seed (either certified or local selection) is commonly under 20%. However, plant breeders have been actively working towards producing cultivars better able to cope with climate change. The Drought Tolerant Maize for Africa project [145] has produced new drought tolerant hybrids [148], and drought tolerant germplasm which also has a higher level of tolerance to higher temperatures is also available [149]. Similarly, the Program for Africa's Seed Systems is seeking to promote the development of seed systems that deliver improved, locally adapted crop varieties to small-scale farmers and the uptake and use of released cultivars [150]. The Millennium Seed Bank, in collaboration with the Global Crop Diversity Trust, is collecting, protecting, and preparing the wild relatives of the world's most important food crops in a form that breeders can readily use to produce cultivars adapted to future climatic conditions that farmers in the developing world will soon be encountering [151]. However, the question remains, can seed of cultivars adapted to climate change be produced in (i) developed countries at a price which farmers in developing countries could afford to pay or (ii) in developing countries in sufficient quantities to meet local demand? In 2013, African countries imported around 40,000 tons of field crop and vegetable seed while countries in Asia imported 79,000 tons [2], only minor contributions to the total demand. For many countries in Africa for example, the annual seed demand exceeds production, and over 80% of seeds which are available to the farming community are not certified and are of unknown quality [150]. Farmers' access to quality seed of a diverse range of adapted cultivars is still impeded by insufficient and inefficient seed production and distribution systems, poor seed quality assurance, inadequate seed policies, and seed price [15,119,150]. Many of the obstacles within seed production can be overcome [119], but the time which will be required to do so is cause for concern. The challenges for seed production which exist presently in developing countries will increase with changing climates [15].

9. CONCLUSIONS

Are there options for adaptation of seed production in response to climate change? There is no escaping the conclusion that the ability of the world's seed industry to supply the quantities of quality seed required for agricultural production will be impaired unless the seed industry undergoes change. Any increases in seed yield in response to elevated CO_2 will be negated by temperature and water stress, and high temperatures at critical times during seed development will reduce seed quality. The genetic quality of cultivars in open pollinated and F1 hybrid crops is also likely to be reduced. Heat stress in particular can induce a breakdown of male sterility and self-incompatibility mechanisms. This results in self-pollinations and the undesired consequences of a reduction in heterozygosity, loss of heterosis, and potential inbreeding depression during seed production. New cultivars with the ability to produce under abiotic stress (including water stress, elevated temperatures, and increased pest and disease pressure) are a must, but while they will be able to be readily utilized in the developed world, their use in developing countries will have many obstacles to overcome. Seed companies in the formal system of developed countries will be more readily able to meet this new demand than those in the informal system of developing countries. Seed crop management options include moving seed production either in latitude (north or south) or in elevation (highlands and mountainous regions) in order to reduce the frequency of abiotic stress and altering sowing date to avoid high temperate stress during seed development [22]. The ability of seed producing countries to do so will depend on their distance from the equator in both hemispheres. There are major challenges ahead for the seed industry. How effectively they can be mitigated will have a major impact on the need to achieve food security internationally.

AUTHOR CONTRIBUTIONS

John Hampton prepared the initial draft of the manuscript. Tony Conner, Birte Boelt, Tom Chastain and Phil Rolston commented on the draft and provided amendments which were incorporated into the final version. John Hampton edited the manuscript.

CONFLICTS OF INTEREST

The authors declare no conflict of interest.

REFERENCES

1. Bruins, M. The evolution and contribution of plant breeding to global agriculture. In Proceedings of the Second World Seed Conference, Rome, Italy, 8–10 September 2009; pp. 18–31.
2. ISF. International Seed Federation website. Available online: https:// www.worldseed.org (accessed on 9 March 2016).

3. Le Buanec, B. Evolution of the seed industry during the past three decades. Seed Test. Int. 2007, 134, 6–10.
4. Hampton, J.G. Maintaining capacity in seed technology and seed testing. In Proceedings of the Second World Seed Conference, Rome, Italy, 8–10 September 2009; pp. 187–194.
5. Le Buanec, B. Opening address. In Proceedings of the Second World Seed Conference, Rome, Italy, 8–10 September 2009; pp. 13–16.
6. Ceccarelli, S.; Grando, S.; Maatougui, M.; Michael, M.; Slash, M.; Haghparast, R.; Rahmanian, M.; Taheri, A.; Al-Yassin, A.; Benbelkacem, A.; et al. Plant breeding and climate changes. J. Agric. Sci. 2010, 148, 627–637.
7. Kurukalasuriya, P.; Rosenthal, S. Climate Change and Agriculture: A Review of Impacts and Adaptations; World Bank: Washington, DC, USA, 2013.
8. Eitzinger, J.; Orlandini, S.; Stefanski, R.; Naylor, R.E.L. Climate change and agriculture: Introductory editorial. J. Agric. Sci. 2010, 148, 499–500.
9. Nelson, G.C.; Rosegrant, M.W.; Koo, J.; Robertson, R.; Sulser, T.; Zhu, T.; Ringler, C.; Msangi, S.; Palazzo, A.; Batka, M.; et al. Climate Change: Impact on Agriculture and Costs of Adaptation; International Food Policy Research Institute: Washington, DC, USA, 2009.
10. Challinor, A.J.; Watson, J.; Lobell, D.B.; Howden, S.M.; Smith, D.R.; Chhetri, N. A meta-analysis of crop yield under climate change and adaptation. Nat. Clim. Chang. 2014, 4, 287–291.
11. Jaggard, K.W.; Qi, A.; Ober, E.S. Possible change to arable crop yields by 2050. Philos. Trans. R. Soc. B Biol. Sci. 2010, 365, 2835–2851.
12. Ainsworth, E.A.; Rogers, A.; Leakey, A.D.B. Targets for crop biotechnology in a future high-CO2 and high-O2 world. Plant Physiol. 2008, 147, 13–19.
13. Ainsworth, E.A.; Beier, C.; Calfapietra, C.; Ceulemans, R.; Durand-Tardif, M.; Farquhar, G.D.; Godbold, D.L.; Hendrey, G.R.; Hickler, T.; Kaduk, J.; et al. Next generation of elevated CO2 experiments with crops: A critical investment for feeding the future world. Plant Cell Environ. 2008, 31, 1317–1324.
14. Maity, A.; Pramanik, P. Climate change and seed quality: An alarming issue in crop husbandry. Curr. Sci. 2013, 105, 1336–1338.
15. Singh, R.P.; Prasad, P.V.V.; Reddy, K.R. Impacts of changing climate and climate variability on seed production and seed industry. Adv. Agron. 2013, 118, 49–110.
16. Hatfield, J.L.; Boote, K.J.; Kimball, B.A.; Ziska, L.H.; Izaurralde, R.C.; Ort, D.; Thomson, A.M.; Wolfe, D. Climate impacts on agriculture: Implications for crop production. Agron. J. 2011, 103, 351–370.
17. Qaderi, M.M.; Reid, D.M. Growth and physiological responses of canola (Brassica napus) to UV-B and CO2 under controlled environment conditions. Physiol. Plant. 2005, 125, 247–259.
18. Baker, H.G.; Allen, L.H., Jr.; Boote, K.J.; Jones, P.; Jones, J.W. Response of soybean to air temperature and carbon dioxide concentration. Crop Sci. 1989, 29, 98–105.
19. Prasad, P.V.V.; Boote, K.J.; Hartwell, L.; Allen, J.R.; Thomas, J.M.G. Super-optimal temperatures are detrimental to peanut (Arachis hypogaea L.) reproductive processes and yield at both ambient and elevated carbon dioxide. Glob. Chang. Biol. 2003, 9, 1775–1787.
20. Olesen, J.E.; Bindi, M. Co-sequences of climate change for European agricultural productivity, land use and policy. Eur. J. Agron. 2002, 16, 239–262.
21. Vanaja, M.; Reddy, P.R.; Lakshmi, N.J.; Maheswari, M.; Vagheera, P.; Ratnakumar, P.; Jyothi, M.; Yadav, S.K.; Venkateswarulu, B. Effect of elevated CO2 concentration on growth and yield of blackgram (Vigna mungo L. Hepper)—A rainfed pulse crop. Plant Soil Environ. 2007, 53, 81–88.
22. Hampton, J.G.; Boelt, B.; Rolston, M.P.; Chastain, T.G. Effects of elevated CO2 and temperature on seed quality. J. Agric. Sci. 2013, 151, 154–162.
23. Farrar, J.F.; Williams, M.L. The effects of increased atmospheric carbon-dioxide and temperature on carbon partitioning, source-sink relations and respiration. Plant Cell Environ. 1991, 14, 819–830.

24. Dornbos, D.L.J.; Mullen, R.E. Influence of stress during soybean seed fill on seed weight, germination and seedling growth rate. Can. J. Plant Sci. 1991, 71, 373–383.
25. Wang, J.; Gan, Y.T.; Clarke, F.; McDonald, C.L. Response of chickpea yield to high temperature stress during reproductive development. Crop Sci. 2006, 46, 2171–2178.
26. Madan, P.; Jagadish, S.V.K.; Craufurd, P.Q.; Fitzgerald, M.; Lafarge, T.; Wheeler, T.R. Effect of elevated CO2 and high temperature on seed-set and grain quality of rice. J. Environ. Bot. 2012, 63, 3843–3852.
27. Fairey, D.T.; Hampton, J.G. Forage Seed Production I. Temperate Species; CAB International: Wallingford, UK, 1998.
28. Deckmyn, G.; Caeyenberghs, E.; Ceulemans, R. Reduced UV-B in greenhouses decreases white clover response to enhanced CO2. Environ. Exp. Bot. 2001, 46, 109–117.
29. Edwards, G.R.; Clark, H.; Newton, P.C.D. The effects of elevated CO2 on seed production and seedling recruitment in a sheep-grazed pasture. Oecologia 2001, 127, 383–394.
30. Wagner, J.; Luscher, A.; Hillebrand, C.; Kobald, B.; Spitaler, N.; Larcher, W. Sexual reproduction of Lolium perenne L. and Trifolium repens L. under free air CO2 enrichment (FACE) at two levels of nitrogen application. Plant Cell Environ.2001, 24, 957–965.
31. Maw, B.R.; Jones, C.S.; Newton, P.D.; Hatier, J.-H.B. Elevated atmospheric CO2 alters heading date of perennial ryegrass. Proc. N. Z. Grassl. Assn. 2014, 76, 217–220.
32. Huxman, T.E.; Hamerlynck, E.P.; Smith, S.D. Reproductive allocation and seed production in Bromus madritensis spp. rubens at elevated CO2. Funct. Ecol. 1999, 13, 769–777.
33. Smith, S.D.; Huxman, T.E.; Zitzer, S.F.; Charlet, T.N.; Housman, D.C.; Coleman, J.S.; Fenstermaker, L.K.; Seemann, J.R.; Nowak, R.W. Elevated CO2 increases productivity and invasive species success in an arid ecosystem. Nature2000, 408, 79–82.
34. Thurig, B.; Korner, C.; Stocklin, J. Seed production and seed quality in a calcareous grassland in elevated CO2. Glob. Chang. Biol. 2003, 9, 873–884.
35. Hikosaka, K.; Kinugasa, T.; Oikawa, S.; Onoda, Y.; Hirose, T. Effects of elevated CO2 concentration on seed production in C3 annual plants. J. Exp. Bot. 2010, 64, 1523–1530.
36. Miyagi, K.M.; Kinugasa, T.; Hikosaka, K.; Hirose, T. Elevated CO2 concentration, nitrogen use, and seed production in annual plants. Glob. Chang. Biol. 2007, 13, 2161–2170.
37. Rogers, A.; Ainsworth, E.A.; Leckey, A.D.B. Will elevated carbon dioxide concentration amplify the benefits of nitrogen fixation in legumes? Plant Physiol. 2009, 151, 1009–1016.
38. Dodd, M.B. Elevated CO2 impacts on grazed pasture: Long-term lessons from the New Zealand FACE. In Proceedings of the 22nd International Grassland Congress, Sydney, Australia, 15–19 September 2013; Michalk, D.L., Millar, G.D., Badgery, W.B., Broadfoot, K.M., Eds.; New South Wales Department of Primary Industry: Sydney, Australia, 2013; pp. 1279–1284.
39. West, J.B.; Hillerislambers, J.; Lee, T.D.; Hobbie, S.E.; Reich, P.B. Legume species identity and soil nitrogen supply determine symbiotic nitrogen-fixation responses to elevated atmospheric CO2. New Phytol. 2005, 167, 523–466.
40. Thomas, R.G. Reproductive development. In White Clover; Baker, M.J., Williams, W.M., Eds.; CAB International: Wallingford, UK, 1987; pp. 63–124.
41. Thomas, R.G. The influence of environment on seed production capacity in white clover (Trifolium repens L.). I. Controlled environment studies. Aust. J. Agric. Res. 1961, 12, 227–238.
42. Thomas, R.G. Growth of the white clover plant in relation to seed production. In Herbage Seed Production; Lancashire, J.A., Ed.; Grassland Research and Practice Series No. I; New Zealand Grassland Association: Palmerston North, New Zealand, 1980; pp. 56–63.

43. Norris, I.B. Flowering of contrasting white clover varieties in relation to temperature in controlled environments. Ann. Bot. 1985, 56, 317–322.
44. Langer, R.H.M. How Grasses Grow, 2nd ed.; The Institute of Biology's Studies in Biology, No. 34; Edward Arnold (Publishers) Ltd.: London, UK, 1979.
45. Alberda, T. The effects of cutting, light intensity and night temperature on growth and soluble carbohydrate content of Lolium perenne L. Plant Soil 1957, 8, 199–230.
46. Aamlid, T.S.; Heide, O.M.; Boelt, B. Primary and secondary induction requirements for flowering of contrasting European varieties of Lolium perenne. Ann. Bot. 2000, 86, 1087–1095.
47. Purvis, O.N.; Gregory, F.G. Devernalization by high temperature. Nature 1945, 155, 113–114.
48. Wilson, J.R. Influence of time of tiller origin and of nitrogen level on floral initiation and ear emergence of four pasture species. N. Z. J. Agric. Res. 1959, 2, 915–932.
49. Ryle, G.J.A. Effects of daylength and temperature on ear size in S.24 perennial ryegrass. Ann. Appl. Biol. 1965, 55, 107–114.
50. Rolston, M.P.; Rowarth, J.S; Young, W.C., III; Mueller-Warrant, G.W. Grass seed crop management. In Herbage Seed Production. Temperate Species; Fairey, D.T., Hampton, J.G., Eds.; CAB International: Wallingford, UK, 1997; pp. 105–126.
51. Hampton, J.G.; Hebblethwaite, P.D. The effects of the environment at anthesis on the seed yield and yield components of perennial ryegrass (Lolium perenne L.). cv. S24. J. Appl. Seed Prod. 1983, 1, 21–22.
52. Idso, S.B.; Kimball, B.A. Growth response of carrot and radish to atmospheric CO2 enrichment. Environ. Exp. Bot.1989, 29, 135–139.
53. Mortensen, L.M. Effects of elevated CO2 concentration on growth and yield of eight vegetable species in a cool climate. Sci. Hortic. 1994, 58, 177–185.
54. Wheeler, T.R.; Morison, J.I.L.; Ellis, R.H.; Hadley, P. The effects of CO2, temperature and their interaction on the growth and yield of carrot (Daucus carota L.). Plant Cell Environ. 1994, 17, 1275–1284.
55. Usuda, H.; Shimogawara, K. The effects of increased atmospheric carbon dioxide on growth, carbohydrates and photosynthesis in radish, Raphanus sativus. Plant Cell Physiol. 1998, 39, 1–7.
56. Curtis, P.S.; Snow, A.A.; Miller, A.S. Genotype-specific effects of elevated CO2 on fecundity in wild radish (Raphanus raphanistrum). Oecologia 1994, 97, 100–105.
57. Case, A.L.; Curtis, P.S.; Snow, A.A. Heritable variation in stomatal response to elevated CO2 in wild radish, Raphanus raphanistrum (Brassicaceae). Am. J. Bot. 1998, 85, 253–258.
58. Quagliotti, L. Effects of different temperatures on stalk development, flowering habit, and sex expression in the carrot Daucas carota L. Euphytica 1967, 16, 83–103.
59. Gray, D.; Steckel, J.R.A.; Dearman, J.; Brocklehurst, P.A. Some effects of temperature during seed development on carrot (Daucus carota) seed growth and quality. Ann. Appl. Biol. 1988, 112, 367–376.
60. Sheen, T.F. Effect of day and night temperature variation and of high temperature on devernalization in radish. Acta Hortic. 2000, 514, 157–162.
61. Weibe, J.H.; Habegger, R.; Liebig, H.P. Quantification of vernalization and devernalisation effects for kohl rabi (Brassica oleracea convar. acephala var. gongylodes L.). Sci. Hortic. 1992, 50, 11–20.
62. Zenkteler, E.; Samordakiewicz, S.; Kaluzewicz, A.; Knaflewski, M. Effect of devernalization on the transition from vegetative to prefloral phase of the broccoli (Brassica oleracea var. italica cv. 'Fiesta') shoot meristem. Acta Agrobot.2012, 65, 29–36.
63. El Murabaa, A.I.M. Effects of high temperature on incompatibility in radish. Euphytica 1957, 6, 268–270.
64. Kumar, P.R.; Yadav, S.K.; Sharma, S.R.; Lal, S.K.; Jha, D.N. Impact of climate change on seed production of cabbage in North Western Himilayas. World J. Agric. Sci. 2009, 5, 18–26.

65. Frenck, G.; van der Linden, L.; Mikkelsen, T.N.; Brix, H.; Jorgensen, R.B. Increased [CO2] does not compensate for negative effects on yield caused by higher temperatures and [O3] in Brassica napus L. Eur. J. Agron. 2011, 35, 127–134.
66. Osborne, J.L.; Awmack, C.S.; Clark, S.J.; Williams, I.H.; Mills, V.C. Nectar and flower production in Vicia faba L. (field bean) at ambient and elevated carbon dioxide. Apidologia 1997, 28, 43–55.
67. Porter, J.R. Rising temperatures are likely to reduce crop yields. Nature 2005, 436, 174.
68. Prasad, P.V.V.; Boote, K.J.; Hartwell, L.; Allen, J.R.; Thomas, J.M.G. Effects of elevated temperature and carbon dioxide on seed-set and yield of kidney bean (Phaseolus vulgaris L.). Glob. Chang. Biol. 2002, 8, 710–721.
69. May, P.; Liao, W.; Wu, Y.; Shuai, B.; McCombie, W.R.; Zhang, M.Q.; Liu, Q.A. The effects of carbon dioxide and temperature on microRNA expression in Arabidopsis development. Nat. Commun. 2013, 4, 2145.
70. Zinn, K.E.; Tunc-Ozdemir, M.; Harper, J.F. Temperature stress and plant sexual reproduction: Uncovering the weakest links. J. Exp. Bot. 2010, 61, 1959–1968.
71. Barnabas, B.; Jager, K.; Feher, A. The effect of drought and heat stress on reproductive processes in cereals. Plant Cell Environ. 2008, 31, 11–38.
72. Rusterholz, H.P.; Erhardt, A. Effects of elevated CO2 on flowering phenology and nectar production of nectar plants important for butterflies of calcareous grasslands. Oecologia 1998, 113, 341–349.
73. Hoover, S.E.R.; Ladley, J.J.; Shchepetkina, A.A.; Tisch, M.; Gieseg, S.P.; Tylianakis, J.M. Warming, CO2, and nitrogen deposition interactively affect a plant-pollinator mutualism. Ecol. Lett. 2012, 11, 1351–1363.
74. Baker, H.G.; Baker, I. Floral nectar sugar constituents in relation to pollinator type. In Handbook of Experimental Pollination Biology; Jones, C., Little, R., Eds.; Van Nostrand Reinhold: New York, NY, USA, 1983; pp. 117–141.
75. Hall, A.E. Breeding for adaptation to drought and heat in cowpea. Eur. J. Agron. 2004, 21, 447–454.
76. Hedhly, A.; Hormaza, J.I.; Herrero, M. Global warming and sexual plant reproduction. Trends Plant Sci. 2008, 14, 30–36.
77. Koti, S.; Reddy, K.R.; Reddy, V.R.; Kakani, V.G.; Zhao, D. Interactive effects of carbon dioxide, temperature, and ultraviolet-B radiation on soybean (Glycine max L.) flower and pollen morphology, pollen production, germination and tube lengths. J. Exp. Bot. 2005, 56, 725–736.
78. Prasad, P.V.V.; Boote, K.J.; Hartwell Allen, L., Jr. Adverse high temperature effects on pollen viability, seed set, seed yield and harvest index of grain-sorghum [Sorghum bicolor (L.) Moench] are more severe at elevated carbon dioxide due to higher tissue temperatures. Agric. For. Meteorol. 2006, 139, 237–251.
79. Marshall, D.L.; Tyler, A.P.; Abrahamson, N.J.; Avritt, J.J.; Barnes, M.G.; Larkin, L.L.; Medeiros, J.S.; Reynolds, J.; Shaner, M.G.M.; Simpson, H.L.; et al. Pollen performance of Raphanus sativus (Brassicaceae) declines in response to elevated CO2. Sex. Plant Reprod. 2010, 23, 325–336.
80. Conroy, J.; Hocking, P. Nitrogen nutrition of C3 plants at elevated atmospheric CO2 concentrations. Physiol. Plant.1993, 89, 570–576.
81. Boote, K.J.; Allen, L.H.; Prasad, P.V.V.; Baker, J.T.; Gesch, R.W.; Snyder, A.M.; Pan, D.; Thomas, J.M.G. Elevated temperature and CO2 impacts on pollination, reproductive growth, and yield of several globally important crops. J. Agric. Meteorol. 2005, 60, 469–474.
82. Klein, A.M.; Vaissiere, B.E.; Cane, J.H.; Steffan-Dewenter, I.; Cunningham, S.A.; Kremen, C.; Tscharntke, T. Importance of pollinators in changing landscapes for world crops. Proc. R. Soc. B Biol. Sci. 2007, 274, 303–313.
83. Schweiger, O.; Biesmeijer, J.C.; Bommarco, R.; Hickler, T.; Hulme, P.E.; Klotz, S.; Kuhn, I.; Moora, M.; Neilsen, A.; Ohlemuller, R.; et al. Multiple stressors on biotic interactions: How climate change and alien species interact to affect pollination. Biol. Rev. 2010, 85, 777–795.

84. Dietlein, D.G. Measurement of carbon dioxide levels in honey bee colonies using a thermistor bridge. Am. Bee J. 1985, 125, 773–774.
85. Ribbands, C.R. Changes in the behaviour of honey-bees following their recovery from anaesthesia. Exp. Biol. 1950, 27, 302–310.
86. Ebadi, R.; Gary, N.E.; Lorenzen, K. Effects of carbon dioxide and low temperature narcosis on honeybees, Apis mellifera. Environ. Entomol. 1980, 9, 144–147.
87. Buhler, R.; Lanzrein, B.; Wille, H. Influence of temperature and carbon dioxide concentration on juvenile hormone titre and dependent parameters of adult worker honey bees (Apis mellifera L.). J. Insect Physiol. 1983, 29, 885–893.
88. Willmer, P.G.; Stone, G.N. Behavioral, ecological, and physiological determinants of the activity patterns of bees. Adv. Stud. Behav. 2004, 34, 347–466.
89. Kjohl, M.; Nielsen, A.; Stenseth, N.C. Potential Effects of Climate Change on Crop Pollination; FAO: Rome, Italy, 2011.
90. Deutsch, C.A.; Tewksbury, J.J.; Huey, R.B.; Sheldon, K.S.; Ghalambor, C.K.; Haak, D.C.; Martin, P.R. Impacts of climate warming on terrestrial ecotherms across latitude. Proc. Natl. Acad Sci. USA 2008, 105, 6668–6672.
91. Simmonds, N.W. Principles of Crop Improvement; Longman: London, UK, 1979.
92. Conner, A.J.; Field, R.J. Herbicide-resistant crops: A new approach to an old problem or a radical new tool? In Herbicide-Resistant Crops and Pastures in Australian Farming Systems; McLean, G.D., Evans, G., Eds.; Bureau of Resource Sciences: Canberra, Australia, 1995; pp. 53–71.
93. Nieuwhof, M. Effect of temperature on the expression of male sterility in Brussels sprouts (Brassica oleracea L. var. gemmifera DC). Euphytica 1986, 17, 265–273.
94. Townsend, C.E. Self-compatibility studies with diploid alsike clover, Trifolium hybridum L., III. Response to temperature. Crop Sci. 1968, 8, 269–272.
95. Kendall, W.A.; Taylor, N.L. Effect of temperature on pseudo-self-compatibility in Trifolium pratense L. Theor. Appl. Genet. 1969, 39, 123–126.
96. Chen, C.C.; Gibson, P.B. Effect of temperature on pollen-tube growth in Trifolium repens after cross- and self-pollination. Crop Sci. 1973, 13, 563–566.
97. Wilkins, P.W.; Thorogood, D. Breakdown of self-incompatibility in perennial ryegrass at high temperature and its uses in breeding. Euphytica 1992, 64, 65–69.
98. Young, L.W.; Wilen, R.W.; Bonham-Smith, P.C. High temperature stress of Brassica napus during flowering reduces micro-and megagametophyte fertility, induces fruit abortion, and disrupts seed production. J. Exp. Bot. 2004, 55, 485–495.
99. Gibson, L.R.; Paulsen, G.M. Yield components of wheat grown under high temperature stress during reproductive growth. Crop Sci. 1999, 39, 1841–1846.
100. Ziska, L.H.; Bunce, J.A. Influence of increasing carbon dioxide concentration on the photosynthetic and growth stimulation of selected C4 crops and weeds. Photosynth. Res. 1997, 54, 199–208.
101. Huxman, T.E.; Hamerlynck, E.P.; Jordan, D.S.; Salsman, K.L.; Smith, S.D. The effects of parental CO2 environment on seed quality and subsequent seedling performance. Oecologia 1998, 114, 202–208.
102. Powell, A.A. Seed vigour and its assessment. In Handbook of Seed Science and Technology; Basra, A.S., Ed.; Food Products Press: New York, NY, USA, 2006; pp. 603–648.
103. Halmer, P. Storage management. In The Encylopedia of Seeds: Science, Technology and Uses; Black, M., Bewley, J.D., Halmer, P., Eds.; CAB International: Wallingford, UK, 2006; pp. 659–662.
104. Hill, M.J. The characteristics of grain and seeds and their reaction to climate. In The Drying and Storage of Grain and Herbage Seeds; Hill, M.J., Ed.; Foundation for Arable Research: Lincoln, New Zealand, 1999; pp. 3–19.
105. Probert, R.J.; Hay, F.R. Keeping seeds alive. In Seed Technology and Its Biological Basis; Black, M., Bewley, J.D., Eds.; Sheffield Academic Press: Sheffield, UK, 2000; pp. 375–410.

106. Fenemore, P.; Lauren, D.; Hill, M.J. Pests and diseases: Their role and significance in affecting seed quality. In The Drying and Storage of Grain and Herbage Seeds; Hill, M.J., Ed.; Foundation for Arable Research: Lincoln, New Zealand, 1999; pp. 93–116.
107. Harrington, J.F. Seed storage longevity. In Seed Biology; Kozlowski, T.T., Ed.; Academic Press: New York, NY, USA, 1972; Volume 3, pp. 145–245.
108. Fischer, G.; Shah, M.; Tubiello, N.; van Velhuizen, H. Socio-economic and climate change impacts on agriculture: An integrated assessment, 1990–2080. Philos. Trans. R. Soc. B 2005, 360, 2067–2083.
109. Hampton, J.G.; Rolston, M.P.; Pyke, N.B.; Green, W. Ensuring the long term viability of the New Zealand seed industry. Agron. N. Z. 2012, 42, 129–140.
110. Olesen, J.E.; Trnka, M.; Kersebaum, K.C.; Skjelvag, A.O.; Seguin, B.; Peltonen-Saino, P.; Rossi, F.; Kozyra, J.; Micale, F. Impacts and adaptation of European crop production systems to climate change. Eur. J. Agron. 2011, 34, 96–112.
111. Kalra, N.; Chakraborty, D.; Sharma, A.; Rai, H.K.; Jolly, M.; Chander, S.; Kumar, P.R.; Bhadraray, S.; Barman, D.; Mittal, R.B.; et al. Effect of increasing temperature on yield of some winter crops in northwest India. Curr. Sci. 2008, 94, 82–88.
112. Kuchirak, C.J. Contribution of planting date trends to increased maize yields in the Central United States. Agron. J.2008, 100, 328–336.
113. Lanning, S.P.; Kephart, K.; Carlson, G.R.; Eckhoff, J.E.; Stougaard, R.N.; Wichman, D.M.; Martin, J.M.; Talbert, L.E. Climate change and agronomic performance of hard red spring wheat from 1950 to 2007. Crop Sci. 2010, 50, 835–841.
114. Hammer, G.; Singh, V.; Nguyen, C.; van Oosterom, E.; McLean, G.; Zheng, B.; Chapman, S.; Jordan, D. Sowing Date Effects on Yield and High Temperature Risk in Sorghum. 2011. Available online: http://www.grdc.com.au/director/events/researchupdates?item_id=3E6A04D500 1209 (accessed on 16 November 2011).
115. Rahman, M.M.; Hampton, J.G.; Hill, M.J. Soybean seed yield as affected by time of sowing in a cool temperature environment. J. New Seeds 2005, 7, 1–15.
116. Board, J.E.; Hall, U. Premature flowering in soybean yield reductions at nonoptimal planting dates as influenced by temperature and photoperiod. Agron. J. 1984, 76, 700–704.
117. Ridge, P.E.; Pye, D.L. The effects of temperature and frost at flowering on the yield of peas grown in a Mediterranean environment. Field Crop. Res. 1985, 12, 339–346.
118. Parry, M.L. (Ed.) Assessment of Potential Effects and Adaptations for Climate Change in Europe; The Europe ACACIA Project, Jackson Environment Institute, University of East Anglia: Norwich, UK, 2000.
119. Cairns, J.E.; Hellin, J.; Sonder, K.; Araus, J.L.; MacRobert, J.F.; Thierfelder, C.; Prasanna, B.M. Adapting maize production to climate change in sub-Sarharan Africa. Food Secur. 2013, 5, 345–360.
120. Prasanna, B.M.; Chaikam, V.; Mahuku, G. Doubled Haploid Technology in Maize Breeding: Theory and Practice; CIMMYT: Mexico City, Mexico, 2012.
121. Varshney, R.K.; Bansal, K.C.; Aggarwal, P.K.; Datta, S.; Craufurd, P.Q. Agricultural biotechnology for crop improvement in a variable climate: Hope or hype? Trends Plant Sci. 2011, 16, 363–371.
122. Ortiz, R. Crop genetic engineering under global climate change. Ann. Arid Zone 2008, 47, 1–12.
123. Wahid, A.; Gelani, S.; Ashraf, M.; Foolad, M.R. Heat tolerance in plants: An overview. Environ. Exp. Bot. 2007, 61, 199–223.
124. Tewolde, H.; Fernandez, C.J.; Erickson, C.A. Wheat cultivars adapted to post-heading high temperature stress. J. Agron. Crop Sci. 2006, 192, 111–120.
125. Salem, M.A.; Kakani, V.G.; Koti, S.; Reddy, K.R. Pollen-based screening of soybean genotypes for high temperature. Crop Sci. 2007, 47, 219–231.
126. Bannayan, M.; Tojosoler, C.M.; Garcia, Y.; Garcia, A.; Guerra, L.C.; Hoogenboom, G. Interactive effects of elevated CO2 and temperature on growth and development of a short-and long-season peanut cultivar. Clim. Chang. 2009, 93, 389–406.

127. Craufurd, P.Q.; Prasad, P.V.V.; Summerfield, R.J. Dry matter production and rate of change of harvest index at high temperature in peanut. Crop Sci. 2002, 42, 146–151.
128. Porch, T.G.; Jahn, M. Effect of high-temperature stress on microsporogenesis in heat-sensitive and heat-tolerant genotypes of Phaseolus vulgaris. Plant Cell Environ. 2001, 24, 723–731.
129. Ahmed, F.E.; Hall, A.E.; Madore, M.A. Interactive effects of high temperature and elevated carbon dioxide concentration on cowpea (Vigna unguiculata L. Walp.). Plant Cell Environ. 1993, 16, 835–842.
130. Kakani, V.G.; Reddy, K.R.; Koti, S.; Wallace, T.P.; Prasad, P.V.V.; Reddy, V.R.; Zhao, D. Differences in in vitro pollen germination and pollen tube growth of cotton cultivars in response to high temperature. Ann. Bot. 2005, 96, 59–67.
131. Dyck, J.A.; Matus-Cadiz, M.A.; Hucl, P.; Talbert, L.; Hunt, T.; Duboc, J.P.; Noss, H.; Clayton, G.; Dodd, J.; Quick, J. Agronomic performance of hard red spring wheat lines sensitive and insensitive to photoperiodism. Crop. Sci. 2004, 44, 1976–1981.
132. Blake, N.K.; Lanning, S.P.; Martin, J.M.; Doyle, M.; Sherman, J.D.; Naruoka, Y.; Talbert, L.E. Effect of variation for major growth habit genes on maturity and yield in five spring wheat populations. Crop Sci. 2009, 49, 1211–1220.
133. Habash, D.Z.; Kehel, Z.; Nachit, M. Genomic approaches for designing durum wheat ready for climate change with a focus on drought. J. Exp. Bot. 2009, 60, 2805–2815.
134. Bita, C.E.; Gerats, T. Plant tolerance to high temperature in a changing environment: Scientific fundamentals and production of heat stress-tolerant crops. Front. Plant Sci. 2013, 4, 1–18.
135. Hong, S.W.; Lee, U.; Vierling, E. Arabidopsis hot mutants define multiple functions required for acclimation to high temperature. Plant Physiol. 2003, 132, 757–767.
136. Roy, S.J.; Tucker, E.J.R.; Tester, M. Genetic analysis of abiotic stress tolerance in crops. Curr. Opin. Plant Biol. 2011, 14, 232–239.
137. Paliwal, R.; Roder, M.S.; Kumar, U.; Srivastava, J.; Joshi, A.K. QTL mapping of terminal heat tolerance in hexaploid wheat (T. aestivum L.). Theor. Appl. Genet. 2012, 125, 561–575.
138. Giorno, F.; Wolters-Arts, M.; Mariani, C.; Rieu, I. Ensuring reproduction at high temperatures: The heat stress response during anther and pollen development. Plants 2013, 2, 489–506.
139. Manderscheid, R.; Weigel, J. Photosynthetic and growth responses of old and modern spring wheat cultivars to atmospheric CO2 enrichment. Agric. Ecosyst. Environ. 1997, 64, 65–73.
140. Ziska, L.H.; Morris, C.F.; Goins, E.W. Quantitative and qualitative evaluation of selected wheat varieties released since 1903 to increasing atmospheric carbon dioxide: Can yield sensitivity to carbon dioxide be a factor in wheat performance? Glob. Chang. Biol. 2004, 10, 1810–1819.
141. Collins, N.C.; Tardieu, F.; Tuberosa, R. Quantitative trait loci and crop performance under abiotic stress: Where do we stand? Plant Physiol. 2008, 147, 469–486.
142. Reif, J.C.; Zhang, P.; Dreisigacker, S.; Warburton, M.L.; van Ginkel, M.; Hoisington, D.; Bohn, M.; Melchinger, A.E. Wheat genetic diversity trends during domestication and breeding. Theor. Appl. Genet. 2005, 110, 859–864.
143. Von Korff, M.; Radovic, S.; Choumane, W.; Stamati, K.; Udupa, S.M.; Grandol, S.; Ceccarelli, S.; MacKay, I.; Powell, W.; Baum, M.; et al. Asymetric allele-specific expression in relation to developmental variations and drought stress in barley hybrids. Plant J. 2009, 59, 14–26.
144. Carena, M.J.; Bergman, N.; Riveland, N.; Eriksmoen, E.; Halvorson, M. Breeding maize for higher yield and quality under drought stress. Maydica 2009, 54, 287–296.
145. Cairns, J.E.; Sonder, K.; Zaidi, P.H.; Verhulst, N.; Mahuku, G.; Babu, R.; Nair, S.K.; Das, B.; Govaerts, B.; Vinayan, M.T.; et al. Maize production in a changing climate. Adv. Agron. 2012, 144, 51–58.
146. Westengen, O.T.; Brysting, A.K. Crop adaptation to climate change in the semi-arid zone in Tanzania: The role of genetic resources and seed systems. Agric. Food Secur. 2014, 3, 3.

147. Pariyar, D.; Hampton, J.G. Base Line Survey—Cool Season Crop Improvement Programme, Nepal; New Zealand Ministry of Foreign Affairs and Trade: Kathmandu, Nepal, 2015.

148. Setimila, P.; MacRobert, J.; Atlin, G.N.; Magorokosho, C; Tarekegne, A.; Makumbi, D.; Taye, G. Performance of elite maize varieties tested in on-farm trials in eastern and southern Africa. In Proceedings of the ASA, CSSA, and SSSA International Annual Meetings, Cincinnati, OH, USA, 21–24 October 2012.

149. Cairns, J.E.; Crossa, C.; Zaidi, P.H.; Grudloyma, P.; Sanchez, C.; Araus, J.L.; Makumbi, D.; Magorokosho, C.; Bänziger, M.; Menkir, A.; et al. Identification of drought, heat, and combined drought and heat tolerance donors in maize (Zea mays L.). Crop Sci. 2013, 53, 1335–1346.

150. Barnett, C.; Chisuo, M.; Pinto, Y. Country case studies on the pass value chain strategy/approach and its impact/effect on small holder farmer yields in Africa. In East Africa Synthesis Report (Kenya, Tanzania, Uganda); Agriculture Learning and Impact Network: London, UK, 2011.

151. Adapting Agriculture to Climate Change. Royal Botanic Gardens, Kew, UK. Available online: http://www.kew.org/science-conservation/reserawch-data/science-directory/projects/ada (accessed on 29 February 2016).

CHAPTER 5

Postharvesting Techniques and Maintenance of Seed Quality

Zubaida Yousaf*, Nadia Saleh, Asba Ramazan and Arusa Aftab

Department of Botany, Lahore College for Women University, Lahore, Pakistan

ABSTRACT

Healthy seeds and grains are the demanding enterprise of the recent era for the production of high yield in the next season. The seeds must be stored for the maintenance of high-yielding crop. During storage, major losses of seeds are caused by various biological and nonbiological factors. There is a need to examine reasonable factors of these crop losses, which ultimately affect the market value and quality of the seed. The quality of seeds can be maintained by using careful postharvest handling techniques. There is need to establish the well-suited methods to assess the losses during the process and to use the best technique to minimize the loss and to ensure the quality and safety of the crop. The target is to achieve the high-quality seeds of the national and international standards that could meet the demand of the supplier. This chapter emphasizes on the aspects and postharvest techniques that are used to maintain seed quality. A comprehensive review of the better, economical, convenient, and productive methods is provided, focused on the needs of developing countries but also with relevance in more industrialized countries.

Keywords: crop maintenance, handling techniques, storage losses, postharvest

1. INTRODUCTION

Seed industry survives for the sake of better quality of seeds and postharvest storage techniques. It is a seed quality program that seeks the high achievement and gives the quality assurance. Seed quality is one of the most demanding enterprises of present days. The majority of the seeds especially cereals and legumes by small-scale farmers are produced and stored on the farms. The major problem is the damage due to biological factors such as molds, and insects can be alleviated by implicating effective storage techniques [1–4]. It is the foremost priority of the farmers to prevent the crop losses and storage of seeds/grains. Farmers often purchase the new seeds/grains from the market to produce the next crop for higher yield due to the risk of crop loss. There is a need to develop the effective strategies for seed storage that can give the confirmation of better crop yield and reduce the risks of storage losses [5].

1.1. Valuable Seeds: Achieving Seed Quality

The value of the seeds allows the farmers to produce high-yielding crops from good quality healthy seeds. Seed quality is the measure of the potential to produce desired quality, healthy, and high-yielding crops at low planting rates [5]. Seed quality can neither be obtained automatically nor by a permanent process. There is nature's pressure against the quality of the seeds. Efforts are being made to produce best quality seeds for the farmers. Any handling or production stage can reduce the quality of the seed. To alleviate these losses, careful technical management is required for all seed operations [6].

1.2. Principles of Seed Storage

Seeds when stored in natural environment or ambient temperature respond readily to temperature, available oxygen, and relative humidity. Metabolic activities, age, and longevity of seeds can be manipulating by controlling the humidity temperature and oxygen [7]. Reduction of seed moisture content up to an appropriate limit is prerequisite for storage as seed could be damaged because of desiccation. Seeds can be stored for a longer period due to lower level of humidity. According to the thumb rule, the life of the seed doubles by decreasing the moisture content to 1% in case if seed moisture content is between 5 and 14%. Higher moisture content is more affected by higher temperature so seeds need to be stored in cool location. The life of the seed doubles by decreasing the temperature to 5°C and is applicable between 0 and 50°C [8].

Oxygen level can be controlled by hermetic storage in a sealed container which reduces the physiological aging of the grains as well as reduces the physical damage due to insects and fungal growth [8].

2. POSTHARVEST TECHNIQUES OF SEED STORAGE

2.1. Drying

Physiological maturity attained by the cereal and legumes at moisture content between 35 and 45% is crop dependent. Temperature affects the storage of seeds at moisture content between 10 and 14%. For high-quality yield of crops, timely harvesting and drying are necessary. Biologically active seeds deteriorate readily under most of the circumstances due to fungal contamination and attack of insects and other pests. The main purpose of drying is to reduce the respiration in seeds [6, 9]. The process also impedes qualitative damage due to fungi and other insect pests. Drying can itself affect the quality of the seeds. Extensive drying under very high temperature can damage the seed. In summer season, simple drying methods are used through exposure to sun and adequate wind. The alternative drying methods have been devised for high-yielding varieties and improved farming practices and irrigation to deal with increased production or harvesting in wet season in multi-cropping [10, 11].

2.2. Sun Drying

Sun drying of seeds is the drying method in the tropical developing countries. The method is employed when crop is ready for the harvest. Some seeds like maize can be sun dried although during drying, crops get sensitive to insect infestation, rodent or birds' attack, and mold damage. A threshed seed drying by spreading on sheets or tray is a common phenomenon but has a risk of soil or stone contamination. For example, paddy at rice mills is at large quantity [6]. Rice is dried on especially build drying floor that allow easy run off of rain water. The seeds are dispersed in thin layer and turned at regular intervals to facilitate drying and covered at night with the help of sheets. There are some disadvantages of the process that temperature is an uncontrollable factor. In paddy rice, high temperature can cause stress or cracks in the seeds which lead to high level of damage during milling. Yield can be contaminated by dust, atmospheric contamination or insect infestation, or human or animal disturbance [6, 9].

2.3. Solar Drying

It is the modification of sun drying in which sun rays are collected in an especially designed unit for air removal in adequate ventilation system. The unit has 20–30° higher temperature than open drying, and less time is consumed in the process [12]. In solar dryers, solar collector is used to heat air which then allowed to pass to the seed beds. It comprises two basic designs: natural convection dryers use thermal gradients and forced convection dryers force the air through solar collectors and seed layers. These dryers are suitable for farm use. The former design of the Asian Institute of Technology in Bangkok has been used as blueprint for several convection dryers and comprises a drying bin, a solar chimney, and a solar collector [13]. The solar collector consists of black polythene sheet or layer of burnt paddy husk; it is covered with clear polythene sheet. Perforated platform presents in drying bin. The disadvantages of the process are as follows: high structural profile, stability problem in windy condition, and the need of replacement of polythene sheets at regular intervals [6].

2.3.1. Mechanical Dryers

The same principle of drying is used by mechanical dryers as forced convection solar dryer; the dryer forced the air through the seed bed and the air is heated with the help of a flat plate instead of conventional means. In modern automated storage system, drying takes place at one of the two points either in prestorage dryers (prior to loading seeds in freestanding loading) or in store dryer (after loading in final storage compartment) [9]. In prestorage dryers, ambient air is used in continuous flow dryers, and heat is generated by thermostatically controlled furnace which is powered by electricity, diesel, or gas. Heat may be supplied by direct or indirect way. Indirect way is preferred due to the separate outlet for the combustion product not through seeds. In the batch dryers, seeds are fed into properly defined batches, whereas in continuous flow dryers, the grains are flowed into the system and recovered at desired moisture content [6].

2.3.2. Tray Dryers

Tray dryers are batch dryers of flat beds. The seeds are dispersed on the mesh tray at the depth of 600–700 mm, and warm dried air is passed through seeds to sufficiently dry them [6, 14].

2.4. Radial Drying Bin

Radial drying bin comprises two vertical metal mesh cylinders, one inside the other. Seeds are loaded between these two cylinders, and air is blown to the inner cylinder and passed from the inner to the outer mesh cylinder. By reversing air through seeds, air can be sucked from the central cylinder. However, there is the risk of overdrying of seeds in the inner cylinder which are in direct contact with the hot air. Air is wetter and cooler at the leaving side toward outside [6].

2.5. Continuous Flow Dryers

Moisture content of the seeds can be removed by sucking or blowing hot air by top-to-bottom passage through the system. Bin is present at the top of the drying section with cooling system at the base. Seed beds can be horizontal, vertical, or inclined. Seeds are moved by conveyors, scrapers, vibration, or gravity. The degree of drying depends upon the speed, size, and rate of flow of outlet conveyor of the dryer. Continuous flow dryer is varied by relative direction of air stream and seed flow [15]. Several continuous flow dry techniques are described below.

2.5.1. Cross Flow

The seed passes through the two perforated sheets downward to the column by allowing horizontal passage of air through the seeds. The advantage of the dryer is that the moisture gradient can be defined at any stage of drying seeds [6].

2.5.2. Counterflow

A round bin is used to unload seeds with the upward flow of air. Little evaporative cooling takes place when hottest air passes through the driest seeds.

2.5.3. Concurrent Flow

In concurrent flow, wettest seeds are exposed to the hottest air during passage of air through seed bed. High temperature improves the efficiency of the dryer and cools the seeds by moisture evaporation [6, 16–18].

2.5.4. Cross Flow

Cross-flow dryers have been used widely in recent years, but mixed-flow dryers have advantages over cross-flow dryers. Mixed flow dryers, combination of concurrent, counter and cross flow dryers has great advantage of efficient fuel consumption. But reduction in output because of uneven flow of seeds leading to uneven drying is one biggest hindrance in adaption of mixed flow drying [6].

2.5.5. Tower (Mixed Flow)

They consist of tall rectangular bins for storage, and triangular ducts are present along the width of the dryer at horizontal position. Half of the ducts are used for induction of warm air and removal of damped and cooled air is done by remaining ducts. It has multiple directions of seed flow and air flow [16–18].

2.5.6. Louvered Bed Dryer

The seeds are passed through different types of batches; the hot air is blown to the seeds to dry them. The dryers work on the principle of cross-flow dryer. The speech and depth of the drying beds depends upon the degree of drying. The two basic designs of the dryers are conveyor dryer and cascade dryer. The cascade dryers are gravity fed cross-flow dryers. The seed depth is controlled by roller dams and speed is controlled by output elevators. There can be incorporated changes to vary the length of the dryer. In conveyor dryer, air is blown to seeds through inclined louvered bed and the seed flow is controlled by variable speed, roller chain conveyor and heavy duty. These dryer can be one directional, two directional, or multidirectional; the variation in direction assists in removal of waste material and reduction in size of the dryer [6].

3. IN STORE DRYING

This is an alternative method of drying in which seeds are load into bulks floor storage or in bin and then they are dried in stores [19].

3.1. Bulk on Floor Storage

They consist of especially strengthened wall which can bear the weight of seeds. The seeds are loaded in uniform depth. At one side of the building, fan is present for the aeration purpose and the plenum chamber runs along the center of the store or walls with perforated lateral ducts, below or above the floor level under the bulk of seeds [6].

3.2. In Bin Drying

This type of drying comprises one or more bins for drying purpose and other bins for storage. The dryers reduce the chance of physical damage due to the lesser handling. The shallow layer of seed along the bins for drying reduces time consumption and makes the process safer. The semi-dried batches free the space for incoming seeds and consist of lateral ventilated system or ventilated floors about 0.5 m above the base [6, 20].

3.3. Bag Dryers

The drying in bags is difficult because there is not proper insurance of passage of air through the seeds. In sack platform dryers air is blown through the floor of air duct whereas in fan blowers heated air is blown from the floor apertures and sacks are placed on them. Larger bags are stacked in the center of tunnel in moisture extraction unit. Through the air ducts, hot air is blown with the help of a fan. One must be careful of correct dimension to avoid uneven drying. However this system is not appropriate for even drying of seeds because of short circuit over certain areas. [6].

4. STORAGE LOSSES

It is reported that a total of 30% crop losses take place due to harvest. However, this is the "worst case" figure to cite for the crops in priority area of development. Storage losses cannot be predicted before harvest. There can be many biological, climate, handling, harvesting, storage, and distribution and social and cultural factors which can cause crop losses. By careful handling, 50% postharvest and storage losses can be disseminated. There is no such method to predict the exact figures of such losses. Efforts have been made to determine the reliable baseline methods to figure out crop loss activities. A methodology has been made to estimate the standardized postharvest losses by crop activities [21]. Loss assessment and loss reduction programs have been prompted by the Food and Agricultural Organization (FAO) of the United Nations. The focus of these programs was to obviate the reduction of crop losses of staple foods. The methodology for assessment of seed loss during harvest was summarized [22], although there was no universal method to assess storage losses. Sampling procedures, varied handling and storage produce, and irregular movement of batches make the loss assessment phenomena non-generalized especially for perishable commodities. Care should be taken in designing a methodology with positive aims to make it acceptable, economical, and meaningful. The methods are varied for perishable commodities due to their different nature whereas relatively uniform for cereal seeds. To find out standard moisture content and dry matter, the weight loss of undamaged and damaged seeds can be compared. The purpose of storage is to avoid both biological and financial losses of yield. To obviate these losses, we must have known the root causes of these deprivations [9].

4.1. Loss and Damage

The term damage and loss can be confusing sometimes when used synonymously. Loss is defined as the measurable qualitative or quantitative decline in foodstuff. Superficial deprivation of commodities is defined as the damage, in which physical spoilage results in loss of commodity. Damaged commodity can either be used but loss is a permanent decline. [6, 7].

4.2. Categories of Storage Loss

Categories of qualitative or quantitative storage loss have economic impacts. Physical weight or volume loss which is considered as quantitative loss can be figured out readily. Quality losses can be measured by the simple judgment of commodity and comparing with standard quality items. Commodities can be rejected by changes in taste, texture, appearance, loss of nutritional value, and the presence of contaminants. The following categories can be listed to demonstrate the storage losses of crops [23].

The primary cause of losses directly affects the stored crops. They may be biological, chemical, or biochemical in nature.

4.3. Biological Losses

Biological factors responsible for crop deprivation are rodents, insects, birds and microbes (fungi and bacteria). Microbes grow on the crops and cause the weight loss, crop spoilage, and other defects which reduce the market demand of the yield. Development of infestation can make trouble if produce is stored for longer period. The damage or loss caused during storage in the cereal seeds by birds, rodents, and the condition can be harsher by microbial (fungi and bacteria) attack in the field. There will be quality loss, if the disease is superficial; it can be quantitative loss if infections penetrate into deeper layers of seeds. In case of superficial disease, it is possible to remove damaged area and use the rest of the portion [14, 24, 25]. *Chemical losses* involve loss of flavor, color, texture, and nutritional value due to pesticides and chemical reactions [26]. *Biochemical losses* can be softening, discoloration, and bad flavor due to enzyme-activated reactions. *Mechanical losses* are due to bruising, breakage, processing, and injury during handling or harvesting. *Physical losses* are related to climate condition such as low or high temperature, improper storage atmosphere, and high humidity. Physical factors can also mediate chemical and biochemical losses [7]. *Physiological losses* involve weight losses due to respiratory heat loss. Susceptibility of infection and damage by pathogen increases during wilting, senescence, ripening, and wilting. Microbiological and biological factors are important in seed, whereas mechanical, physiological, and microbial factors cause losses in perishable crops. The factors that encourage primary crop losses are secondary losses by inappropriate handling of equipment, technology, and control. The factors are lack of harvesting equipment, skills, packaging, handling, adequate containers, appropriate transport, drying and storage conditions, proper processing technique, and adequate management [6, 9].

4.4. Weight Loss

Weight loss is not necessarily the sign of loss of crops. Reduction in moisture content can be a reason of weight loss. Shrinkage factor is a tool to recognize for commercial transactions. Moisture loss can lead to economic loss if it is not taken into account by grading for price control. Weight loss can be due to feeding of birds, insects, rodents, and microorganisms. Weight loss can be measured by comparing the weight before and after storage in the sack. There can be also an increase in moisture content due to water production in seed by insect infestation and can lead to an increase in weight. If insect infestation increase the moisture content of seed or insect may consume seed, leaving behind dust, weight loss cannot be detected easily [27]. To detect these losses, a useful mass of infested and noninfested seeds is ground into flour, and their weight is compared. A decrease in flour yield of infested mass than sound mass will be observed. Be aware of malpractices to compensate weight loss by adulteration of stones, earth, or sand and water. So there is a need to assess not only moisture changes but also the quantity of foreign matter present in the yield [23, 28].

4.5. Quality Loss

Quality is an important trait considered by the consumers, and local traders have different criteria for assessment according to circumstances. Biochemical factors (acidity, sugars, flavor, and smell) involve size, shape, and appearance. Quality loss can also be due to contamination and foreign matter (insect fragments, rodent hair, excreta, weed seeds, earth, glass, stones, and parts of plants) content. Contaminants that are difficult to remove comprise soluble excretion of insects, pesticides, oils, toxins produced from fungal infections, and pathogenic organisms spread by rodents. Loss potential will be increased by raising the standard rules by consumers [24, 29].

4.6. Nutritional Loss

The loss depends upon the qualitative and quantitative loss of nutritional value to the human population which affects the nutritional status of that population. This is mainly caused by feeding of pest on selective part of seed. *Plodia* and *Ephestia* feed selectively on the seed embryo and remove the vitamin and protein content. Many pests feed on bran of cereal seeds as a result reduced vitamin content. *Liposcelis* spp. selectively feeds on bran of rice and embryo (Pike, 1994). Weevil is an endosperm feeder and declines the carbohydrate content [6, 30].

4.7. Loss of Seed Viability

The loss of seed viability is linked with. The reasons of the losses may be temperature, excessive respiration, moisture content, infestation, light, and infestation-controlling methods. Insects that are selective to attack the embryo cause great losses in germination as compared to others. Seed loss can be detected by standard germination tests [31].

4.8. Commercial Losses

Commercial losses are either due to direct consequences (foregoing factors) or indirect consequences (cost of preventive action or equipment). There may be a loss of goodwill, monetary loss, and loss caused by legal action. Intercountry trade might be affected by commercial loss. Losses can be rapidly reduced by experience and knowledge. Postharvest losses not necessarily take place due to inappropriate storage. There may be biological, physical, or mechanical factors in deterioration of cereal seeds. There is a need to broaden the intervention techniques to deliver high-quality product from the field to the market. For example, after the Tanzania outbreak of pest in the maize crop, Somalia and Malawi refused to take the maize due to insect spreading [32].

4.9. Temperature-Dependent Injury

High-temperature solar radiation on exposure deteriorates the fresh commodities rapidly. By increase of temperature, respiration also increases; there should be proper ventilation and cooling of the crops to obviate the problem. In similar way, there can be crop injury due to low temperature between 0 and −2°C. However some crops of tropical and subtropical origin showed tolerance for chilling injuries at 12–14°C. Chilling injuries (skin pitting, discoloration, uneven or abnormal ripening, and sensitivity to rapid deterioration) are apparent when commodity is removed from the environment [7, 29].

5. ASSESSMENT OF LOSSES

5.1. Assessment of Seed Losses in Storage

The agents causing seed loss during storage are insects, rodents, and molds. In past this issue gained attention of many scientist but still there is a need to find some more effective technique to avoid seed loss due to insects. Insects can cause both qualitative and quantitative loss by boring or feeding on seeds; weight loss has got more attention [6].

5.2. Weight Losses Caused by Insects

The assessment is made through collection of samples of seed and various intervals after storage and comparing the samples to observe the changes in subsequent samples. Assessing quantity loss with subsequent samples at different intervals will be used to estimate storage losses at different occasions. Sample collection from each batch of seed and their quantity loss are estimated accordingly. Sample should be collected without disturbing the pattern of infestation in the bulk stores. Three samples must be collected when subsequent regular sampling is not possible: first, at time of storage; next, in the midway of storage period; and last, at the final few weeks of seed storage.

The pattern of used seed and quantity loss is noted [22, 27].

5.3. Methods of Estimation of Weight Loss

There are two methods used for estimation of weight loss by the insects, when subsequent sampling is possible: volumetric method and thousand grain mass (TGM) method.

In this case, when subsequent sampling is impossible, count and weight and converted percentage methods are used [6].

5.4. Volumetric Method

Volumetric method is commonly known as the standard volume weight (SVW) or bulk density method. This is used to measure the bulk density of clean sample by the use of equipment. SVW is figured out from the sample of the seeds at the beginning of the storage period, and losses are determined. This method strictly calculates the weight loss by grain boring of insects and moisture difference over a subsequent time period in standard volume container. In stand volumetric method to determine accurate ratio for moisture content and dry weight of seed, moisture can be taken as constant term and the crop as dry matter. However, volume and frictional characters can also be affected by changes in moisture content. There is direct relation between moisture content and volume of the sample therefore seed should be packed loosely. To make the moisture constant, it is necessary to calculate the standard volume of dry matter at different moisture contents. The process is time-consuming and needs care and equipped laboratory [33]. Another factor affect the volume of sample is weight of insecticidal dust, adheres to the seed surface increase the volume of seed and frictional character. Sieving can be a useful phenomenon to remove dust. However, volumetric phenomena are less helpful due to overassessment of losses [6, 34].

5.5. Thousand Grain Mass Method

This method differs from volumetric method in comparison with fixed number of seeds instead of fixed volume. This means that weight of seeds is multiplied by thousand and corrected by a dry matter. It is calculated by weighing and counting the seeds in a given sample. Standard reading is made by taking measurement at the beginning of storage of seeds, and then subsequent readings are compared with baseline reading [35].

5.6. Count and Weight Method

The method, sometimes called gravimetric method, is applied where the baseline readings of seed storage are not obtained in the start of the season. A sample of 1000 seeds and minimal medium are used in this estimation. Weight and number of seeds in each sample fraction is determined after isolation of damaged seeds. The results are then calculated by putting the values in the following equation:

$$\frac{(U \times N_d) - (D \times N_a)}{U\,(N_d + N_a)} \times 100 = \%\ \text{weight loss} \tag{1}$$

where U = weight of undamaged seeds, D = weight of damaged seeds, N_a = number of undamaged seeds, and N_d = number of damaged seeds.

In this method, moisture content of the separate fraction is unnecessary for a single sample, and differences of assumption are likely small. The method does consider hidden infestation in damaged category, and it considers seed infestation that is random by the insect which is not necessarily to be true [36]. The method can cause misleading results for low level of infestation and multiple infestations in large seeds. The method is useful in field level for quick estimation at extremes. To overcome the biased estimation, several refinements have been made. For example, different-sized seeds have different moisture contents and can have hidden infestation. These seeds are graded according to size and categories before counting and weighing [22]; severely attacked grains are separated, and reading of hidden grains is taken after emergence of infestation [37, 38]. There is another way to know the hidden infestation by dissecting seeds, but the method is tedious and has a chance of change in moisture content of seeds, for the sake of calculations that are needed to be made on dry matter.

5.7. A Count and Weight Method Modification

In this modification of count and weight method, highly infected grains have also been taken into account. To avoid the underestimation of losses, the number of infected seeds is counted for sum of the seeds in whole sample [39]. To assess the weight loss, the following equation is used:

$$\frac{[TND(D + U)] + FW(N_dU - N_aD)}{[TND(D + U)U] + [FW(N_d + N_a)U]} \times 100 \tag{2}$$

where U is the weight of undamaged seeds, D is the weight of damaged seeds, N_a is the number of undamaged seeds, and N_d is the number of damaged seeds.

The weight loss is calculated by averaging weight loss of two subsamples.

The percentage weight loss is calculated by deviation in Eq. (3):

$$\frac{\text{Undamaged weight (UW)} - \text{final weight (FW)}}{\text{Undamaged weight (UW)}} \times 100$$

(3)

The undamaged weight of the sample (in the absence of missing seeds) is figured out by applying assumption in count and weight method. The average unit weight of undamaged grains in the original sample will be equal to the average unit weight in the remaining undamaged grain subsamples. The undamaged weight of the whole original grain sample is figured out as the product of the total number of seed estimated in the original sample and unit weight of undamaged seeds in the subsamples. The total number of seeds is the sum of the damaged and undamaged seeds. The final step will be the ratio of the final sample weight to the average unit weight of the seeds. The method focuses on missing seeds of the sample instead of underestimating lost samples as in conventional count and weight method [39].

5.8. The Converted Percentage Damage Method

The method is suitable for quick estimation of loss by boring insects without using equipment such as during a rapid field appraisal. Weight loss is assessed by taking percentage damage seed as sample. A study has been made to find out the relationship of damage and weight loss [23]. A conversion factor can be used subsequently to estimate the weight loss of same type to seeds in different samples. The conversion factor is calculated by the following formula by using same count and weight method and so faces same type of errors:

$$\frac{\text{\% damage seeds}}{\text{\% weight loss}} = \text{Conversion factor}$$

(4)

To overcome the error of count and weight method, 10% or more damaged sample seeds are used. The sample size should not be smaller than 500 seeds taken as the percentage of the total seeds present. Rapid loss technique is a rapid weight loss assessment in the field based on the damage and weight loss relationship. Standard graphs are used in this technique to relate percentage damage and weight loss of samples under study [6, 21]. For preparing reference graph, ten samples of 500 seeds are required in the laboratory. Weight loss is estimated by count and weight method, and percentage damage is calculated. The graph is plotted between percentage weight loss and percentage damage. In the field, clean sample of 200 seeds is gathered, percentage damage seed is determined, and loss figure of reference to graph is read off. For a sample of 200 seeds of Bambara or cowpeas, the technique depicts accuracy on ±7%. The method is better than percentage damage method and allows good assessment of weight loss by taking many individual samples in short time period at the field. Visual scales are rapid loss assessment methods. The abovementioned techniques are time-consuming and need appropriate equipment and well-trained technicians. Visual scale has benefit in assessing damage in fields, is ideal for field use and rapidity, and is required only for reference scales and operator unbiased. This technique enables wider coverage in small time and small sample error. On-site loss assessment avoids extra spoilage and anomalous results by double checking on the site before leaving. The visual scale should be calibrated according to the objective of the study, weight loss, and farmer perception of value [6, 21].

5.9. Losses by Vertebra Pests

Vertebra pests such as rodents and birds remove the whole seed from the sample so it is impossible to estimate loss by vertebra pests. The loss can be assessed by comparing the reference percentage of seed loss and average seed weight [40]. Population studies and feeding trials are used to calculate the losses by pests and rodents, but these are often with small accuracy in comparison to efforts expanded [41]. Pests used stored grains only as their part of diet; feeding trial can overestimate the loss of stored seeds. The quantity of seed loss by rodents is questionable. Loss of crops due to rodents goes behind when compared with loss of storage container, personal property, buildings, and potential health risks.

5.10. Weight Loss by Molds

Seeds infected by molds will cause weight loss; the weight loss can be assessed by the same method that is used to calculate weight loss by insect. The moldy seed weight loss increased due to the absorption of moisture so the loss by mold can be compensated. The method is not that good to assess the real loss of seeds, and the seeds may be misled as undamaged seeds due to no apparent indication of infestation on surface. To estimate the weight loss due to mold, damaged seeds are separated from undamaged seeds, and then moldy seeds are discriminated from damaged seeds. The weight loss due to mold will be equal to the weight of mold [21, 27].

5.11. Total Loss in a Season

The above losses are the seed losses at a given point in a time. The picture can be misleading; there must be a relation in the pattern of seed used in a season. In an undisturbed stored crop, if the sample loss is 10% throughout the storage seasons, the total loss will be due to insects. The seeds will loss at different quantities at different intervals of time due to insect exposure at different time lengths throughout the season [27]. The percentage loss of seeds increases gradually with time due to increase in insect infestation. After permitting changes in moisture content, the loss in seeds can be calculated by measuring the weight of seeds at the time when the seeds are still in the store and when taken out from the store. The loss other than insect damage can be obtained by subtracting the loss caused by other insects. The overall losses of the seed after store are considerably lower than the suggested value. Many loss assessment protocols at commercial and farm level have been reported [22]. The key to get the best assessment result is to draw an acceptable methodology for each commodity.

There were small figures for losses in commercial operations and none for cooperative level storage. The condition reflects the buying and selling of seeds in the developing world in short time period. This gives the picture of private sector (market liberation and parastatal marketing) involvement, but there is few information about storage loss. Entrepreneurs could store large quantity of seed for longer time period. However private sector has increased this level for farm storage. Many efforts, time and money was invested to measure the storage losses in farm storage, but the effort was not fruitful like the earlier projects. Moreover, the study should be undertaken with the postharvest sector as a whole, and precise measurements should be avoided. Social survey can be helpful to discover the farmers' problem for loss assessment and appropriate measurement techniques [42].

6. HARVESTING AND MATURITY INDICES

Harvesting and improper handling of commodities can cause bruises and injuries which directly affect their market value and make them unattractive. Injuries give the open space for microbial attack, that cause rotting, respiration increase, and shortage of storage life. Improper harvesting can cause crop loss and severe damage in seeds [9, 43].

6.1. Harvesting and Handling

Harvesting is the first step of postharvest and is the last step of crop production. The method and condition of the harvest affect the further handling, processing, and storage of crops. Premature harvest causes loss of quality of seeds, and due to high water content, they will deteriorate in the store. Overmature crop harvest causes biological and physical losses of crops by consistent wetting and drying of crops [9]. Damp seeds need to be threshed quickly and dried after harvest. Different parts of plant are harvested by different harvesting methods; in case of forage, the whole plant is trimmed off; in case of cereal seeds, partial or part of the plant is threshed and cleaned; and straw or chaff is removed for further processing. Small scale produces performed threshing and harvesting by threshing combines harvesters (equipped through community groups) while in developing countries threshing and harvesting carried out by hand unlikely cause damage or deterioration of crops in store. Large-scale commercial producers use mechanical harvester equipment; their use is limited due to the production of cash crops. Post-harvest risk of crop damage in storage is reduces by manual harvesting Small scale produces performed threshing and harvesting by threshing combines [9, 21]. Conventionally, seeds are beaten with stick or against hard surface (wooden bar, log of wood, stone, and wooden metal or tub) for the sake of threshing. The methods can cause damage or cracks on the seeds, while seeds that are trodden under foot will be a less damaging method for the seeds. Grain heads or ears of sorghum, millet, or wheat are commonly beaten with sticks. However, manual harvesting is laborious, and economical process has the risk of physical damage. Maize cobs are beaten with sticks or shelled by hand, which result in high-level damage. Mechanized threshers are made to reduce the damage of seeds; the models are highly sophisticated.

Seed harvesting is carried out by a combination of various steps of threshing, cleaning, or combine harvester. Large-scale harvesting is carried out by mechanical equipment that is especially designed to harvest cereal seeds [6].

6.2. Storage Structures for Seeds

Seeds/grains are durable crops that usually require simple systems in their storage.

6.3. Farm-Level Storage

For protective outdoor or indoor storage, the seeds must be guarded for physical damage such as high temperature, adverse weather, snow and rain, and biological factors which include microorganisms, birds, rodent, mites, and insects. The process of farm storage is used by various countries to store major part of seeds [44, 45]. The storage structure varies in capacity and has range of 100 kg to few metric tons. Modifications in locally prepared storage structure could be done according to climatic conditions. There are some traditional storage structure. The bins are commonly made of ferro-cement, plywood, metal, and high-density and high-molecular-weight polyethylene. Plywood is the most suitable storage structure and underground structures of different shapes and sizes, which work on the principle of hermetic storage. This builds the higher level of carbon dioxide and low oxygen level which is lethal to pest and microbial attack during seed storage [46]. Traditional methods are cheaper, but they are not effective against pest and microbial attack. Silos or metal bins are also used at farm level to store seeds.

6.4. Bagged Storage

In developing countries, seeds are stored in gunny or woven polypropylene bags in traditional warehouses, whereas silos for bulk storage, seeds elevator, and flat storage structure are used in developed countries [45, 47]. Bag storage is a laborious and costly process and has greater chance of seed spillage and biological losses. There may be water seepage and humidity problem due to inappropriate flooring of warehouses. Bags do not need any fumigation facilities or aeration system. In developing countries, the system will be uneconomical due to small farm size and cheaper manual labor.

6.5. Bulk Storage

There are two types of bulk storage of seeds; it may be vertical (silos or bins) or horizontal (on floor stores). Horizontal stores comprise especially constructed floors of warehouses containing proper ventilation on the floor and walls that are strengthened to withstand the weight of the seeds. Bins and silos are especially designed stores, either round or square, in grouped or freestanding form, and incorporating unloading and loading usually of aeration systems. The alternative forms of bulk storage are belowground or partially belowground storage or enameled, sealed silos, for the storage of high moisture content seeds. The process is appropriate to bulk handling or storage of seeds [48, 49].

6.6. Hermetic Storage

The traditional methods of storage in the natural build of oxygen and lower level of oxygen protect the seeds from biological damage. This traditional storage method is not effective for the seeds that have lesser moisture content, and infestation is less than insects per kilograms of seeds. The controlled atmosphere treatment and fumigation have to be supplemented in hermetic storage [50].

6.7. Outdoor Storage

This is the temporary measure of storage in case of lack of permanent storage. The godowns and silos are built on plinth, and the stacks of seeds are covered with polyethylene covers. The stacks must be properly aerated in a week by raising the cover to the seventh or eighth layer. The cover and plinth (CAP) technique is used commonly for wheat and paddy. However, there is a chance of damage of cover by wind or rain, and effective fumigation cannot be achieved [44].

6.8. Storage Practices For Quality Seeds

Quality of seeds can be maintained by especially designing the small stores to silo or warehouses which play a protective role against adverse temperature conditions, ground water, rain water, pests, and thefts. The structure and contents of the store should be managed [9].

7. MOISTURE CONTROL

A water disposal system and well-designed roof are required (overhung or gutter) to hinder the water running to the store. Drains move the water away from the stores. In large warehouses, side by side joining of stores should be avoided to prevent the water run into the store. Raised floor and proper drainage system protect against the risk of the flood, and water-proof floors and walls guard against ground water. There must be proper ventilation system to control the humidity inside the storage structure [27].

7.1. Protection against Temperature

It is difficult to control temperature in the storage structures; it needs special design elements to control temperature. Temperature can be ascertained by the use of controlled ventilation. In the cold night, temperature can be manipulated by insulated stores. Stores can be built in east-west direction, and using shiny material outside the stores can be an effective strategy to control heat. Thick wall and wide roofs (for shade) can further reduce the heat of the storage structure. Heater and refrigerators can be installed to modify temperature in stores; the equipment will perform well in case of insulated stores. However in these stores, degree of insulation depends upon atmospheric condition outside of it. [6, 51].

8. MODIFIED ATMOSPHERE STORAGE (MAS) AND CONTROLLED ATMOSPHERE STORAGE (CAS)

In modified atmospheric condition, the gases are added or removed from surrounding to make a controlled atmospheric composition around seeds, which differs from air (78.08% nitrogen, 20.95% oxygen, and 0.03% carbon dioxide). This includes the reduction of oxygen and elevation of carbon dioxide concentration. CAS and MAS are different in level of control although CAS is more exact. The process is used to facilitate whole-store fumigation [52, 53].

8.1. Protection against Theft

Security guard or fence may be needed to guard the store. Lights are installed to illuminate the stores [6].

8.2. Protection from Birds and Rodents

Bird and rodent contact can be hindered by avoiding the free spaces for the access to the seeds. Rodent guards are commonly tight-fitting shelves, or conventional methods could be tying of thorn bushes to the stores as barrier. Walls of the stores should be smooth, and roof should be rodent proof to avoid their contact inside the storage structure [54].

8.3. Protection against Insect

The small size insects can easily enters the storage buildings therefore it is difficult to make insect-proof walls and roofs because they can enter from even small cracks. For protection against insects, smooth walls, controlled ventilation, cleaned stores, floor without cracks, and curved floor wall can be effective. For protection against termites, termite-resistant material should be used (concrete, stones, fire bricks, and termite-resistant wood) [55].

9. TRANSPORT

The transport of commodities from field to storage houses can produce some injury, which can later cause deterioration of produce. This main focus is to keep the produce dry and moisture-free. The risk of insect infestation of seeds from the contaminated container has insect residual. Mechanical injury is produced from vibration during transport, poor condition of vehicle and roads, poor driving, insecure container stacking, the use of unsuitable container, and careless handling. Overheating is produced from the sun or vehicle engine that results in moisture loss of produce, and this leads to natural breakdown and decay [6].

10. QUALITY AND SAFETY

The quality of a product defines the class, degree, excellence, or superiority of the crop. The quality of the durable commodity is the combination of characteristics, properties, and attributes that give the commodity value as the food or a source of next crop yield. Foreign material or high moisture content can affect the marketing quality of the crop. High moisture may encourage the shrinkage or biological and biochemical damage in seeds. Low moisture in pulses and paddy rice can break or damage the seeds. The discolored and broken seeds lower the marketing quality and increase the chance to insects and microorganism attack [14, 27]. The main objective of a farmer is to produce apparently good product with few visual defects, and product must score high on yield, disease resistance, ease of harvest, and shipping quality and achieve the national and international quality standards. For the buyer or consumer, the appearance is more important; they are eagerly concerned in healthy seed and lifelong storage. The product should be assured safe for the distribution to suppliers and market [7].

11. POSTHARVEST QUALITY ASSESSMENT TECHNIQUES

There are two methods to assess the quality of crop after harvest such as analytical or objective and subjective or sensory method [56]. There are limitations for subjective method in quality evaluation of hazardous material [57]. Objective method is destructive or nondestructive in nature. Manual sorting of commodities is outdated, costly, unreliable, subjective, laborious, and slow. Food depicts complex and dynamic behavior with respect to varietal and time dependency. Nondestructive methods classified according to physical principles are visual inspection, x-ray imaging and computed tomography, near-infrared spectroscopy, ultrasonic testing, thermographic testing, electromagnetic testing, liquid penetrant testing, magnetic particle testing, acoustic emission testing, infrared and thermal testing, magnetic resonance imaging, electronic nose, etc. [58].

Quality assessment or grading has the following benefits:

i. It helps to facilitate the purchase of an unseen product.
ii. Quality and safety improvement incentives.
iii. Grading makes the market information meaningful.
iv. It helps in facilitating quality and price comparison.
v. It reduces the deception and fraudulent marketing risk.
vi. Diverse marketing mechanism can be enabled through the process such as commodity exchange, future trading, inventory credit, credit letters, and facilitating resolution of disputes regarding composition or quality [6].

12. FUTURE CHALLENGES FOR POSTHARVEST TECHNOLOGY

In increasing urban population and shifting lifestyle in developing countries, demand for food has increased many times. Seed is a unit which not only is used as food but also responsible for production of the next generation. The most attention-demanding perspectives in seed biology are postharvest management and quality maintenance of seed. Although in past years, great success has been achieved in development of novel packaging, storage and transport systems, pests, and seed-borne disease control for market access. However, to achieve future goals, further research and technology development should be dedicated to explore genetic aspects of quality traits like stress resistance, resistance to postharvest diseases, and pest management. Researchers should try to work on integrated approaches for postharvest management of seed. Nanotechnology is an emerging field which is leading to tremendous achievements in crop sciences. Seed biologist should also try to explore this field so that seed could be preserving for longer time without altering the genetics.

13. CONCLUSIONS

Seed quality should be maintained for the better quality crops. Maintenance of the seeds is the major problem in the developing countries these days. There is a need to establish better, economical, convenient, and productive methods for postharvest handling and seed storage. Researchers should focus on translating knowledge into agricultural outputs. The chapter deals with the value of healthy seeds, factors affecting the seed quality, postharvest techniques for the seed storage, and techniques and safety measures for their quality assessment for the maintenance of good quality seeds to meet the international standards emphasizing on the needs of developing countries. This could be achieved by opting more sensitive and advance technologies with special emphasis on seed response to environmental and maternal factors.

REFERENCES

1. Delouche JC, Caldwell WP. Seed vigor and vigor tests. Proceeding of Association of Official Seed Analyst. 1960;50:124–129.
2. Delouche JC. Planting seed quality. In: Proceeding of 1969 Beltwide Cotton Prod.-Mech. Conference. New Orleans. January 1969. p. 16–18.
3. Grabe DF. Agronomic significance of seed deterioration. Agronomy Abstract. 1965;57:40.
4. Woodstock LW. A respiration test for corn seed vigor. Proceeding of Association of Official Seed Analyst. 1966;56:95–98.
5. Gregg BR, Billups GL. Seed Conditioning Technology Part A. Science Publishers. USA. 2010, Vol. 2, p. 1–2.
6. Boxall RA, Brice JR, Taylor SJ, Bancroft RD. Crop post-harvest, science and technology, volume 1, principles and practice. In: Golob P, Farrell G, Orchard E (Eds) Crop Post- Harvest: Science and Technology. Blackwell Science Ltd, Oxford, UK. 2002. p. 141–204.
7. Mohammed M. Manual on post-harvest management strategies to reduce losses of perishable crops. UWI/CTA/NAMDEVCO Workshop on Strategies to Reduce Post-harvest Losses in Perishable Commodities at NAMDEVCO Conference Facility. Piarco, Trinidad. February 24–25, 2014.
8. Harrington JF. Seed storage and longevity. Cited in: Kozlowski TT, Seed Biology. 1972;3:145–245.
9. Kiaya V. Post-harvest Losses and Strategies to Reduce Them. New York: Action Contre la Faim (ACF International). 2014.
10. López J, Uribe E, Vega-Gálvez A, Miranda M, Vergara J, Gonzalez E, Di Scala K. Effect of air temperature on drying kinetics, vitamin C, antioxidant activity, total phenolic content, non-enzymatic browning and firmness of blueberries variety O Neil. Food and Bioprocess Technology. 2010;3:772–777.
11. Mujumdar AS, Law CL. Drying technology: Trends and applications in post-harvest processing. Food and Bioprocess Technology. 2010;3:843–852.
12. Chen XD, Mujumdaer AS (Eds) Drying Technologies in Food Processing. Oxford: Blackwell. 2008.
13. Boothumjinda S, Exell RHB, Rongtawng S, Kaewnikom W. Field tests of solar rice dryers in Thailand. In: Proc ISES Solar World Forum. Perth, Oxford: Australia (Parkville ISES), Pergamon Press. 1983. p. 1258–1263.
14. Kitinoja L, Gorny JR. Post-harvest Technology for Small-scale Produce Marketers Economic Opportunities, Quality and Food Safety. Post-harvest Horticulture Series. Department of Pomology, University of California. Davis, USA. 1999. p. 21.

15. Radajewski W, Jolly P, Abawi GY. Optimization of solar grain drying in a continuous flow dryer. Journal of Agricultural Engineering Research, 1987;38:127–144.
16. Colliver DG, Brook RC, Peart RM. Optimal management procedure for solar grain drying. American Society of Agricultural Engineers. Chicago. 1978;12:78–35.
17. Heid WG. Solar-assisted Combination Drying: An Economic Evaluation. Agricultural Economic Report, Economics, Statistics and Co-operative Services, US Department of Agricultural, No. AER-453. 1980.
18. Hellickson MA. Cost/performance goals for agricultural solar collectors. American Society of Agricultural Engineers. 1979;54:79–40.
19. Chung DS. Lee C. Physical and thermal properties of grains. In: Champ BR and Highley E (Eds) Preserving Grain Quality by Aeration and In-store Drying. Proceedings of an International Seminar. Kuala Lumpur, Malaysia. Canberra, Australia: ACIAR. October 9–11, 1986.
20. Bartosik RE. A model-based fan and burner control strategy for the in-bin drying and conditioning of corn. 2005.
21. Harris KL, Lindblad CJ. Post-harvest Grain Loss Assessment Methods. Minnesota: America Association of Cereal Chemist. 1978. p. 193.
22. Boxall RA. A Critical Review of the Methodology for Assessing Farm-level Grain Losses after Harvest. Slough, UK: Tropical Products Institute G191. 1986.
23. Boxall RA. Storage losses. In: Golob P, Farrell G, Orchard JE (Eds) Crop Post-harvest: Science and Technology Volume 1: Principles and Practice. Oxford: Blackwell Sciences Ltd. 2002. p. 143–169.
24. Kader AA. Post-harvest Technology of Horticultural Crops. 3rd ed. Oakland: University California Agriculture and Natural Resources, Publ. 3311. 2002.
25. Gross K, Wang CY, Saltveit ME. The Commercial Storage of Fruit, Vegetables and Florist and Nursery Stocks. USDA Agriculture Handbook 66. 2002. http://www.ba.ars.usda.gov/hb66/index.html.
26. Atanda SA, Pessu PO, Agoda S, Isong IU, Ikotun I. The concepts and problems of post-harvest food losses in perishable crops. African Journal of Food Science. 2011;5:603–6013.
27. Boxall RA. Post-harvest losses to insects – A world overview. International Bioterioration and Biodegradation. 2001,48:137–152. Retrieved from www.scopus.com.
28. Grolleaud M. Post-harvest Losses: Discovering the Full Story. Rome: FAO. 1997. Available from: http://www.fao.org/docrep/004/ac301e/AC301e04.htm#3.2.1%20Rice
29. Lipinski B, Hanson C, Lomax J, Kitinoja L, Waite R, Searchinger T. Reducing Food Loss and Waste. World Resources Institute Working Paper. Washington, DC: 2013.
30. Grolleaud M. Post-harvest Losses: Discovering the Full Story. Overview of the Phenomenon of Losses during the Post-harvest System. Rome, Italy: FAO, Agro Industries and Post-Harvest Management Service (AGSI). 2002.
31. ISTA. International rules for seed testing. Proceedings of the International Seed Testing Association. 1966;31:49–91.
32. Tyler PS, Golob P, Compton J, Bickersteth S. Study on Phytosanitary Requirements to Promote Maize Trade in Eastern and Southern African Countries at Risk from the Larger Grain Borer. Report for the European Commission. Chatham, UK: Natural Resources Institute. 1990.
33. Adams JM, Schulten GGM, Harris KL, Lindblad CJ. Post-harvest Grain Loss Assessment Methods. American Association of Cereal Chemists. USA. 1978 p. 83–95.
34. FAO I. WFP. The State of Food Insecurity in the World. Food and agriculture organization of the united nations. Rome. 2013.
35. Reed C. The precision and accuracy of the standard volume weight method of estimating dry weight losses in wheat, grain sorghum and maize, and a comparison with the thousand grain mass method in wheat containing fine material. Journal of Stored Products Research. 1987;23:223–231.

36. Adams JM, Harman GW. The Evaluation of Losses in Maize Stored in a Selection of Small Farms in Zambia with Particular Reference to the Development of Methodology. Slough, UK: Tropical Products Institute G109. 1977.
37. Ratnadass A, Fleurat-Lessard F. Improvement in the practical methods of assessment of losses caused by insects in grain stored at the village level in tropical Africa. In: Fleurat-Lessard F, Ducom P (Eds) Proceedings of the 5th International Working Conference on Stored Product Protection, Bordeaux. Bordeaux, France: INRA. 1991.
38. Ratnadass A, Berté S, Diarra D, Cisé B. Insect losses in sorghum stored in selected Malian villages, with particular emphasis on varietal differences in grain resistance. In: Highley E, Wright EJ, Banks HJ and Champ BR (Eds) Proceedings of the 6th International Working Conference on Stored Product Protection, Canberra. Wallingford, UK: CAB International. 1994.
39. Compton JAF, Floyd S, Ofosu A, Agbo B. The modified count and weigh method: An improved procedure for assessing weight loss in stored maize cobs. Journal of Stored Products Research. 1998;34:277–285.
40. Boxall RA, Gillett R. Farmer Level Storage Losses in Eastern Nepal. Slough, UK: Tropical Products Institute G157. 1982.
41. Hernandez A. Drummond DC. A study of rodent damage to food in some Cuban warehouses and the cost of preventing it. Journal of Stored Products Research. 1984;20:83–86.
42. Goletti F, Wolff C. The Impact of Post-harvest Research. Washington, DC: International Food Policy Research Institute, Markets and Structural Studies Division Discussion Paper, 29. 1999.
43. FAO. Global Food Losses and Waste: Extent, Causes and Prevention. . Food and agriculture organization of the united nations Rome. 2011.
44. Semple RL. Post-harvest technology: Storage development and application in developing Asian countries. In: Semple RL, Hicks PA, Lozare JV, Castermans A (Eds) Grain Storage Systems in Selected Asian Countries. REAPASIA. People Republic of China. 1992. p. 9–59.
45. McFarlane JA. Storage methods in relation to post-harvest losses in cereals. Insect Science Application. 1988;9:747–754.
46. Shejbal J, de Bioslambert JN. Modified atmosphere storage of grains. In: Multon JL (Ed). Preservation and Storage of Grains, Seeds and Their By-Products: Cereals, Oilseeds, Pulses, and Animal Feed. New York: Lavoisier. 1988. p. 749–777.
47. Kennedy L, Devereau AD. Observations of large-scale outdoor maize storage in jute and woven polypropylene sacks in Zimbabwe. In: Highley E, Wright EJ, Banks HJ, Champ BR (Eds) Stored Product Protection: Proceedings of the 6th International Working Conference on Stored-Product Protection. Wallingford: CAB International. 1994. p. 290–295.
48. Multon JL. Spoilage mechanisms of grains and seeds in the post-harvest ecosystem, the resulting losses and strategies for the defence of stocks. In: Multon JL (Ed.) Preservation and Storage of Grains, Seeds and Their By-Products: Cereals, Oilseeds, Pulses and Animal Feed. New York: Lavoisier. 1988. p. 3–63.
49. Bailey JE. Whole grain storage. In: Sauer DB (Ed). Storage of Cereal Grains and Their Products. St. Paul, MN: American Association of Cereal Chemistry. 1992. p. 157–182.
50. Alvindia DG, Caliboso FM, Sabio GC, Regpala AR. Modified atmosphere storage of bagged maize outdoors using flexible liners: A preliminary report. In: Highley E, Wright EJ, Banks HJ, Champ BR (Eds) Stored Product Protection: Proceedings of the 6th International Working Conference on Stored-Product Protection. Wallingford: CAB International. 1994. p. 22–26.
51. Longstaff BC. Temperature manipulation and the management of insecticide resistance in stored grain pests: A simulation study for the rice weevil, Sitophilus oryzae. Ecological Modelling. 1988;42:303–313.

52. Banks HJ, Annis PC. Comparative advantages of high CO2 and low O2 types of controlled atmospheres for grain storage. In: Calderon M, Barkai-Golan R (Ed.) Food Preservation by Modified Atmospheres. Boca Raton: CRC Press. 1990. p. 94–122.
53. Paster N, Calderon M, Menesherov M, Barak V, Mora M. Application of biogenerated modified atmospheres for insect control in small grain bins. Tropical Science. 1991;32:355–358.
54. Smith RH. Rodents and birds as invaders of stored-grain ecosystems. In: Jayas DS, White NDG, Muir WE (Eds) Stored-Grain Ecosystems. New York: Marcel Dekker. 1995. p. 289–323.
55. Hagstrum DW, Flinn PW, Shuman D. Acoustical monitoring of stored-grain insects: An automated system. In: Highley E, Wright EJ, Banks HJ, Champ BR (Ed.) Stored Product Protection: Proceedings of the 6th International Working Conference on Stored-Product Protection. Wallingford: CAB International. 1994. p. 403–405.
56. Ranganna S. Handbook of Canning and Aseptic Packaging. Tata McGraw-Hill Publishing Company. Dehli, India. 2000.
57. Jha SN, Garg R. Non-destructive prediction of quality of intact apple using near infrared spectroscopy. Journal of Food Science and Technology. 2010;47: 207–213.
58. Siddiqui MW. Post-harvest Biology and Technology of Horticultural Crops: Principles and Practices for Quality Maintenance. CRC Press. USA. 2015.

CHAPTER 6

Resistance Levels to Root Rot and Angular Leaf Spot Diseases in Selected High Iron Bean Genotypes

Mukamuhirwa F[1,2*], Mukankusi MC[3], Tusiime G[1], Butar L[2], Musoni A[2], Ngaboyisonga C[2], Gahakwa D[2], Gibson P[1] and Kelly K[1]

[1]*Department of Agricultural production, Makerere University, Kampala, Uganda*

[2]*Rwanda Agriculture Board, Kigali, Rwanda*

[3]*CIAT Africa, National Agricultural Research laboratories-Kawanda, Kampala, Uganda*

ABSTRACT

Common bean production is constrained by different diseases the major ones being, Angular Leaf Spot (ALS), bean root rot, anthracnose, Common Bright Bacteria (CBB), Bean Common Cosaic Virus (BCMV) and Bean Common Mosaic Necrotic Virus, (BCMNV). The aim of this study was to identify new and better sources of broad resistance to both bean ALS and Root Rot diseases among nutritional bean varieties. Fifty seven varieties were planted in the screen house of CIAT Africa based at Kawanda Agricultural Research Laboratories Institute (KARL). Virulent inocula actually used at CIAT were used to test these genotypes. Different varieties were resistant to specific isolates but interestingly, only ACC 714 contained broad resistance to both Andean and MesoAmerican isolates of bean Angular Leaf Spot as well as Fusarium root rot and Pythium root rot at mean, median and mode basis. Since different nutritional bean varieties have varying levels of resistance to different pathogens, it may be possible to pyramid these resistance genes into appropriate background so as to provide durable resistance in biofortified bean genotypes higher in iron and zinc content.

Keywords: Nutritional bean varieties; Pathogens; Resistance; Iron content

INTRODUCTION

Biofortification, a practice of enhancing contents of minerals with nutritional significance in food products is regarded as one of the cheap approaches to improve human nutrition [1,2]. Traditionally, crops have been mostly improved for agronomic traits and to a lesser extent for pest and disease resistance. Biofortification is only a recent practice. The most limiting micronutrients in the diets of the rural and urban poor in Rwanda and Uganda are Fe and Zn, resulting into anemia and depressed immunity, respectively [3]. Efforts have been made in this study to breed for increased Fe and Zn in common beans in Rwanda and Uganda. However, successful deployment of high Fe and Zn common beans in both countries will require that such varieties are high yielding but also resistant to some of the most important diseases. The interaction of disease borne pathogens with the crop-bio system complicates the demonstration of superior genotypes across environments, and thus it results into scale or rank shift of trait performance. The bean root rot (*Pythium sp, Fusarium solani fsp. phaseoli, Rhizoctonia solani, Macrophomina phaseoli* and *Sclerotium rolfsii*) and angular leaf spot (*Phaeoisariopsis griseola*) are currently regarded the most important bean diseases in Uganda and Rwanda [4]. The plant diseases lead to food deficit and food insecurity. In order to reduce the yield losses due to disease, it is important to define the diseases contributing to reduced yield, accurately estimate the severity of disease and propose the possible solutions. Therefore, this study was carried out to determine the levels of resistance to these two diseases in common bean genotypes bred for high Fe and Zn contents.

MATERIALS AND METHODS

Research Sites and Plant Materials

This study was carried out in the CIAT Africa screen house based at Kawanda Agricultural Research Laboratories Institute (KARL), Uganda. It was carried out from June to October 2012. 57 bean genotypes were used in the study. These genotypes were regional nutrition nursery breeding lines; advanced G × E stable lines; susceptible checks for root rot and ALS; low Fe check (CAL 96); and resistance checks for root rot (MLB49-89A/ RWR719) and ALS (MEX54/ BAT332) (**TABLE 1**).

Table 1: Plant materials used in the study of characterization of resistance of selected high iron beans to root rot and Angular Leaf Spot diseases.

No.	Genotypes	Origin	No.	Genotypes	Origin
1	NGWIN × CAB2/2/3/1/1	Rwanda	30	NUA 69	CIAT
2	NGWIN × CAB2 × (RWV3317)	Rwanda	31	KAB06F2.8-12	CIAT
3	MAC 42	Rwanda	32	RWR 2154	Rwanda
4	RWV 3316	Rwanda	33	RWR 2245	Rwanda
5	NUV 219-1	CIAT	34	KAB06F2.8-36	CIAT
6	CAB 2	Rwanda	35	KAB06F8.8-35	CIAT
7	RWV 2359	Rwanda	36	CODMLB 001	CIAT
8	Garukurare	Rwanda	37	HM 21-7	CIAT
9	Kivuzo	Rwanda	38	Ngwaku-Ngwaku.	RDC
10	RWV1129	Rwanda	39	NUA 45	CIAT
11	Ndimirakaguja volubile	Rwanda	40	NUA 59	CIAT
12	Icyana 2	Rwanda	41	NUA 56	CIAT
13	RWV 2361	Rwanda	42	NUA 35	CIAT
14	MAC 44	Rwanda	43	Gitanga	Rwanda
15	RWV 3006	Rwanda	44	Zebra	CIAT
16	RWV 2887	Rwanda	45	ACC 714	CIAT
17	MAC 74	Rwanda	46	CODMLB 033	CIAT
18	Agronome 2	Rwanda	47	ROBA I	CIAT
19	VRA 4	CIAT	48	SMC 21	CIAT
20	Rugandura	Rwanda	49	SEMC 16	CIAT
21	RWV 2070	Rwanda	50	SMC 18	CIAT
22	Kiangara	DRC	51	SEMC 17	CIAT
23	VCB 81013	CIAT	52	GLP 2	CIAT
24	Gasirida	Rwanda	53	CAL 96 (check. low Fe, root rot susceptible)	Uganda
25	MEX54/ BAT332 resistant check	CIAT	54	DOR 500 (low Fe check	Uganda
26	MIB 456-High Fe Universal Check	CIAT	55	Maharagi Soya	Rwanda
27	Decelaya 1-Low Fe Check	Rwanda	56	MBC32	CIAT
28	KAB06F2.8-27	CIAT	57	Nyiramogorori2	Rwanda
29	NUA 99	CIAT			

Preparation Of Pathogen Inocula, Application And Disease Evaluation

FUSARIUM SOLANI FSP. PHASEOLI: Inoculum for *Fusarium solani* fsp. *phaseoli* was prepared from isolate FSP-3, the most virulent isolate for *Fusarium* root rot obtained from infected bean fields in the *Fusarium* root rot hot spot in south-western Uganda [5]. *Fusarium* inoculum was isolated and prepared following CIAT's laboratory training manual. The inoculum was reactivated by sub culturing it on a fresh PDA culture media. Sorghum grains were used as a medium for fungal inoculum multiplication. Approximately 400 ml of water to every 300 g of sorghum grains were placed in polyethylene bags, sterilized and allowed to cool for 12 h. A disc of agar bearing *Fusarium* spp. culture was incubated in the polyethylene bags over the sterilized sorghum grains in a sterile environment at a room temperature for 14 days to allow uniform growth. After incubation, *Fusarium* inoculum was mixed with the loam sandy soil previously sterilized by steaming on firewood for four hours and left overnight to cool. Sterile soil and inoculum were mixed in a ratio of 1:8 and put in a wooden flat tray and left to stabilize for seven days. In each tray, was planted 5 test varieties, each in 2 rows. A susceptible check (Cal 96) and a resistant check (MLB49-89A) were includes in each tray. Disease assessment was done twenty one days after planting by carefully uprooting all the seedlings planted per variety taking care not to damage roots and hypocotyls, and washing with clean tap water. For each variety, twenty plants were evaluated per replicate. FRR severity was assessed by scoring disease on roots and hypocotyls and scoring based on a 1 to 9 disease score [6]. Acccording to this scale, 1=no visible symptoms, 3=light discoloration either without necrotic lesions or with approximately 10% of the hypocotyls and root tissues covered with lesions; 5=approximately 25% of hypocotyls and root tissues covered with lesions but tissues remain firm with deterioration of the root system; 7=approximately 50% of hypocotyls and root tissues covered with lesions combined with considerable softening, rotting and reduction of root system, 9=approximately 75% or more of the hypocotyls and root tissues affected with advanced stages of rotting combined with severe reduction in the root system. The experiment was conducted in RCBD design with 3 replications. Data were subjected to ANOVA using Genstat [7].

PYTHIUM SP: Inoculum for Pythium was prepared from *P. ultimum* isolate MS 61 from long term storage at CIAT, Kawanda. The inoculum was reactivated by sub-culturing it on a fresh PDA culture media. Finger millet grains were used as a medium for fungal growth. Approximately 200 ml of water to every 300 g of millet grains were placed in polyethylene bags, double sterilized and allowed to cool for 12 h. A disc of agar bearing *Pythium spp* culture was incubated in the polyethylene bags over the sterilized finger millet grains in a sterile environment at a room temperature for 14 days to allow uniform growth.

After incubation, Pythium inoculum was mixed with the loam sandy soil previously sterilized earlier as for *Fusarium*. Inoculum and soil were mixed in a 1:8 ratio, put in wooden flat trays and left to stabilize for seven days. Test varieties were planted in these trays as done for *Fusarium solani* above. Again, CAL 96 and RWR719 were included as susceptible and resistant checks respectively. Three weeks after planting, plants were uprooted, washed carefully and immediately evaluated for damage. Disease severity was scored using the 1 to 9 CIAT scale. Twenty plants per replication were evaluated. The experiment was laid out as a RCBD with 3 replications. Data were subjected to ANOVA using Genstat 14th edition.

PHAEOISARIOPSIS GRISEOLA: For this pathogen, isolates KAK3, a virulent Andean isolate and isolate 2A a virulent Meso-American isolate were used for testing the 57 genotypes. Isolates were prepared for inoculation following the CIAT laboratory training manual procedures.

Forest black soil, lake sandy and decomposed farm yard manure were mixed in a ratio of 3:1:1 (currently used by CIAT) and was used in the screening. Test beans were planted in 5 L buckets and arranged in a RCBD with 3 replicates. Twenty-one (21) days after planting, bean leaves were inoculated using a hand sprayer, by misting inoculum onto lower surfaces of leaves when beans had developed two trifoliate leaves. Plants were thereafter covered with polyethylene bags maintained that way for three days.

On the appearance of ALS symptoms, the disease severity was assessed every three days for seven times. Four to five plants were evaluated per variety per replication. For the Andean isolate, genotypes CAL96 and MEX54 were included in the trial as susceptible and resistant checks respectively. MCM5001 and BAT332 were used as susceptible and resistant checks respectively for Meso-American isolate. The disease was rated according to the CIAT 1 to 9 scales CIAT [8] and the data were subjected to ANOVA using Genstat 14th edition.

Data Analysis

Area under Disease Progress Curve (AUDPC) based on mean, was calculated for ALS while the mean mode, median and disease index [9] were used for Pythium and *Fusarium* root rot.

The AUDPC value for each genotype was calculated by trapezoidal integration [10] and is given by: AUDPC=Σ ((Xi+Xi+1)/2(ti+1-ti)) in which: X_i and X_{i+1} is the disease severity for two consecutive assessments, and t_{i+1}-t_i the interval between two consecutive assessments [11]. The disease index for each variety was calculated according to Kobriger et al. [9] as: Disease Index=(Σ(disease class × number of plants in class)/ ((total plants) × 9) × 100) Statistical Analysis was performed by the ANOVA statistical procedure of Genstat GenStat 14th Edition.

RESULTS

Augural Leaf Spot (Als)

Tested varieties had a significant differences on resistance to ALS for both Andean and Mesoamerican isolates (P<0.001) (**TABLE 2**).

Table 2: ALS Anden and MesoAmerican isolates mean squares.

	Andean Isolate		MesoAmerican Isolate	
Source	DF	MS	DF	MS
Rep	2	707.77	2	701.30
Genotype	51	784.48***	55	1227.20 ***
Residual	95	70.05	101	364.20
Total	148	324.86	158	668.80
CV		13.49		29.07
Mean		62.02		65.64
LSD		13.57		30.91

***Significant at $p \leq 0.001$.

However, genotypes with the same letter were not significantly different (**TABLE 3**). Fourteen genotype including MAC 42, NUV 219-1, MAC 44, VRA 4, RWV 2070, KAB06F2.8-27, NUA 99, NUA 69, KAB06F2.8-12, RWR 2245, NUA 59, ACC 714, CODMLB 033, MBC32, and resistant checks (Mex54/BAT332) were identified as resistant to both Andean and Meso-American bean angular leaf spot isolates.

Table 3: AUDPCvalues for Andean and MesoAmerican isolates on tested genotypes.

Genotypes	AUDPC for Andean isolate	Group	AUDPC for Meso-American isolate	Group
Ngwin × CAB2/2/3/1/1	75.0	mno	63.2	Bcdefghijklmno
Ngwin × CAB2(RWV3317)	75.2	mno	81.6	Jklmnopqrst
MAC 42	29.0	ab	40.4	Abcd
RWV 3316	75.6	no	93.2	Opqrst
NUV 219-1	51.7	efgh	51.3	Abcdefghij
CAB 2	-	-	39.5	Abc
RWV 2359	76.6	o	51.4	Abcdefghij
Garukurare	75.4	mno	64.6	Bcdefghijklmno
Kivuzo	73.7	lmno	103.5	Rst
RWV1129	75.4	mno	53.5	Abcdefghij
Ndimirakaguja vol	75.6	no	78.4	Hijklmnopqrst
Icyana 2	73.0	klmno	87.8	Mnopqrst
RWV 2361	73.0	klmno	74.4	Fghijklmnopqr
MAC 44	60.5	fghijkl	54.0	Abcdefghijk
RWV 3006	64.6	hijklmno	44.8	Abcdefg
RWV 2887	76.0	o	91.6	Nopqrst
MAC 74	65.5	ijklmno	50.6	Abcdefghi
Agronome 2	72.7	klmno	87.0	Mnopqrst
VRA 4	35.2	abcd	42.9	Abcde
Rugandura	60.5	fghijkl	85.5	Lmnopqrst
RWV 2070	23.5	a	35.5	Ab
Kiangara	72.1	klmno	55.5	Abcdefghijkl
VCB 81013	75.0	mno	62.2	Bcdefghijklmn
Gasirida	76.2	o	95.6	Pqrst
MEX54/BAT332 resistant checks	31.3	ab	26.8	A
MIB 456-High Fe Universal Check	76.8	o	98.6	Qrst
Decelaya 1-Low Fe Check	-	-	50.0	Abcdefghi
KAB06F2.8-27	30.5	ab	59.2	Bcdefghijklm
NUA 99	48.4	def	44.0	Abcdef
NUA 69	34.5	abc	45.3	Abcdefg
KAB06F2.8-12	32.5	ab	49.0	Abcdefgh
RWR 2154	73.0	klmno	106.7	T
RWR 2245	56.1	fghij	53.7	Abcdefghij
KAB06F2.8-36	55.4	fghi	105.7	St
KAB06F8.8-35	51.0	efg	78.4	Hijklmnopqrst
CODMLB 001	74.2	mno	72.3	Efghijklmnopq
HM 21-7	71.0	klmno	52.5	Abcdefghij
Ngwaku-Ngwaku	73.6	lmno	54.2	Abcdefghijk
NUA 45	-	-	-	-
NUA 59	41.0	bcde	52.0	Abcdefghij
NUA 56	59.6	fghijk	70.9	Defghijklmnopq
NUA 35	67.5	ijklmno	62.8	Bcdefghijklmno
Gitanga	69.4	jklmno	67.8	Cdefghijklmnopq
Zebra	74.2	mno	106.9	T
ACC 714	31.5	ab	49.4	Abcdefgh
CODMLB 033	35.1	abcd	43.7	Abcdef
Roba 1	75.2	mno	41.0	Abcd
SMC 21	76.2	o	65.9	Bcdefghijklmnop
SEMC 16	-	-	55.7	Abcdefghijkl
SMC 18	75.5	mno	73.2	Efghijklmnopqr
SEMC 17	62.2	ghijklmn	75.0	Ghijklmnopqrs
GLP 2	-	-	85.0	Klmnopqrst
CAL 96-Low Fe check susceptible check	75.1	mno	90.7	Nopqrst
DOR 500-Low Universal check	74.2	mno	70.7	Defghijklmnopq
Maharagi soya	66.9	ijklmno	50.2	Abcdefghi
MBC32	47.2	cdef	48.9	Abcdefgh
Nyiramogorori2	75.0	mno	80.9	Ijklmnopqrst
LSD	13.6		30.9	

Root Rot

A high and significant (P<0.001) difference in resistance to *Fusarium* root rot and Pythium root rot isolates was observed among the tested genotype (**TABLE 4**). However, the means, modes and medians for both *Fusarium* and Pythium root rot isolates were not significantly different (**TABLE 5**). Seven genotype; Rugandura, Kiangara, Decelaya 1, Gitanga, Zebra, ACC 714,

Nyiramogorori2 were identified as resistant varieties to both *Fusarium* root rot and Pythium root rot.

Table 4: Means squares for both beansFusarium and Pythium root rot.

		Fusarium root rot isolate			Pythium root rot isolate			
		Mean	Median	Mode		Mean	Mode	Median
Source of variation	DF	MS	MS	MS	DF	MS	MS	MS
Reps	2	10.247 **	25.158 **	50.518 **	2	28.08 ***	36.21 ***	38.838 ***
Genotype	54	8.485***	15.768 ***	17.717 ***	54	13.48***	24.01***	22.62 ***
Residual	106	2.043	5.626	7.533	101	1.731	5.015	4.49
Total	162	4.291	9.248	11.458	157	6.107	11.946	11.163
CV		31.458	55.560	57.988		24.607	41.28	40.53
Mean		4.544	4.269	4.733		5.347	5.42	5.23
LSD		2.314	3.840	4.443		2.131	3.63	3.43

**, ***significant at P ≤ 0.01 and P ≤ 0.001 respectively.

Table 5: Means, modes and medians values for both beans Fusarium and Pythium root rot.

Genotype	Fusarium root rot				Pythium root rot			
	Mean	Median	Mode	DI	Mean	Median	Mode	DI
Ngwin x CAB2/2/3/1/1	5.54	6.66	6.66	65	6.72	8.99	8.99	76
Ngwin × CAB2 × (RWV3317)	6.73	6.99	6.66	73	8.44	8.99	8.99	94
MAC 42	3.79	2.16	1.99	42	6.02	6.66	6.66	66
RWV 3316	4.22	2.49	4.33	47	2.83	1.99	1.99	30
NUV 219-1	4.91	4.33	4.33	54	5.06	2.32	4.32	57
CAB 2	6.05	4.83	6.66	59	4.99	5.92	5.91	69
RWV 2359	7.91	9.19	8.98	88	8.07	8.99	8.99	89
Garukurare	3.93	3.66	4.66	43	2.64	1.99	1.99	29
Kivuzo	5.71	6.66	6.66	62	3.74	2.16	1.99	42
RWV1129	8.51	8.99	8.99	96	8.99	8.99	8.99	100
Ndimirakaguja vol	3.66	2.33	1.99	40	3.06	2.32	1.99	34
Icyana 2	3.48	2.33	4.33	39	2.46	1.99	1.99	28
RWV 2361	4.72	4.33	4.33	51	3.25	2.16	2.32	36
MAC 44	5.47	6.99	6.66	62	6.56	6.99	8.99	75
RWV 3006	3.87	2.66	4.66	40	4.69	4.66	4.32	44
RWV 2887	4.99	4.66	6.66	56	7.14	8.99	8.99	79
MAC 74	4.12	2.33	4.33	47	7.14	8.99	8.99	81
Agronome 2	5.66	5.71	5.99	48	9.14	9.07	9.09	100
VRA 4	4.80	4.49	4.33	55	6.13	5.16	4.99	71
Rugandura	2.00	1.99	1.99	22	2.02	1.99	1.99	22
RWV 2070	7.61	8.99	8.99	84	5.92	5.66	6.66	64
Kiangara	2.03	1.99	1.99	23	2.83	1.99	1.99	31
VCB 81013	3.63	4.33	4.33	41	2.36	1.99	1.99	27
Gasirida	2.99	1.99	1.99	33	5.85	4.99	4.66	63
MLB49-89A/ RWR719 Resistant check	2.13	1.99	1.99	24	2.01	1.99	1.99	22
MIB 456-High Fe Universal Check	2.39	1.99	1.99	27	3.48	2.32	2.32	38
Decelaya 1-Low Fe Check	3.03	1.99	1.99	37	2.58	2.77	2.73	22
KAB06F2.8-27	5.58	4.83	6.66	64	5.19	6.66	6.66	56
NUA 99	6.05	6.99	8.99	67	7.06	6.99	6.66	79
NUA 69	5.56	4.66	6.66	62	7.10	8.99	8.99	78

KAB06F2.8-12	4.52	3.83	2.33	50	6.84	6.99	6.66	73
RWR 2154	4.95	4.66	6.66	58	7.65	8.99	8.99	85
RWR 2245	5.84	4.33	4.33	68	3.57	1.99	1.99	40
KAB06F2.8-36	5.23	4.66	6.66	59	8.91	8.99	8.99	98
KAB06F8.8-35	4.73	3.33	4.33	50	6.17	4.66	4.32	71
CODMLB 001	3.51	2.66	2.33	39	7.42	8.99	8.99	81
HM 21-7	3.49	4.33	4.33	48	4.20	2.32	2.32	47
Ngwaku-Ngwaku	3.11	2.33	2.33	32	4.74	5.92	5.91	58
NUA 45								
NUA 59	6.91	6.99	8.99	70	7.39	8.99	8.99	81
NUA 56	7.96	8.99	8.99	89	8.56	8.99	8.99	95
NUA 35	6.86	8.99	8.99	76	8.48	8.60	8.62	98
Gitanga	2.00	1.99	1.99	22	1.99	1.99	1.99	22
Zebra	2.65	1.99	1.99	28	3.21	1.99	1.99	36
ACC 714	2.00	1.99	1.99	22	2.18	1.99	1.99	24
CODMLB 033	3.89	2.33	4.33	43	5.87	4.32	4.32	62
Roba 1	2.33	2.33	2.33	26	3.26	2.32	4.32	37
SMC 21	3.66	1.99	1.99	41	6.40	6.99	6.99	67
SEMC 16	4.76	4.49	4.33	52	5.24	5.82	6.99	63
SMC 18	3.68	2.33	1.99	41	5.58	4.32	4.32	53
SEMC 17	4.96	4.66	4.33	55	6.61	8.99	6.99	75
GLP 2								
CAL 96-Low Fe check susceptible check	7.56	8.99	8.99	86	8.61	8.99	8.99	96
DOR 500-Low Universal check	3.22	2.33	4.33	36	5.57	4.32	4.32	63
Maharagi soya	3.54	1.99	1.99	40	3.36	2.32	2.32	37
MBC32	5.45	6.66	6.66	61	6.60	6.16	8.99	73
Nyiramogoron2	2.03	1.99	1.99	23	2.01	1.99	1.99	22

DI=Disease Index.

DISCUSSION

In this study, genotypes that make up the regional nutritional nursery were screened to identify new and better sources of resistance to both bean angular leaf spot and root rot diseases. Genotypes exhibited different reactions to the different diseases. Only ACC 714 exhibited resistance to both Andean and MesoAmerican isolates of bean angular leaf spot and also to *Fusarium*and Pythium root rot. This genotype is highly recommended for both nutritional and multi-resistance breeding program. The results also indicate that resistance to different pathogens can be pyramided into appropriate background to provide broad spectrum resistance in iron and zinc biofortified bean genotypes.

The results of this study showed that some of the nutritional bean genotypes have good levels of resistance to a particular pathogen. In 2007, Wagara and Kimani [12] reported that some on the nutrient rich bean varieties evaluated in Kenya possessed good level of resistance to major diseases occurring in farmer fields. For example, they reported that

Kiangara had high to moderate resistance to major biotic constraints in Kenya. In the current study, Kiangara exhibited an intermediate level of resistance to both ALS and root rot. Results suggest that selecting biofortified beans for broad resistance to diseases in plant breeding programs is possible; and that this can result in significant amounts of genetic resources for multiple purposes with minimal resources. Thus, our results support the fact that breeding for higher iron as well as high zinc content and multiple resistance in common beans could contribute significantly to improving the life status of individuals dependent on beans as staple foods [13].

CONCLUSION AND RECOMMENDATIONS

The study revealed variations among the screened nutritional bean varieties according to resistance to different pathogens. This implied the potential for utilization of some of these varieties to pyramid useful disease resistance and high Fe and Zn content quantitative trait loci into appropriate background to provide durable resistance in bean genotypes higher in iron and zinc content. The variety ACC 714 was attributed to a broad resistance among 57 genotypes screened for bean angular leaf spot (*Psuedocercospora griseola*) and bean root rot (*Pythium ultimum* and *Fusarium* solani fsp. phaseoli) under inoculation in the screenhouse. Efforts should be made to promote selection of genotypes that combine high capacity to accumulate high iron and zinc content in their seed and high level of resistance to major stresses.

ACKNOWLEDGEMENTS

We are grateful to Rwanda Agriculture Board, Alliance for Green Revolution in Africa, Makerere University, CIAT Africa, Harvest Plus project for the support to this work.

REFERENCES

1. Harvest Plus (2011) Breeding crops for better nutrition.
2. Pfeiffer WH, McClafferty B (2007)Biofortification: Breeding micronutrient dense crop. In: Kang MS, Priyadarshan PM (eds.) Breeding Major food Staples for the 21stCentry,pp: 61-91.
3. Strang B, Lane A (2009) Investing in the future: A united call to action on vitamin and mineral deficiencies Global Report, pp: 5-6.

4. Alliance for Green Revolution in Africa (AGRA) (2008) Uganda bean varieties show resistance to diseases.
5. Mukankusi CM (2008) Improving resistance to Fusarium Root rot (Fusariumsolani (Mart) Sacc. f.sp. phaseoli (Burkholder) WC Snyder & HN Hans) in common bean (Phaseolus vulgaris L.). PhD Thesis, University ok Kwazulu Natal, Pietermaritzburg, South Africa, p: 200.
6. AbawiGS, Pastor-Corrale MA (1990) Root rot of beans in Latin America and Africa: Diagnosis, research methodologies and management strategies. CIAT Publication No 35, Cali Colombia, p: 114.
7. Payne RW, Murray DA, Harding SA, Baird DB, Sourtar DM (2007) Introduction to GenStat for Windows.10th edition. VSN International, Hemel Hempstead, UK.
8. Van SA, Pastor-Corrales MA (1987) Standard System for the Evaluation of Bean Germplasm. Cali, Colombia, p: 54.
9. Kobriger KM, HagedornDJ, Stevenson WR (1998) Analysis of Snap bean root rot potential of Wilsconsin fields. Extension publications, A3242.
10. Durham KM, Navabi A, PaulsKP (2011) Evaluation of Common Bacterial Bright Resistance in a resistant intercross population of common bean. The University of Guelph, Ontorio, Canada.
11. Bergamin FA, CarneiroSMTPG, Godoy CV, Amorim L, Berger RD, et al. (1997) Angular leaf spot of Phaseolus beans: Relationships between disease, healthy leaf area and yield. Phytopathology 87: 507.
12. Wagara IN, Kimani PM (2007) Resistance of nutrient rich bean varieties to major biotic constraints in Kenya. African Crop Science Conference Proceedings 18: 2087-2090.
13. Welch RM, House WA,Bebe S, Cheng Z (2000) Genetic Selection for Enhanced Bioavailable Levels of Iron in Bean (Phaseolus vulgaris L.) Seeds. J Agric Food Chem 48: 3576-3580.

CHAPTER 7

Formulation, Characterization and Properties of Hemp Seed Oil and Its Emulsions

Veronika Mikulcová [1], Věra Kašpárková [1,2,*], Petr Humpolíček [2,3] and Leona Buňková [4]

[1]*Department of Fat, Surfactant and Cosmetics Technology, Faculty of Technology, Tomas Bata University in Zlin, nam. T. G. Masaryka 5555, 760 01 Zlin, Czech Republic*

[2]*Centre of Polymer Systems, Tomas Bata University in Zlin, nam. T. G. Masaryka 5555, 760 01 Zlin, Czech Republic*

[3]*Polymer Centre, Faculty of Technology, Tomas Bata University in Zlin, 760 01 Zlin, Czech Republic*

[4]*Department of Environmental Protection and Engineering, Faculty of Technology, Tomas Bata University in Zlin, nam. T. G. Masaryka 5555, 760 01 Zlin, Czech Republic*

ABSTRACT

The formulation, characterization, and anticipated antibacterial properties of hemp seed oil and its emulsions were investigated. The oil obtained from the seeds of *Cannabis sativa* L. in refined and unrefined form was characterized using iodine, saponification, acid values, and gas chromatography, and was employed for the preparation of stable oil-in-water emulsions. The emulsions were prepared using pairs of non-ionic surfactants (Tween, Span). The effects of the emulsification method (spontaneous emulsification vs. high-intensity stirring), hydrophilic lipophilic balance (HLB), type and concentration of surfactant, and oil type on the size and distribution of the emulsion particles were investigated. It was found that the ability to form stable emulsions with small, initial particle sizes is primarily dependent on the given method of preparation and the HLB value. The most efficient method of emulsification that afforded the best emulsions with the smallest particles (151 ± 1 nm) comprised the high-energy method, and emulsions stable over the long-term were observed at HBL 9 with 10 wt % concentration of surfactants. Under high-intensity emulsification, refined and unrefined oils performed similarly. The oils as well as their emulsions were tested against the growth of selected bacteria using the disk diffusion and broth microdilution methods. The antibacterial effect of hemp seed oil was documented against *Micrococcus luteus* and *Staphylococcus aureus* subsp. *aureus.* The formulated emulsions did not exhibit the antibacterial activity that had been anticipated.

Keywords: hemp seed oil; emulsion; particle size; fatty acid composition; antibacterial activity

1. INTRODUCTION

Hemp seed oil, obtained from the seeds of *Cannabis sativa* L., is known for its nutritive, health-enhancing properties and bioactivity. Compared to other vegetable oils, it is an especially rich source of both n-3 and n-6 essential fatty acids, namely linoleic acid (18:2 n-6, at 55 wt %) and alpha-linolenic acid (18:3 n-3, at 20 wt %). The content of gamma-linolenic acid (18:3 n-6) equals approximately 1–4 wt %, while that of stearidonic acids (18:4 n-3) ranges from 0.5–2 wt % [1]. Of no less importance is that hemp seed oil contains a moderate to high amount of tocopherols and tocotrienols (100 to 150 mg per 100 g of oil), phytosterols, phospholipids, carotenes, and minerals [2]. Interestingly, the aforementioned beneficial properties of hemp seed oil offer numerous potential applications, e.g., as components in functional foodstuffs and in treating various health problems. Here, the lowering of high cholesterol and high blood pressure can be named as examples [3,4]. The health benefits of hemp seed oil are attributed mainly to its desirable n-3 and n-6 fatty acid ratio, 3:1, which is suggested as being optimal for human nutrition [5,6]. It has been shown that unbalanced intake of n-3 and n-6 fatty acids is associated with many diseases, e.g., diabetes, cardiovascular diseases, and cancer. The presence of gamma-linolenic acid in the oil, which is deficient in the average Western diet, is also noteworthy [6]. This unique composition of hemp seed oil differs from the other common seed oils and offers opportunities for the development of special nutritional formulations [7,8,9]. However, besides the well-documented health and nutrition effects, the antimicrobial and specifically the antibacterial effects of hemp seed oil still remain questionable [10,11].

Although many studies have focused on preparing oil-in-water (O/W) emulsions consisting of either mineral or synthetic oils, less attention has been paid to their formation in the presence of vegetable oils as a dispersed phase. Moreover, most studies and applications of vegetable oil-based emulsions involve one of the following commonly used oils; palm, soybean, rapeseed, sunflower, coconut, palm-kernel, cottonseed, groundnut, and olive [4,12,13]. However, with the development of the nutraceutical and functional foods market, more attention is now devoted to non-traditional vegetable oils and their encapsulation in emulsions. Examples include argan oil [14], pomegranate seed oil [15], and grapeseed and sesame oils [16]. In this context, the application of hemp seed oil in emulsions, which might further enhance its beneficial properties, has not been fully exploited. Due to reduced sizes of droplets and high surface to volume ratio, loading of the oil in nanoemulsions may improve the bioavailability of present unsaturated fatty acids [7], help to protect the oil against oxidation and interaction with other ingredients, and contribute as a dietary source of natural antioxidants for disease prevention and health promotion [9,17]. Only a few studies have actually dealt with hemp seed oil emulsions (water-in-oil) while, as far as the authors are aware, even less research has concentrated on oil-in-water emulsions [18,19].

The past few years have also witnessed a widening in the utilization of oils from plant sources into non-food areas, such as the pharmaceutical and cosmetics industries, due to their qualities of being non-toxic, biodegradable, and environmentally friendly. This is reflected in current research on the exploration of vegetable-oil-based emulsions [20,21]. The aim of this paper was to prepare stable oil-in-water emulsions based on hemp seed oil, and determine the influence of emulsion composition and preparation methods on their characteristics, including stability. In this regard, the quality of two brands of oil extracted from hemp seeds was investigated to describe differences between the properties of refined and unrefined oil when utilized in emulsions. Research was also carried out on the possible bioactivity of the hemp seed oils and their emulsions as pertaining to their antibacterial properties against common pathogens.

2. RESULTS AND DISCUSSION

Prior to preparing the emulsions, basic characterization of the hemp seed oils was carried out in order to elucidate differences between their refined and unrefined types. The iodine values of 155.8 ± 1.9 and 167.4 ± 0.6 g iodine/100 g of oil were determined for refined and unrefined oil, respectively. Correspondingly, the saponification values of 197.6 ± 4.5 and 202.2 ± 3.9 mg KOH/g of oil, and acid values of 1.7 ± 0.6 and 0.7 ± 0.0 mg KOH/g of oil were measured. Comparing the obtained results with the data published by Anwar et al. [22] revealed reasonably good compliance. The published iodine values were quite similar to those herein, ranging from 154 to 165 g iodine/100 g oil and the saponification values were slightly lower (184–190 mg KOH/g of oil). The fatty acid composition of the oils is provided in Table 1. The main differences between the contents of particular fatty acids pertained to oleic and alpha-linolenic acids. Regarding the content of alpha-linolenic acid in unrefined and refined oils, values of 20.3 ± 0.03 wt % and 16.7 ± 0.04 wt %, respectively, were measured. The content of oleic acid was 12.1 ± 0.03 wt % and 9.0 ± 0.1 wt % for the unrefined and refined samples, respectively. However, the variations observed did not obviously deviate from the fatty acid contents reported for a range of hemp seed oils by different authors [22,23]. It is therefore apparent that oil composition prevailingly depends on its origin and/or the extraction procedure used to obtain it. Unrefined hemp seed oil also possessed trace concentrations of eicosadienoic and myristic acids.

Table 1. Composition of fatty acids (g/100 g) in unrefined and refined hemp seed oils determined by gas chromatography (GC).

Fatty Acid	Concentration (g/100 g)	
	Unrefined Oil	Refined Oil
Myristic	0.04 ± 0.03	n.d.
Palmitic	5.9 ± 0.27	6.2 ± 0.13
Palmitoleic	0.1 ± 0.00	0.1 ± 0.02
Stearic	2.2 ± 0.04	2.4 ± 0.07
Oleic	9.0 ± 0.15	12.1 ± 0.03
Linoleic	55.3 ± 0.12	57.3 ± 0.03
gamma-linolenic	4.4 ± 0.03	3.0 ± 0.02
alpha-linolenic	20.3 ± 0.03	16.7 ± 0.04
Arachidic	1.7 ± 0.04	1.0 ± 0.04
Eikosanoic	0.7 ± 0.05	0.8 ± 0.00
Eikosenoic	0.4 ± 0.02	0.4 ± 0.02
Eikosadienoic	0.1 ± 0.01	n.d.

n.d.: not determined.

2.1. Formulation of Emulsions

The O/W emulsions were prepared by two different emulsification procedures; these also involved varying the types and concentrations of emulsifiers, with the hydrophilic-lipophilic balance (HLB) ranging from 6 to 10, as well as utilizing two types of hemp seed oil. Key characteristics of an emulsion are the size and distribution of the emulsion particles, as their changes indicate the stability of the formulation. As expected, it was discerned that particle size and distribution were notably influenced by the method of preparation. This is illustrated in Figure 1, which shows the particle size of emulsions and methods for preparing the same, relating to the samples based on Span 85/Tween 85.

As can be seen, the diameters of the particles present in all emulsions prepared by the high-energy method were significantly smaller ($p \leq 0.001$), regardless of HLB values, than the emulsions produced by the low-energy EIP (emulsion inversion point) method. This observation is in agreement with a study published by Ostertag et al. [24]. In this study, only relatively large droplets (>600 nm) were formulated by the EIP method using long-chain triglyceride oils (such as olive, grape seed, sesame, peanut, and canola oils). Similar findings were reported by Gullapalli and Sheth [25], who found that non-ionic emulsifiers reduced the particle size of hydrocarbon-in-water emulsions more effectively than triglyceride-in-water emulsions. Recently, exotic vegetable oil-in-water nanoemulsions with the addition of ethoxylated and acetylated lanolin have been obtained by using a low energy EIP method. It has been shown that the lanolin derivative addition caused alterations of droplet size and conductivity of the systems, however the droplet size remained still within the nanometer range (20–200 nm) [26]. Lane et al. [27] reported on flaxseed and algae nanoemulsions formulated with combinations of Tween 40 and lecithin. Stable emulsions were prepared up to 50 wt % of oil content with a droplet size of 192 nm for flaxseed-oil and 182 nm for emulsions loaded with algae oil. In the study of Krasodomska and Jungnickel [28], various seed oils (apple, strawberry, and raspberry) have been used as components of the oil phase in O/W emulsions. The best emulsion contained 4 and 5 wt % of seed oil together with other components of the oil phase and the O/W ratio was 20/80. The emulsification procedure also affected the particle size distribution of the emulsions, and homogenization via the high-energy method brought about emulsions with narrower distribution, in comparison with the EIP method. This is shown in Figure 2 for systems consisting of the Tween 80/Span 80 mixture (5 wt %) at HLB 9, where only the main particle population was observed when the high-energy method was utilized, while EIP afforded bimodal size distribution with the occurrence of two main droplet populations.

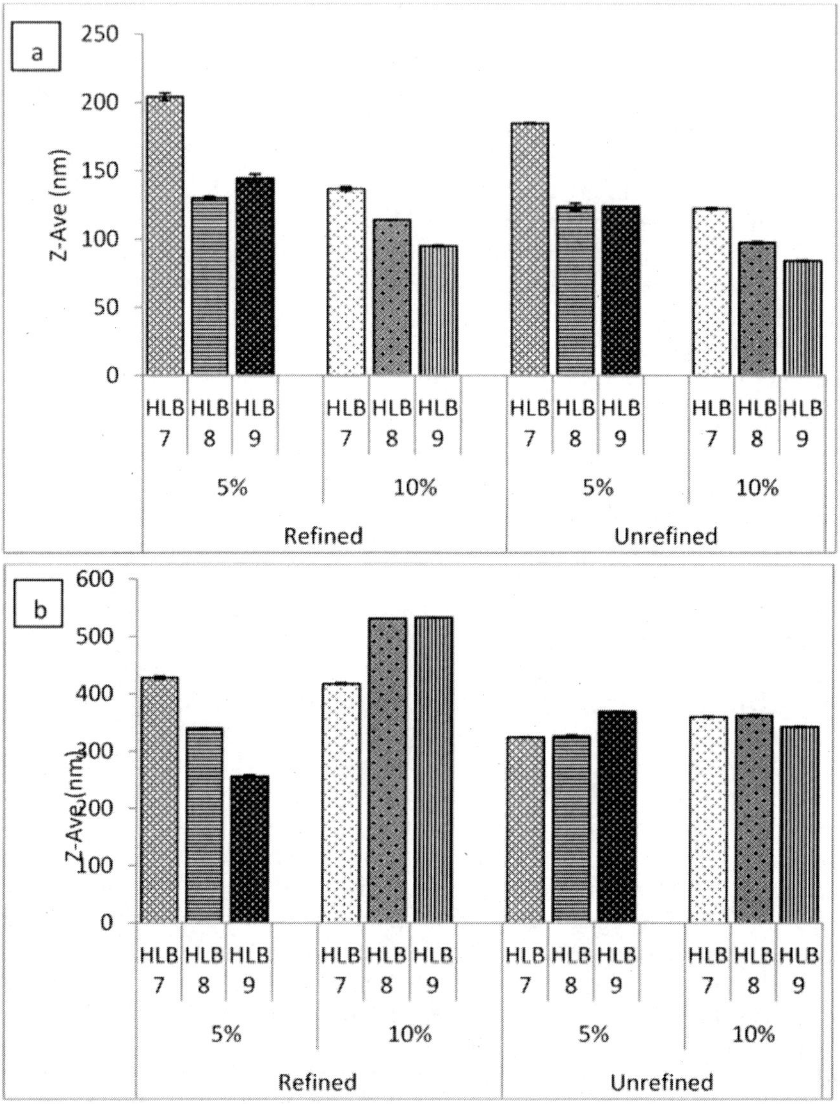

Figure 1. Effect of oil type (refined vs. unrefined), hydrophilic lipophilic balance (HLB), and surfactant concentration on the particle size of emulsions prepared by (**a**) high-energy method and (**b**) low-energy method. Tween 85/Span 85 ratios in the emulsions were 1.30 (HLB 7), 2.07 (HLB 8), and 3.60 (HLB 9).

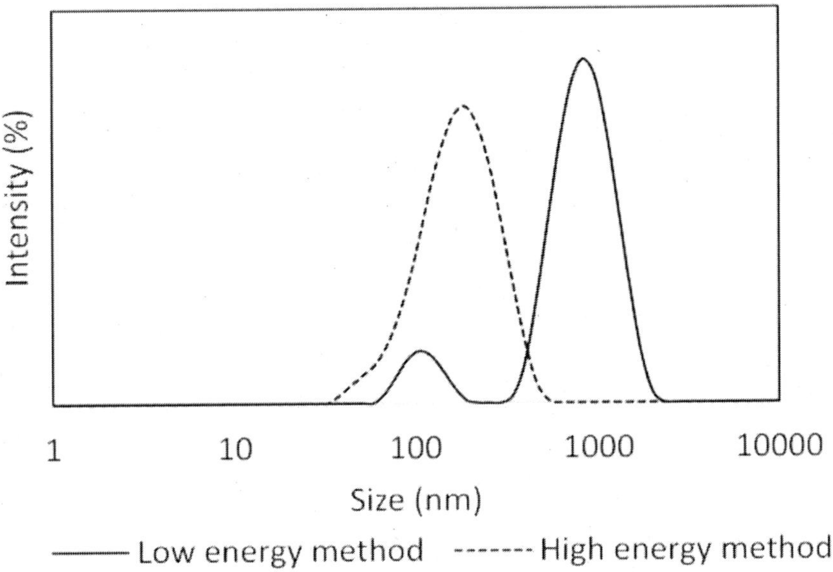

Figure 2. Particle size distribution recorded after emulsification with low-energy vs. high-energy conditions at emulsions with 5 wt % Tween 80/Span 80 and HLB 9. The Tween 80/Span 80 ratio was 0.78.

On average, emulsions prepared using the high-energy method contained considerably smaller droplets than systems prepared with the EIP technique. Immediately after preparation, the former of the two systems showed particles from 151 ± 1 nm to 209 ± 5 nm, whilst the particle sizes of the latter were significantly higher ($p \leq 0.001$) and ranged from 502 ± 22 nm to 1050 ± 29 nm. Correspondingly, the polydispersity index (PDI) of the high-energy samples was approximately 0.18 ± 0.01 to 0.26 ± 0.01, whereas the PDI of the emulsions prepared by EIP ranged from 0.26 ± 0.09 to 0.75 ± 0.44. It is known that the PDI is an estimate of the width of the droplet distribution in samples. The PDI values of approximately 0.1 correspond to the polydispersity of monodisperse standards and values greater than 0.7 indicate that the sample has a very broad size distribution. The PDI values measured on the studied emulsions therefore clearly show that the high-energy method provides significantly narrower droplet distributions than the EIP technique.

2.2. Influence of HLB and Oil Type

Finding the optimum HLB value required for the successful encapsulation of the hemp seed oil was based on the premise that at optimum HLB, the mean particle sizes of the emulsion droplets are at their minimum. This factor also influences, to a large extent, the stability of the emulsions produced. Another possible procedure to determine optimum HLB encompasses visual observation of the emulsions, and the system with minimal creaming and phase separation is deemed to possess the optimum HLB [29]. The values chosen for preparing the emulsions ranged from 6–10 for the pairs of Tween 80/Span 80 and 7–9 for Tween 85/Span 80. Figure 3 depicts the impact of the HLB value on the particle sizes of the emulsions, revealing the evolution of droplet size concurrent with changes in HLB for Tween 80/Span 80 under differing preparation conditions. This figure shows that the optimal HLB for hemp seed oil lies between 8 and 9, and is identified with affording the minimum average particle size, irrespective of the method used for producing the emulsions. In the case of the EIP method, a U-shaped curve is clearly visible, showing the strong correlation that exists between particle size and HLB. Unfortunately, despite finding the optimum HLB value for this system, it was not possible to formulate stable emulsions with EIP and emulsions prepared by this low-energy method became unstable or broke down within several minutes of preparation. On the contrary, for analogous emulsions prepared by the high-energy method, merely negligible changes in droplet size alongside changes in HLB were observed. Furthermore, the emulsions were reasonably stable, not exhibiting any sign of destabilization. In the literature, the aforementioned U-shape dependence of particle size on HLB is much less reported for the high-energy methods than the low-energy methods (such as EIP) [30]. Figure 3 also highlights the significant impact of HLB and the method of emulsion preparation on the PDI, which was systematically higher in emulsions prepared by EIP. Hence, the low-energy method utilized is far more sensitive to the proper choice of emulsion composition than the high-energy method. Of all the samples, the smallest particles (84 ± 1 to 122 ± 2 nm) were achieved at any given HLB with the Tween 85/Span 85-based emulsions (10 wt %) prepared by the high-energy method.

The above results demonstrated that emulsions with fine droplets could primarily be produced by the high-energy approach, but the question remained as to whether the type of oil (refined vs. unrefined) actually influenced the particle sizes of emulsions formulated through both approaches. Figure 1a clearly shows the similar behaviour exhibited by the Tween 85/Span 85 emulsions over the range of the HLB tested, irrespective of the type of oil used when applying the high-energy method. This contrasted with emulsions prepared by the EIP method (Figure 1b), which tended to vary and behave differently, even if they shared the exact same formulation. Here, a statistically significant difference ($p \leq 0.01$) was observed for particle sizes of the emulsions formulated with refined and unrefined oils, respectively. It was also seen that higher concentrations of surfactant provided more uniform systems across the HLB range examined and enabled the preparation of emulsions with significantly smaller particles ($p \leq 0.01$).

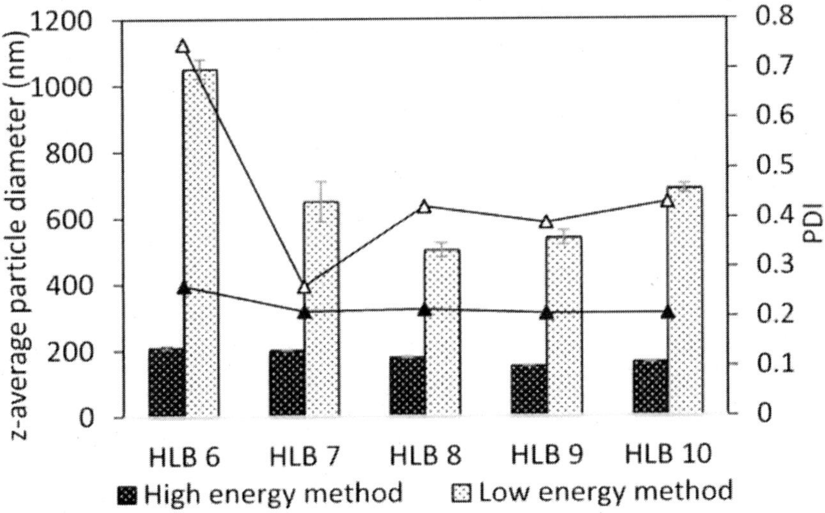

Figure 3. The influence of the HLB on particle sizes (columns) and polydispersity index (lines) of the nanoemulsions (5 wt % Tween 80/Span 80) prepared by the low-energy and high-energy methods using refined oil.

2.3. Emulsion Stability

The long-term stability of the hemp seed oil-in-water emulsions was assessed over a period of a maximum of 230 days by recording changes in their visual appearance during storage at 25 and4 °C. The best stability discerned for emulsions formed by the EIP method, regarding the visual assessment, was demonstrated by emulsions containing 10 wt % Tween 85/Span 85 at HLB 7. These emulsions were stable for 8 days and such early break-down was attributed to their relatively large sizes of particles (590 to 780 nm). However, emulsions prepared on the Ultraturrax device (IKA, Staufen, Germany), stabilized by Tween 80/Span 80, were stable for 24 days of storage, with no phase separation or creaming observed. Later, when creaming of the emulsions was recorded, no significant changes were observed in particle size, as measured by dynamic light scattering (DLS). The greatest stability was shown by Tween 80/Span 80-based systems at HBL 9 and 10 wt % surfactant (stable for 230 days), and the Tween 85/Span 85 system (200 days), stored at both temperatures. This long-term stability might be associated with the small, initial droplet size observed after preparation. According to study conducted by Rao and McClements [31], lemon oil nanoemulsions with initially smaller droplets were more stable against coalescence, aggregation, as well as flocculation. Similar findings were also reported in several other papers. For example, in recent works studying the effect of various processing factors on the formulation of vegetable oil-based nanoemulsions [32,33], the smaller droplets resulted in their better long-term stability, which also correlated well with values of droplet sizes predicted by using mathematical modelling.

Variations in mean particle sizes during the storage of emulsions formulated with Tween 85/Span 85 (10 wt %) at HLB 9 are illustrated in Figure 4, which documents only minor changes, regardless of the storage temperature. This finding is consistent with the study of Rebolleda et al. [34], in which wheat bran oil nanoemulsions stabilized by mixtures of Span 80/Tween 80 showed only negligible changes in the droplet size of the emulsions during storage.

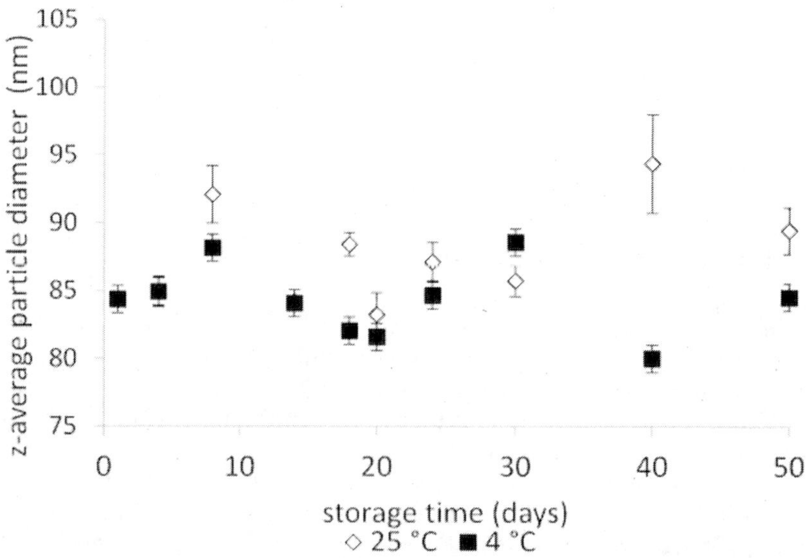

Figure 4. Changes in the z-average diameter of emulsion droplets as a function of storage time and temperature (4 and 25 °C) determined for emulsions with Tween 85/Span 85 (10 wt %) at HLB 9.

2.4. Antibacterial Activity

Both hemp seed oils and their emulsions were screened with respect to their antibacterial effects. Table 2 displays the results from the disc diffusion method as recorded for both oils, revealing their weak antibacterial activity against selected species. Using this method, the antibacterial activity was determined by the measurement of the diameter of the inhibitory zone (mm) formed around the discs soaked with the oil and the size of the disc was subtracted from the size of the inhibitory zone. The unrefined oil possessed inhibition to all strains, with inhibition zones ranging from 0.3 to 3.3 mm, and the effect was mainly dependent on the type of bacteria. These results were further supported by a broth microdilution method (data not presented). A previous report on the antimicrobial activity of hemp seed oil was published by Leizer et al. [11], who observed some bioactivity during primary screening. In past years, the antibacterial activity of essential oils extracted from *C. sativa* was, however, more extensively studied [35,36].

Table 2. Inhibitory effect of hempseed oils (mean ± SD, $n = 3$) against most common pathogenic bacteria, expressed as diameter of the inhibition zone in mm (diameter of the disc was subtracted from the total size of inhibition zone). Means within a line with the same superscript differ significantly ($p \geq 0.005$).

	Bacterial Strain	Inhibition Zone Size (Ø ± σ) mm	
		Unrefined Oil	Refined Oil
G+	*Bacillus cereus* CCM 2010	2.3 ± 0.6 [a]	0.0 ± 0.0 [a]
	Bacillus subtilis subsp. *subtilis* CCM 2216	2.3 ± 1.8 [a]	0.0 ± 0.0 [a]
	Micrococcus luteus CCM 732	3.3 ± 1.8	2.7 ± 1.2
	Staphylococcus aureus subsp. *aureus* CCM 3953	3.0 ± 0.0	3.0 ± 1.2
G−	*Citrobacter freundii* CCM 7187	2.3 ± 0.6 [a]	0.0 ± 0.0 [a]
	Enterococcus faecalis CCM 4224	2.3 ± 0.6	2.7 ± 0.6
	Escherichia coli CCM 3954	0.3 ± 0.6	0.0 ± 0.0
	Salmonella enterica subsp. *enterica* ser. Enteritidis CCM 4420	3.0 ± 1.8	2.0 ± 1.2
	Serratia marcescens subsp. *marcescens* CCM 303	2.7 ± 0.6	0.7 ± 1.2
	Pseudomonas aeruginosa CCM 395	1.7 ± 0.6	2.3 ± 0.6

In order to understand the difference in antibacterial activity of both types of oil, the results from antibacterial testing were compared to the fatty acid analyses. As stated earlier, the antibacterial action of fatty acids is usually recognized for long-chain unsaturated fatty acids, including oleic acid, linoleic acid, and linolenic acid [37]. The higher content of alpha-linolenic acid in the unrefined oil could explain the increase in antibacterial activity, compared with findings for the refined oil. Another possible explanation might pertain to the fact that during the refining process, minor components such as tocopherols and tetrahydrocannabinol are removed, which may contribute to the antibacterial activity of unrefined hemp seed oils [38].

As reported earlier [39], gram-positive bacteria were seen to be more sensitive to the unrefined oil than gram-negative strains, as a result of differences in the composition of the bacterial cell wall. In accordance with this fact, *E. coli* was the most resistant species, and *M. luteus* and *S. aureus* proved to be most sensitive to the oils tested. However, statistical analysis did not prove any significant difference between the group of gram positive and gram negative strains regarding the effect of refined and unrefined oils, respectively. Though the observed antibacterial activity of hemp seed oil was weak, it can be regarded as an added value to the main positive characteristics of hemp seed oil, namely the content of n-3 and n-6 fatty acids. Additionally, an investigation was conducted on the potential antibacterial effects of the formulated emulsions. Previously, a synergistic effect of antibacterial substances encapsulated in emulsions or nanoemulsions has been observed, as reported by Ghosh et al. [40]. The enhanced activity of emulsions against microorganisms is primarily explained by the presence of non-ionic surfactants in the formulations. Several studies [30,41] reported on increased antibacterial activity when non-ionic Tween 80 made up part of the formulation. Therefore, a similar effect was expected for the hemp seed oil emulsions studied herein. Nevertheless, the disk diffusion and broth dilution methods did not reveal the rise that had been anticipated in antibacterial activity by emulsions containing oil, relative to pure forms of oils alone.

3. EXPERIMENTAL METHODS

3.1. Materials

Two types of cold-pressed hemp seed oil were employed: an unrefined oil was kindly donated by Míča a Harašta (Prague, Czech Republic), and a refined, commercially available oil was purchased from Cannaderm (Prague, Czech Republic). The non-ionic surfactants Span 80 (Sorbitane monooleate, HLB 4.3), Span 85 (Sorbitane trioleate, HLB 1.8), Tween 80 (Polyoxyethylenesorbitan monooleate, HLB 15), and Tween 85 (Polyoxyethylenesorbitan monooleate, HLB 11) were supplied by Sigma-Aldrich (Steinheim, Germany) and were used without further purification.

3.2. Microorganisms

The test microorganisms, including gram-positive and gram-negative strains, were obtained from the Czech Collection of Microorganisms (CCM, Czech Republic). The bacteria were selected to represent the major spoilage classes. Tests utilized gram-positive *Bacillus subtilis*, subsp. *subtilis* CCM 2216, *Bacillus cereus* CCM 2010, *Enterococcus faecalis* CCM 4224, *Micrococcus luteus* CCM 732, *Staphylococcus aureus*, subsp. *aureus* CCM 3953, and gram-negative *Citrobacter freundii* CCM 7187, *Escherichia coli* CCM 3954, *Proteus vulgaris* CCM 1799, *Pseudomonas aeruginosa* CCM 3955, and *Serratia marcescens*, subsp. *marcescens* CCM 303. All strains were maintained on nutrient agar (5 g L^{-1} peptone, 5 g L^{-1} NaCl, 1.5 g L^{-1} beef extract, 1.5 g L^{-1} yeast extract, 15 g L^{-1} agar; from Hi-Media Laboratories Bombay, India) and were sub-cultured onto fresh media every two weeks. The initial test inocula of the microorganisms were prepared from 24 h cultures. Each bacterial suspension was adjusted by dilution with a nutrient broth to 5×10^8 CFU mL^{-1}.

3.3. Characterization of Hemp Seed Oil

Basic characteristics in terms of iodine, saponification, and acid value were determined using respective methods described elsewhere [42]. Fatty acid methyl esters (FAME) were prepared by transesterification with KOH in methanol. In brief, oil was mixed with methanolic potassium hydroxide (1 M), boiled for 30 min and cooled to room temperature. Hexane and aqueous sodium chloride solution were added, and the organic, upper layer was separated for gas chromatography (GC) analysis. The composition of fatty acids was determined using GC on a Shimadzu GC-14A device equipped with an flame ionization detector (FID). A capillary DB-WAX column (30 m × 0.25 mm, Agilent, Santa Clara, CA, USA) was used. The temperature program employed was as follows: the column temperature was programmed at 110 °C and maintained for 3 min, then the temperature was raised to 220 °C at a rate of 15 °C/min, and an isothermal step followed at this temperature for 10 min. The temperatures for the injector and detector were set at 225 °C and 230 °C, respectively. Identification of the fatty acids present was carried out using the FAME SUPELCO 37 Component FAME Mix (Sigma Aldrich, Steinheim, Germany) standard. The content of the respective FAME in oil was expressed in percent by applying an internal normalization procedure.

3.4. Preparation of Emulsions

Two different methods were carried out to prepare the emulsions: high-energy emulsification and low-energy, phase-inversion emulsification (EIP). The oil-to-water (O/W) ratio of 5/95 (*w/w*) was employed. Suitable pairs of Spans and Tweens (Table 3) at the amounts of 5 and 10 wt % were used to formulate emulsions with a hydrophilic-lipophilic balance (HLB) ranging from 6 to 10. The required HLB was calculated using the following formula:

$$HLB = w_1 \times HLB_1 + w_2 \times HLB_2 \quad (1)$$

wherein w_1 and w_2 represent weight fractions of the emulsifiers utilized with HLB_1 and HLB_2, respectively. High-energy emulsification was performed on an Ultra Turrax T 25 device (IKA, Staufen, Germany). Appropriate amounts of hemp seed oil (5 wt %), a suitable pair of emulsifiers (5 or 10 wt %), and distilled water (add to 100 wt %) were heated in a test tube to 70 °C and homogenized immediately at 13,400 rpm for 15 min. The low-energy, phase-inversion procedure went as follows. Both the water and oil phases were heated to 70 °C and maintained at this temperature. The aqueous phase with a dissolved, water-soluble surfactant (Tween) was added drop-wise to the oil phase, which contained the oil-soluble surfactant (Span) and hemp seed oil. A constant stirring rate of 1050 rpm was utilized over a duration of 30 min. Homogenization was performed using an RZR Heidolph homogenizer (Heidolph Instruments GmbH & Co. KG, Schwabach, Germany).

Table 3. Mixtures on non-ionic surfactants used for obtaining the chosen HLB values; design of the study used for the preparation of hemp seed oil emulsions. Amounts of surfactants are given for their 5 wt % contents in 100 g of emulsions.

Amount of Surfactants (g)				HLB of the Mixture (*Tween/Span Ratio*)
Tween 80 (HLB 15)	Span 80 (HLB 4.3)	Tween 85 (HLB 11)	Span 85 (HLB 1.8)	
0.794	4.206	-	-	6 (*0.19*)
1.262	3.738	-	-	7 (*0.34*)
-	-	2.826	2.174	7 (*1.30*)
1.729	3.271	-	-	8 (*0.52*)
-	-	3.370	1.630	8 (*2.07*)
2.196	2.804	-	-	9 (*0.78*)
-	-	3.913	1.087	9 (*3.60*)
2.664	2.336	-	-	10 (*1.14*)

3.5. Particle Size Measurements

Particle size, particle size distribution, and the polydispersity index (PDI) were determined by dynamic light scattering (DLS) on a Zetasizer Nano ZS instrument (Malvern Instruments, Malvern, UK). Measurements for the hydrodynamic radii of emulsion droplets were expressed as intensity-weighted z-average diameters (nm). Analyses were carried out at a scattering angle of 90° at the temperature of 25 °C. The stability of the emulsions was evaluated at regular time intervals by measuring the particle size under different storage conditions (25 and 4 °C).

3.6. Antibacterial Testing

The oil samples and their emulsions were screened for the antibacterial activity they exhibited against common pathogenic bacteria by utilizing the disk diffusion and broth microdilution methods.

Disk diffusion method: Suspensions of each tested microorganism (100 μL) were spread on Mueller-Hinton sterile agar plates (Hi-Media Laboratories, Bombay, India). Sterile paper discs of 6 mm diameter, soaked with 5 μL of sample (hemp seed oil or emulsions), were placed on surfaces of the agar plates. As a reference, sample discs soaked with emulsions absent of hemp seed oil were utilized. Antibacterial activity was evaluated by measuring the diameter of the inhibition zone in mm after 24 h of incubation at 30 °C (*Bacillus subtilis* subsp. *subtilis*; *Pseudomonas aeruginosa*; and *Bacillus cereus*) or 37 °C (the remaining bacteria) and expressed in mm. In order to calculate the inhibition zone, the diameter of the paper disc was subtracted from the diameter of the inhibition zone.

Broth microdilution method: Under sterile conditions, 20 μL of bacterial suspension and 200 μL of a sample (containing hemp seed oil or a prepared emulsion) were pipetted into each well of the 96-well sterile microplate. Nutrient agar inoculated with bacterial suspension was employed as a positive reference. Emulsions, absent of hemp seed oil were used as a negative reference. Plates were then incubated for 30 min at 30 °C. After incubation, 100 μL of each individual suspension present in a respective well was spread over the surface of an agar plate and incubated again for 24 h at either 30 °C or 37 °C, depending on the bacteria used.

3.7. Statistical Analysis

The sizes of emulsion particles (DLS), iodine, saponification, and acid values as well as fatty acid composition (GC) of the hemp-seed oils were analysed at least in triplicate (n = 3); means and standard deviations were calculated in accordance with the Dean-Dixon method. Correspondingly, antibacterial testing (disk diffusion and broth microdilution methods) was conducted in triplicate and the Dean-Dixon method was utilized to calculate the means and standard deviations. The T-test was applied to determine statistical differences between the individual samples (Statistica, StatSoft, Inc., Tulsa, OK, USA). The p values of ≤0.05 were considered statistically significant.

4. CONCLUSIONS

This study has shown that emulsions of bioactive hemp seed oil can be formulated by employing pairs of non-ionic surfactants (Span, Tween) under appropriate conditions. It was found that the ability to form stable emulsions of small, initial particle size is primarily dependent on the given method of preparation and the HLB value. The low-energy method was suitable for producing emulsions without a high-energy input, but these systems turned out to be unstable due to the large, initial droplet sizes (502 ± 22 nm to 1050 ± 29 nm). However, high-energy homogenization produced nanoemulsions with fine droplets (151 ± 1 nm to 209 ± 5 nm), also supporting the stability of the emulsions. Regarding the influence of oil type on the formulations, emulsions containing refined and unrefined hemp seed oil performed similarly when using the high-energy method but differed when low-energy emulsification was employed. Testing for antibacterial properties by the disk diffusion and broth dilution methods confirmed the activity of the hemp seed oil utilized herein, although there was no sign of the enhancement that had been anticipated in the capability of the oil to act against bacteria via encapsulation in the emulsions. Regardless of this fact, the results suggest that such formulated emulsions could serve to fortify foodstuffs, as hemp seed oil is an exceptionally rich source of essential fatty acids, and the n-6 to n-3 fatty acid ratio in the oil stands at 3:1, which is considered optimal for human dietary purposes.

ACKNOWLEDGMENTS

This work was supported by the Ministry of Education, Youth and Sports of the Czech Republic—Program NPU I (LO1504). The authors are also grateful to the Czech Science Foundation (17-05095S). The author, Veronika Mikulcová, acknowledges the support of an internal Grant from TBU in Zlín, IGA/CPS/2017/001 financed from funds of specific academic research.

AUTHOR CONTRIBUTIONS

Veronika Mikulcová: was responsible for the preparation and characterization of the emulsions, and she also partially performed the characterization of the hemp seed oil, and conducted and evaluated the results of the microbiological studies; Věra Kašpárková: performed the characterization of the hemp seed oil, designed and supervised the formulation of the emulsions; Petr Humpolíček: analysed the experimental data and conducted the statistical analysis; Leona Buňková: supervised and designed the microbiology studies; All authors contributed to the writing of the manuscript.

CONFLICTS OF INTEREST

The authors declare that there is no conflict of interest.

REFERENCES

1. Callaway, J.C. Hempseed as a nutritional resource: An overview. Euphytica 2004, 140, 65–72.
2. Grotenhermen, F.; Russo, E. Cannabis and Cannabinoids: Pharmacology, Toxicology, and Therapeutic Potential, 1st ed.; Integrative Healing Press: New York, NY, USA, 2002.
3. Jones, K. Nutritional and Medicinal Guide to Hemp Seed; Rainforest Botanical Laboratory: Gibsons, BC, Canada, 1995.
4. Raikos, V.; Neacsu, M.; Morrice, P.; Duthie, G. Physicochemical stability of egg protein-stabilised oil-in-water emulsions supplemented with vegetable powders. Int. J. Food Sci. Technol. 2014, 49, 2433–2440.
5. Dunford, N.T. 2-Hemp and flaxseed oil: Properties and applications for use in food a2-talbot, geoff. In Specialty Oils and Fats in Food and Nutrition, 1st ed.; Woodhead Publishing: Cambridge, UK, 2015; pp. 39–63.

6. Hazekamp, A.; Fischedick, J.T.; Díez, M.L.; Lubbe, A.; Ruhaak, R.L. Chemistry of Cannabis. In Comprehensive Natural Products II, 1st ed.; Elsevier: Oxford, UK, 2010; pp. 1033–1084.
7. Acosta, E. Bioavailability of nanoparticles in nutrient and nutraceutical delivery. Curr. Opin. Colloid Interface Sci. 2009, 14, 3–15.
8. Sun, Y.; Xia, Z.; Zheng, J.; Qiu, P.; Zhang, L.; McClements, D.J.; Xiao, H. Nanoemulsion-based delivery systems for nutraceuticals: Influence of carrier oil type on bioavailability of pterostilbene. J. Funct. Foods 2015, 13, 61–70.
9. Raikos, V.; Ranawana, V. Designing emulsion droplets of foods and beverages to enhance delivery of lipophilic bioactive components—A review of recent advances. Int. J. Food Sci. Technol. 2017, 52, 68–80.
10. Ali, E.M.; Almagboul, A.Z.; Khogali, S.M.; Gergeir, U.M. Antimicrobial activity of Cannabis sativa L. Chin. Med. 2012, 3, 61–64.
11. Leizer, C.; Ribnicky, D.; Poulev, A.; Dushenkov, S.; Raskin, I. The composition of hemp seed oil and its potential as an important source of nutrition. J. Nutr. Funct. Med. Foods 2000, 2, 35–53.
12. Joye, I.J.; McClements, D.J. Biopolymer-based nanoparticles and microparticles: Fabrication, characterization, and application. Curr. Opin. Colloid Interface Sci. 2014, 19, 417–427.
13. Gunstone, F.D. Vegetable Oils in Food Technology: Composition, Properties and Uses, 2nd ed.; Wiley-Blackwell: Oxford, UK, 2011.
14. El-Abbassi, A.; Neves, M.A.; Kobayashi, I.; Hafidi, A.; Nakajima, M. Preparation and characterization of highly stable monodisperse argan oil-in-water emulsions using microchannel emulsification. Eur. J. Lipid Sci. Technol. 2013, 115, 224–231.
15. Goula, A.M.; Adamopoulos, K.G. A method for pomegranate seed application in food industries: Seed oil encapsulation. Food Bioprod. Process 2012, 90, 639–652.
16. Komaiko, J.; McClements, D.J. Low-energy formation of edible nanoemulsions by spontaneous emulsification: Factors influencing particle size. J. Food Eng. 2015, 146, 122–128.
17. Yu, L.L.; Zhou, K.K.; Parry, J. Antioxidant properties of cold-pressed black caraway, carrot, cranberry, and hemp seed oils. Food Chem. 2005, 91, 723–729.
18. Kowalska, M.; Ziomek, M.; Zbikowska, A. Stability of cosmetic emulsion containing different amount of hemp oil. Int. J. Cosmet. Sci. 2015, 37, 408–416.
19. Raikos, V.; Konstantinidi, V.; Duthie, G. Processing and storage effects on the oxidative stability of hemp (Cannabis sativa L.) oil-in-water emulsions. Int. J. Food Sci. Technol. 2015, 50, 2316–2322.
20. Burton, G.; Goo, C.S.; Zhang, Y.Q.; Jun, M.B.G. Use of vegetable oil in water emulsion achieved through ultrasonic atomization as cutting fluids in micro-milling. J. Manuf. Process 2014, 16, 405–413.
21. Lawal, S.A.; Choudhury, I.A.; Nukman, Y. Evaluation of vegetable and mineral oil-in-water emulsion cutting fluids in turning aisi 4340 steel with coated carbide tools. J. Clean. Prod. 2014, 66, 610–618.
22. Anwar, F.; Latif, S.; Ashraf, M. Analytical characterization of hemp (Cannabis sativa) seed oil from different agro-ecological zones of Pakistan. J. Am. Oil Chem. Soc. 2006, 83, 323–329.
23. Teh, S.S.; Birch, J. Physicochemical and quality characteristics of cold-pressed hemp, flax and canola seed oils. J. Food Compos. Anal. 2013, 30, 26–31.
24. Ostertag, F.; Weiss, J.; McClements, D.J. Low-energy formation of edible nanoemulsions: Factors influencing droplet size produced by emulsion phase inversion. J. Colloid Interface Sci. 2012, 388, 95–102.
25. Gullapalli, R.P.; Sheth, B.B. Influence of an optimized non-ionic emulsifier blend on properties of oil-in-water emulsions. Eur. J. Pharm. Biopharm. 1999, 48, 233–238.
26. Pereira, T.; Guerreiro, C.; Maruno, M.; Ferrari, M.; Rocha-Filho, P. Exotic vegetable oils for cosmetic o/w nanoemulsions: In vivo evaluation. Molecules 2016, 21, 248.

27. Lane, K.E.; Li, W.; Smith, C.J.; Derbyshire, E.J. The development of vegetarian omega-3 oil in water nanoemulsions suitable for integration into functional food products. J. Funct. Foods 2016, 23, 306–314.
28. Krasodomska, O.; Jungnickel, C. Viability of fruit seed oil o/w emulsions in personal care products. Colloids Surf. A: Physicochem. Eng. Asp. 2015, 481, 468–475.
29. Florence, A.T.; Attwood, D. Physicochemical Principles of Pharmacy, 5th ed.; Pharmaceutical Press: London, UK, 2011.
30. Sevcikova, P.; Kasparkova, V.; Hauerlandova, I.; Humpolicek, P.; Kucekova, Z.; Bunkova, L. Formulation, antibacterial activity, and cytotoxicity of 1-monoacylglycerol microemulsions. Eur. J. Lipid Sci. Technol. 2014, 116, 448–457.
31. Rao, J.; McClements, D.J. Lemon oil solubilization in mixed surfactant solutions: Rationalizing microemulsion and nanoemulsion formation. Food Hydrocoll. 2012, 26, 268–276.
32. Mehmood, T. Optimization of the canola oil based vitamin e nanoemulsions stabilized by food grade mixed surfactants using response surface methodology. Food Chem. 2015, 183, 1–7.
33. Homayoonfal, M.; Khodaiyan, F.; Mousavi, S.M. Optimization of walnut oil nanoemulsions prepared using ultrasonic emulsification: A response surface method. J. Disper. Sci. Technol. 2014, 35, 685–694.
34. Rebolleda, S.; Sanz, M.T.; Benito, J.M.; Beltrán, S.; Escudero, I.; González San-José, M.L. Formulation and characterisation of wheat bran oil-in-water nanoemulsions. Food Chem. 2015, 167, 16–23.
35. Wasim, K.; Haq, I.; Ashraf, M. Antimicrobial studies of the leaf of Cannabis sativa L. Pak. J. Pharm. Sci. 1995, 8, 29–38. [PubMed]
36. Novak, J.; Zitterl-Eglseer, K.; Deans, S.G.; Franz, C.M. Essential oils of different cultivars of Cannabis sativa L. And their antimicrobial activity. Flavour Fragr. J. 2001, 16, 259–262.
37. Seidel, V.; Taylor, P.W. In vitro activity of extracts and constituents of pelagonium against rapidly growing mycobacteria. Int. J. Antimicrob. Agents 2004, 23, 613–619.
38. Matthaus, B.; Bruhl, L. Virgin hemp seed oil: An interesting niche product. Eur. J. Lipid Sci. Technol. 2008, 110, 655–661.
39. Altieri, C.; Bevilacqua, A.; Cardillo, D.; Sinigaglia, M. Effectiveness of fatty acids and their monoglycerides against gram-negative pathogens. Int. J. Food. Sci. Technol. 2009, 44, 359–366.
40. Ghosh, V.; Mukherjee, A.; Chandrasekaran, N. Eugenol-loaded antimicrobial nanoemulsion preserves fruit juice against, microbial spoilage. Colloid Surf. B 2014, 114, 392–397.
41. Toutain-Kidd, C.M.; Kadivar, S.C.; Bramante, C.T.; Bobin, S.A.; Zegans, M.E. Polysorbate 80 inhibition of pseudomonas aeruginosa biofilm formation and its cleavage by the secreted lipase lipa. Antimicrob. Agents Chemother.2009, 53, 136–145.
42. AOCS. Official Methods and Recommended Practices of the American Oil Chemists' Society, 6th ed.; AOCS Press: Champaign, IL, USA, 2011.

CHAPTER 8

Potassium Fertilization and Physiological Soybean Seed Quality

Everton Vinicius Zambiazzi, Adriano Teodoro Bruzi*, Maria Laene Moreira de Carvalho, Igor Oliveri Soares, Alan Mario Zuffo, Pedro Milanez de Rezende, Diego Henrique Miranda

Department of Agriculture, Federal University of Lavras (UFLA), Lavras, Brazil

ABSTRACT

The purpose of this study was to evaluate the influence of increased application rates of potassium on the physiological quality of seeds from different soybean cultivars. Seeds from two locations (São Gotardo, MG and Lavras, MG, Brazil) were used. After harvest, the seeds were sent to the Central Seed Analysis Laboratory of the Federal University of Lavras. We used a randomized design in a 4 × 6 factorial arrangement of four cultivars and six doses of potassium. All tests were performed with two replicates of 50 seeds (300 seeds per treatment). Germination, emergence under controlled conditions, accelerated aging, electrical conductivity, and tetrazolium tests were performed. Data were subjected to analysis of variance. In soybean seed production fields with high potassium content in the soil, higher application rates of potassium do not increase the physiological quality of seeds.

Keywords: Glycine max L., Seed Vigor, Seed Production

1. INTRODUCTION

The use of high vigor seeds in soybean is critical in establishing the crop and bringing about high yield. In this context, it should be noted that nutrient availability can influence the chemical composition of seeds and therefore their metabolism. Thus, an adequate supply of nutrients can provide for good plant development and produce metabolites for seed development.

The chemical composition of seeds may be influenced by genetic factors, the environment, and, moreover, the nutrients applied to the soil. Potassium fertilization can influence germination and seed vigor since it is an essential element for oil synthesis and transport in soybeans [1] [2] . Similar results were obtained by [3] , who found that seed vigor was significantly affected by potassium fertilization rates.

Some authors show the evident importance of maintaining high potassium concentration in plant tissue and grain; and a higher potassium level in grain is positively associated with yield and physiological quality [4] . In a study by [5] , the author states that the rates of pathogens in seeds increase when less than 120 $kg{\cdot}ha^{-1}$ is applied, indicating that potassium deficiency may predispose seeds to infection by numerous pathogens.

In general, the literature reports that seed quality is significantly higher when potassium is applied at rates greater than 80 $kg{\cdot}ha^{-1}$ [5] . However, results are still contradictory [6] . Thus, the aim of this study was to evaluate the influence of potassium application rates on the physiological quality of seeds from different soybean cultivars.

2. MATERIALS AND METHODS

The study was performed at the Central Seed Analysis Laboratory of the Universidade Federal de Lavras (Federal University of Lavras)—UFLA (Lavras, Minas Gerais [MG], Brazil). We used seeds from Lavras and São Gotardo, MG. Chemical and physical analyses of the field experiments, and rainfall and average temperature results of the sites are shown in Table 1 and Figure 1.

A randomized complete block design with three replications was used in a 4 × 6 factorial arrangement, with four soybean cultivars (TMG 127 RR, TMG 1179 RR, BRS MG 850 GRR, and NA 7255 RR) and six potassium application rates (0, 40, 80, 120, 160, and 200 $kg{\cdot}ha^{-1}$). Potassium chloride (KCl^-) with 60% K^+ was used as a source. Each plot consisted of two 5 m rows at a spacing of 0.5 m.

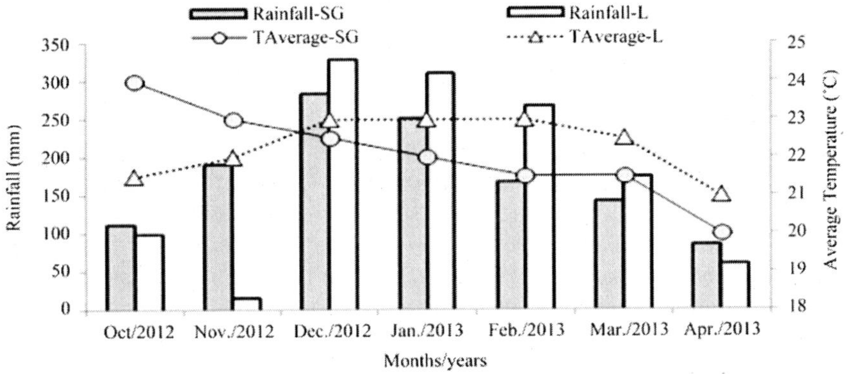

Figure 1. Monthly rainfall and air temperature average in São Gotardo, MG (SG) and Lavras, MG (L) over the 2012/2013 crop season. Source: National Institute of Meteorology (INMET).

Table 1. Chemical and physical composition of a Distroferric typical Oxisol soil (0 - 20 cm) before setting up the experiments in Lavras, MG and São Gotardo, MG in the 2012/2013 crop season.

Site	pH	Ca^{2+}	Mg^{2+}	Al^{3+}	$H^{+} + Al^{3+}$	SB	CEC	P	K	OM	V
	H_2O	cmol$_c$·dm^{-3}						mg·dm^{-3}		dag/kg^{-1}	%
São Gotardo[1]	5.27	3.60	1.10	0.08	7.05	4.90	11.90	28.14	103	4.68	40.94
Lavras[2]	5.90	4.70	1.30	0.00	2.90	6.30	9.20	7.21	118	2.61	68.51
Site	Zn^{2+}	Mn^{2+}	Cu^{2+}	B	Fe^{2+}	S	Clay	Silt	Sand	Textual class	
	mg/dm^3						dag/kg			-	
São Gotardo[1]	17.10	14.50	11.30	0.50	-	-	50	28	22	Clayey	
Lavras[2]	5.31	13.33	0.60	0.33	-	-	64	20	16	Clayey	

$H + Al^{3+}$: potential acidity; SB: sum of bases; CEC: cation exchange capacity at pH 7.0; OM: organic matter; V: base saturation. [1]Experimental area of the Experimental Farm of the Cooperativa Agropecuária do Alto Paranaíba Ltda—COOPADAP. [2]Experimental area of the Centro de Desenvolvimento Científico e Tecnológico em Agropecuária of the Universidade Federal de Lavras.

The experiments were harvested manually and then threshed mechanically. Samples from the plots were homogenized and classified by sieving. For the laboratory tests, we used the seeds contained in the 5.5 and 6.0 mm sieves.

The field experiment laboratory evaluations were performed using a randomized design; all tests were performed with two replicates of 50 seeds per replicate, totaling 300 seeds per treatment. Physiological quality was evaluated by the following tests:

Germination: seeds were distributed on germitest paper towel with a distilled water volume at the ratio of 2.5 times the dry weight of the substrate. They were then placed in a Mangelsdorf model germination chamber at a temperature of 25°C. Evaluations were performed on the 8th day after sowing, according to the criteria set forth in the Regra de Análises de Sementes (Seed Analysis Guidelines) [7] . The mean results were expressed as a percentage of normal seedlings.

Accelerated Aging: in this test we adopted the criteria established by [8] . We used 42 g of seed samples per treatment. The samples were placed on stainless steel screens adapted to plastic gerbox boxes with 40 mL of water at the bottom. The boxes were sealed and taken to a water-jacketed model incubation chamber set at 41°C for 72 h. Then the seeds were subjected to germination tests and the evaluations as described in the methodology above.

Emergence under controlled conditions: the substrate was composed by mixing soil + sand (2:1) and then placed in plastic trays. Upon sowing, the seeds were covered with a layer of substrate (2 - 3 cm). The trays were kept in a plant growth chamber at 25°C with daily watering according to plant needs. As of emergence of the first seedling (cotyledon visible), daily evaluations were performed by computing the number of seedlings until stabilization, with final evaluations at 14 days after sowing. The average percentage at the end of emergence (% E) was considered. To calculate the speed of emergence index (ESI) of the seeds, we adopted the equation suggested by [9] .

Tetrazolium test: the seeds were placed on moistened germitest paper and kept for 16 hours at 25°C. After this period, the seeds were placed fully submerged in 0.075% tetrazolium salt solution in plastic containers, remaining for three hours at 40°C in a water-jacketed model incubation chamber. After this period, the seeds were washed in water and again placed in plastic containers with water. Then viability and vigor were determined by classifying each seed in one of the eight groups described by [10] .

Electrical conductivity: we used the conductivity of mass method and placed 50 previously weighed seeds/ replicates in plastic beakers containing 75 mL of deionized water and then transferred them to an incubator chamber regulated at the temperature of 25°C for 24 hours [11] . After this period, the containers were removed from the incubator chamber and gently stirred, and the electrical conductivity of the solution was measured with the aid of a conductivity meter (MS TECNOPON®-mCA150). The mean values were expressed as $\mu S \cdot cm^{-1} \cdot g^{-1}$.

After collecting and tabulating the data, all the evaluated parameters were analyzed. Mean values were compared according to [12] at 5% probability. Statistical analysis was performed with the aid of the statistical package [13] .

3. RESULTS AND DISCUSSION

From combined analysis, a significant difference was detected in all the parameters evaluated for cultivars. This was expected since the cultivars have different growth habits and genetic backgrounds, leading to such variation. Several reports in the literature show that genotype influences the physiological quality of soybean seeds [6] .

There was a significant effect of the production site on the quality of seed lots. The effect of environment on the expression of physiological quality is often highlighted in the literature [6] [14] and this supports the reports obtained in this study. We also found a significant interaction between cultivars and locations, i.e., the cultivars showed no consistent performance irrespective of environment. These results confirm the reports presented by [15] - [17] for physiological quality in different environments. Significant interactions between cultivars and locations have been frequently reported in the literature for physiological seed quality [6] [16] [18] . In fact, this occurred because the two sites were different in altitude and latitude which are essential factors for seed production.

Additionally, we should also highlight the unpredictable environmental factors, such as precipitation and temperature (Figure 1). Considering the period of the field test, it is evident that in December, January, and February (Lavras) we have a greater accumulation of rainfall and higher average temperature. This fact may be related to the lower vigor and germination percentage observed in this location compared to São Gotardo.

The main objective of this study concerns the influence of potassium levels on seed quality. However, no significant difference was found for the application rates in the tests carried out. This allows us to infer that the physiological quality of soybean seeds, on average, is not dependent on the potassium application rate used.

There are several reports in the literature that confirm this fact. For [2] , increasing potassium application rates did not affect germination and vigor. However, according to [19] , potassium promoted higher germination percentage and vigor.

One possible explanation for the doses used not having an effect on the physiological quality of seeds is the high content of potassium in the soil. There are several reports in the literature showing that the effect of potassium on physiological quality, yield, and chemical composition is due to the natural content of the nutrient in the experimental area. According to [1] , it is necessary to carry out experiments in areas of low natural nutrient content.

The mean values from individual site analyses are shown in Table 2. Clearly, the São Gotardo environment provided high physiological quality. For germination percentage, for example, the São Gotardo environment had 15.5% greater germination than Lavras. In this study, seeds from São Gotardo surpassed those produced in Lavras by 31.11% in the tetrazolium/vigor test. This may be related to the climatic conditions observed in Lavras, especially at harvest time. Excess rainfall provides a favorable microclimate for the occurrence of pathogens and also compromises the physiological quality of seeds.

The average of the cultivars and combined analysis involving all the quality tests shown in Table 2 allow us to infer that BRS MG 850 GRR showed better performance for all tests; however, it did not differ from NA 7255 RR for viability determined by tetrazolium. For germination, the best performance was from BRS MG 850 GRR, at about 81.44%. The other cultivars obtained a germination percentage below that required for the marketing of seeds in Brazil, which is 80% [20] .

One of the factors that contributed to low germination was the high percentage of dead and infected seeds observed in the germination test (Figure 2). Germination percentage was inversely proportional to the percentage of dead and infected seeds observed in seed lots with different cultivars. The seeds are susceptible to contamination from numerous pathogens [21] . The pathogens cause contamination of the seeds, which, because of an increase in the degree of deterioration, resulted in decreased germination and seed vigor [22] .

Estimates of higher magnitude were obtained for emergence percentage compared to germination percentage obtained in the laboratory. BRS MG 850 GRR had the highest emergence rate (90%).

The superiority of the emergence test in relation to the germination test was previously observed by [23] and [24] . According to these same authors, this fact can be explained by the escape mechanism in which the seedling releases the infected seed coat to the ground in order to emerge, while in the germination test in the laboratory under roll paper the seed coat remains associated with the cotyledons and the associated fungi cause seed deterioration.

The emergence speed index (ESI) varied among cultivars, with the highest estimate observed for seedlings of the cultivar BRS MG 850 GRR. The speed index is associated with emergence vigor of soybean seeds because seeds with higher ESI tend to perform better and therefore have higher speed of emergence in the field, better resisting the stresses that may occur during emergence [25] . It should also be noted that seeds with higher ESI develop a closed canopy more quickly, which results in efficient weed control [26] .

Table 2. Mean values of individual and combined analysis per site for germination (G-%), Emergence (E-%), emergence speed index (ESI), accelerated aging (AA-%), electrical conductivity (EC), vigor index (VIG), and seed viability (VIA) tetrazolium, for soybean cultivars evaluated, from Lavras-MG and São Gotardo-MG in the 2012/2013 crop.

Cultivars	G	E	ESI	AA	EC	VIG	VIA
TMG127 RR	48.92 c[1]	73.08 b	45.39 b	53.92 c	114.52 a	76.06 b	85.19 b
NA 7255 RR	56.81 b	76.78 b	49.11 b	54.25 c	109.11 a	80.86 b	88.11 a
BRS MG 850 GRR	81.44 a	90.08 a	59.75 a	81.67 a	72.80 b	86.22 a	91.97 a
TMG 1179 RR	62.42 b	76.25 b	48.06 b	60.92 b	117.18 a	66.25 c	80.06 c
Overall mean ± SE	62.40 ± 2.61	50.58 ± 2.56	79.05 ± 1.48	62.69 ± 3.88	77.35 ± 6.19	86.33 ± 3.35	103.40 ± 2.17
CV (%)	22.00	14.00	11.00	22.00	15.00	10.00	19.00
São Gotardo, MG							
Overall mean	70.15 a	92.51 a	58.28 a	80.74 a	74.70 b	92.90 a	95.99 a
CV (%)	16.00	4.00	8.00	14.00	13.00	4.00	5.00
Lavras, MG							
Overall mean	54.64 b	65.58 b	42.87 b	44.64 b	132.11a	61.79 b	76.68 b
CV (%)	29.00	18.00	20.00	26.00	20.00	25.00	15.00

[1] Values followed by the same letter belong to the same group by the Scott & Knott (1974) test at 5% probability. SE—standard error mean.

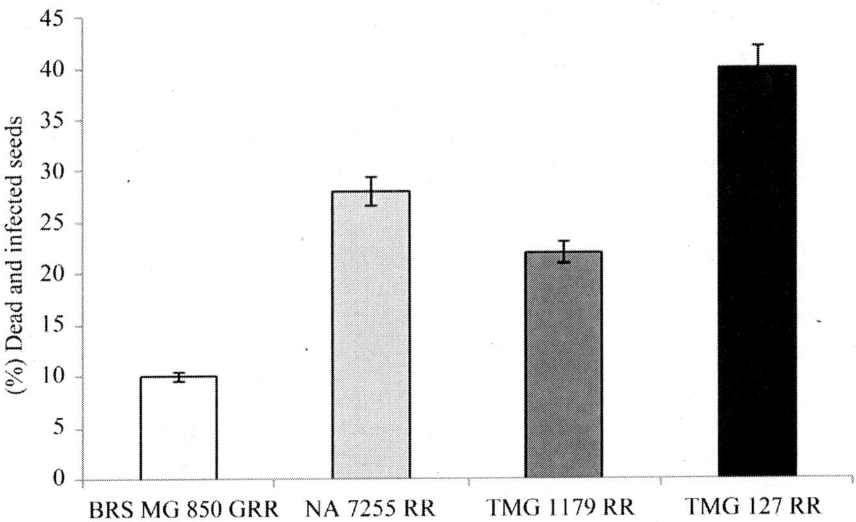

Figure 2. Combined analysis of the total percentage of dead and infected seeds observed in the germination test for each cultivar from Lavras, MG and São Gotardo, MG in the 2012/2013 crop season. Each column represents the mean of three replicates (±standard deviation).

In the accelerated aging test (AA), upon analyzing the mean values of each cultivar, it was observed that the TMG 1179 RR and TMG 127 RR cultivars followed cultivar NA 7255 RR, showing greater sensitivity to conditions of high relative humidity and air temperature.

Accelerated aging is simulation of adverse environmental factors, temperature, and high relative humidity, which are factors seen as causing seed deterioration [8] [27] . Under these conditions, seeds with low physiological quality deteriorate faster than more vigorous seeds, establishing differences in physiological potential [28] . Thus, high vigor seed lots must maintain their viability when subjected to such conditions. Low vigor seeds will bring about reduced viability [29] .

In regard to the results for electrical conductivity, the highest values were observed for the cultivars TMG 1179 RR, TMG 127 RR, and NA 7255 RR, differing from BRS MG 850 GRR. According to [30] soybean seed lots with high vigor may have electrical conductivity values of at most 70 - 80 $\mu S \cdot cm^{-1} \cdot g^{-1}$. In this study, the results obtained show that only BRS MG 850 GRR exhibited values considered ideal for electrical conductivity (72.80 $mS\mu S \cdot cm^{-1} \cdot g^{-1}$), thereby leading to production of seeds with high to medium vigor for the cultivar concerned.

For seed vigor and viability, both determined by the tetrazolium test, best performance was observed for BRS MG 850 GRR with 86.22% and 91.97%, respectively, although the cultivar NA 7255 RR did not differ in regard to viability.

Among the methods of quality control adopted by seed companies in Brazil, the tetrazolium test stands out for soybean due to the speed, accuracy, and large amount of information provided by the test. In addition, in assessing the viability and vigor of seed lots, it provides possible causes responsible for the reduction in soybean quality that can come from mechanical damage, damage caused by moisture, and damage from stink bug attack, which are the problems that most affect the physiological quality of soybean seeds [10] .

It is evident that, on average, the most damage leading to reduction in vigor and viability, assessed by the tetrazolium test, was caused by stink bug attack (48.3%), moisture content (32%), and mechanical factors (19.7%).

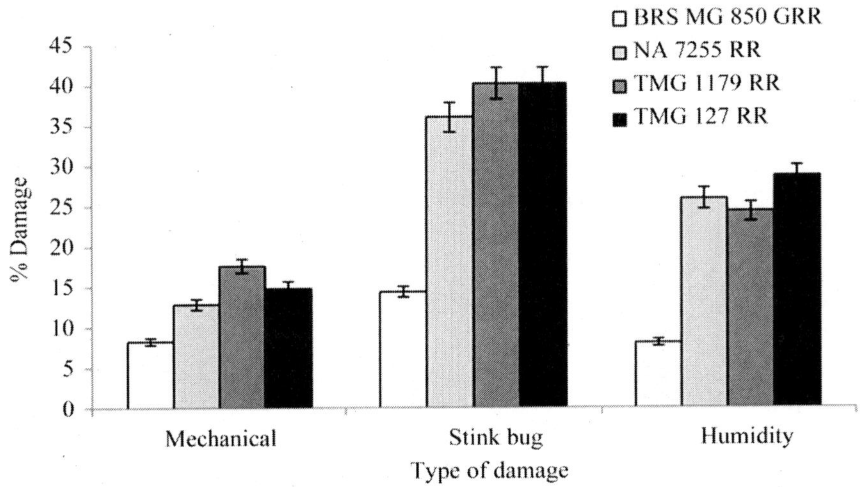

Figure 3. Combined analysis of the damage caused and the respective % of damage observed in the tetrazolium test for each cultivar from Lavras, MG and São Gotardo, MG in the 2012/2013 crop season. Each column represents the mean of three replicates (±standard deviation).

The main damage observed in the tetrazolium test was that caused by the attack of stink bugs. BRS MG 850 GRR showed 14.3% damage caused by this insect; the other cultivars showed damage ranging from 35.9% to 40.2% (Figure 3).

In the incidence of stink bug, the site of the lesion is more important than the number of bites because damage to the hypocotyl prevents germination, while multiple lesions on cotyledons reduce vigor and emergence, but not germination [31] . Stink bugs can also cause delayed maturation from leaf retention, hampering harvest [32] , due to the excessive amount of water present in the plant.

In assessing damage caused by stink bugs in soybean, [33] observed results similar to those found in the present study. They also found correlation between the intensity of the attack of bugs and leaf retention.

The data related to damage caused by moisture show that BRS MG 850 GRR exhibited 8% damage. The other cultivars exhibited values ranging from 24.3% to 28.6%, while the cultivar that showed the greatest magnitude was TMG 127 RR. This may be related to the maturity group of each cultivar. BRS MG 850 GRR had samples taken in a period with less rainfall. Thus, the operation was performed when the seeds had lower moisture content, leading to greater vigor and viability compared with other cultivars.

In respect to mechanical damage, 8.2% injury was noted in BRS MG 850 GRR. The other cultivars showed damage ranging from 12.8% to 17.5%. Such damage may have been aggravated by mechanical threshing, which favors its occurrence. In a study by [17] , evaluating different genotypes in three distinct environments, the authors observed that mechanical damage and damage caused by moisture were responsible for the loss of vigor and viability of seeds in all environments.

4. CONCLUSION

In soybean seed production fields with high potassium content in the soil, the application of differing rates of potassium did not have immediate effects on seed quality.

ACKNOWLEDGEMENTS

Everton Vinicius Zambiazzi, Adriano Teodoro Bruzi, Maria Laene Moreira de Carvalho, Igor Oliveri Soares, Alan Mario Zuffo, Pedro Milanez de Rezende, Diego Henrique Miranda (Conselho Nacional de Desenvolvimento Científico e Tecnológico), the Capes (Coordenação de aperfeiçoamento de pessoal de nível superior), and FAPEMIG (Fundação de Amparo à Pesquisa do Estado de Minas Gerais) for scholarships and the finnacial support, and the Federal University of Lavras for the logistic support.

REFERENCES

1. Pettigrew, W.T. (2008) Potassium Influences on Yield and Quality Production for Maize, Wheat, Soybean and Cotton. Physiologia Plantarum, 133, 670-681.
2. Veiga, A.D., Von Pinho, É.V.D.R., Veiga, A.D., Pereira, P.H.D.A.R., Oliveira, K.C.D. and Von Pinho, R.G. (2010) Influência do Potássio e da Calagem na Composição Química, Qualidade Fisiológica e Na Atividade Enzimática de Sementes de Soja. Ciência e Agrotecnologia, 34, 953-960.
3. Pedroso Neto, J.C. and Rezende, P.M. (2005) Doses e Modos de Aplicação de Potássio na Produtividade de Grãos e Qualidade de Semente de Soj a (Glycine max (L.), Merrill). FAZU em Revista, 27-36.
4. Vyn, T.J.Y.X., Bruulsema T.W., Jackson, C.-J.C., Rajcan, I. and Brouder, S.M. (2002) Potassium Fertilization Effects on Isoflavone Concentrations in Soybean [Glycine max (L.) Merr.]. American Chemical Society, 50, 3501-3506.
5. Sfredo, G.J. (2008) Soja No Brasil: Calagem, Adubação e Nutrição Mineral. Empresa Brasileira de Pesquisa Agrope- cuária, Embrapa Soja.

6. Krueger, K.G.A., Mallarino, A.P. and Mullen, R.E. (2013) Phosphorus and Potassium Fertilization Effects on Soybean Seed Quality and Composition. Crop Science, 53, 602.
7. BRASIL (2009) Ministério da Agricultura, Pecuária e Abastecimento. Regras para Análise de Sementes.
8. Marcos Filho, J. (1999) Teste de Envelhecimento Acelerado. In: Krzyzanowski, F.C., Vieira, R.D. and França Neto, J.B., Eds., Vigor de Sementes: Conceitos e Testes, ABRATES, Londrina, 1-24.
9. Maguire, J.D. (1962) Speed of Germination-Aid in Selection and Evaluation for Seedling Emergence and Vigor. Crop Science, 2, 176-177.
10. França-Neto, J.B., Krzyzanowski, F.C. and Costa, N.P. (1998) O Teste de Tetrazólio em Sementes de Soja.
11. Krzyzanowski, F.C., Vieira, R.D. and Neto, J.B.F. (1999) Vigor de Sementes: Conceitos e Testes. Associação Brasileira de Tecnologia de Sementes, Comitê de Vigor de Sementes.
12. Scott, A.J. and Knott, M.A. (1974) A Cluster Analysis Method for Grouping Means in the Analysis of Variance. Bio- metrics, 30, 507-512.
13. Ferreira, D.F. (2011) Sisvar: A Computer Statistical Analysis System. Ciência e Agrotecnologia, 35, 1039-1042.
14. Schnebly, S.R. and Fehr, W.R. (1993) Effect of Years and Planting Dates on Fatty Acid Composition of Soybean Genotypes. Crop Science, 33, 716-719.
15. Ávila, M.R., Braccini, A.L., Motta, I.S., Scapim, C.A. and Braccini, M.C. (2003) Sowing Seasons and Quality of Soybean Seeds. Scientia Agricola, 60, 245-252.
16. Lima, W.F., Pípolo, A.E., Moreira, J.U., de Carvalho, V.C.G.P., Prete, C.E.C. and Arias, C.A.A. (2008) Interação Ge- nótipo-Ambiente de Soja Convencional e Transgênica Resistente a Glifosato, no Estado do Paraná. Pesquisa Agropecuária Brasileira, 43, 729-736.
17. Gomes, G.D.R., Benin, G., Rosinha, R.C., Galvan, D., Pagliosa, E.S., Pinnow, C., et al. (2012) Produção e Qualidade Fisiológica de Sementes de Soja em Diferentes Ambientes de Cultivo. Semina: Ciências Agrárias, 33, 2593-2604.
18. de Pádua, G.P., Zito, R.K., Rantes, N.E. and Neto, J.D.B.F. (2010) Influência do Tamanho da Semente na Qualidade Fisiológica e na Produtividade da Cultura da Soja. Revista Brasileira de Sementes, 32, 9-16.
19. Mascarenhas, H.A.A., Bullsani, E.A., Miranoa, M.A.C., Braga, N.R. and Carlos, V.N.A.J. (1988) Deficiência de Potá- ssio em Soja no Estado de São Paulo: Melhor Entendimento do Problema e Possíveis Soluções. Informações Agro- nômicas, 42, 1-4.
20. Carraro, I.M. and Peske, S.T. (2005) Uso de Sementes de Soja no Estado do Paraná. Revista Brasileira de Sementes, 27, 75-80.
21. Lopes, K.P., Bruno, R.L.A., Costa, R.F., Bruno, G.B. and Rocha, M.S. (2006) Efeito do Beneficiamento na Qualidade Fisiológica e Sanitária de Sementes do Algodoeiro Herbáceo. Revista Brasileira de Engenharia Ambiental, 10, 426- 435.
22. Binotti, F.F.S., Haga, K.I., Cardoso, E.D., Alves, C.Z., de Sá, M.E. and Arf, O. (2008) Efeito do Período de Envelhe- cimento Acelerado no Teste de Condutividade Elétrica e na Qualidade Fisiológica de Sementes de Feijão. Acta Scientiarum Agronomy, 30, 247-254.
23. Henning, A.A. and Neto, J.B.F. (1980) Problemas na Avaliação da Germinação de Semente de Soja Com Alta In- cidência de Phomopsis sp. Revista Brasileira de Sementes, 2, 9-22.
24. Bizzetto, A. and Homechin, M.R. (1997) Efeito do Período e da Temperatura de Armazenamento na Qualidade Fisiológica e Sanitária de Sementes de Soja com Altos íNdices de Phomopsissojae (Leh.). Revista Brasileira de Se- mentes, 19, 296-303.
25. Dan, L.G.M., Dan, H.A., Barroso, A.L.L. and Braccini, A.D.L. (2010) Qualidade Fisiológica de Sementes de Soja Tratadas com Inseticidas sob Efeito do Armazenamento. Revista Brasileira de Sementes, 32, 131-139.

26. Neto, J.B.F., Krzyzanowski, F.C. and Henning, A.A. (2010) A Importância do Uso de Semente de Soja de Alta Qualidade. Informativo ABRATES, 20, 37-38.
27. Torres, S.B. and Filho, J.M. (2001) Teste de Envelhecimento Acelerado em Sementes de Maxixe (Cucumisanguria L.). Revista Brasileira de Sementes, 23, 108-112.
28. Guedes, R.S., Alves, E.U., de Oliveira, L.S.B., de Andrade, L.A., Gonçalves, E.P. and de Melo, P.A.R.F. (2011) Envel- hecimento Acelerado na Avaliaçãoo da Qualidade Fisiologia de Sementes de Dalbergia nigra (Vell.) Fr. All. Semina: Ciências Agrárias, 32, 443-450.
29. AOSA (1983) Seed Vigor Testing Handbook. East Lansing.
30. Vieira, R.D. and Krzyzanowski, F.C.T. (1999) Teste de Condutividade Elétrica. In: Krzyanowski, F.C., Vieira, R.D. and Neto, J.B.F., Eds., Vigor de Sementes: Conceitos e Testes, ABRATES, Londrina, 1-26.
31. Trumper, E.V. and Edelstein, J.D. (2008) El Complejo de Chinchesfitofagosen Soja: Revision y Avances Enelestudio de su Ecologia y Manejo. [S.l.]: INTA.
32. Sosa-Gomez, D.R. and Moscardi, F. (1995) Retencão Foliar Diferencial em Soja Provocada por Percevejos (Hete- roptera: Pentatomidae). Anais da Sociedade Entomológica do Brasil, 24, 401-404.
33. Belorte, L.C., Ramiro, Z.A., Faria, A.M. and Marino, C.A.B. (2003) Danos causados por percevejos (Hemiptera: Pen- tatomidae) em cinco cultivares de soja (Glycinemax (L.) Merrill, 1917) no município de Araçatuba, SP. Arquivos do Instituto Biológico,70, 169-175.

CHAPTER 9

Evaluating the Role of Seed Treatments in Canola/Oilseed Rape Production: Integrated Pest Management, Pollinator Health, and Biodiversity

Gregory Sekulic [1] and Curtis B. Rempel [1,2,*]

[1]*Canola Council of Canada, 400-167 Lombard Ave, Winnipeg, MB R3B 0T6, Canada*

[2]*Faculty of Agricultural and Food Sciences, University of Manitoba, Ellis Building, Winnipeg, MB R3T 2N2, Canada*

ABSTRACT

The use patterns and role of insecticide seed treatments, with focus on neonicotinoid insecticides, were examined for canola/oilseed rape production in Canada and the EU. Since nearly all planted canola acres in Western Canada and, historically, a majority of planted oilseed acres in the EU, use seed treatments, it is worth examining whether broad use of insecticidal seed treatments (IST) is compatible with principles of integrated pest management (IPM). The neonicotinoid insecticide (NNI) seed treatment (NNI ST) use pattern has risen due to effective control of several early season insect pests, the most destructive being flea beetles (*Phyllotreta sp.*). Negative environmental impact and poor efficacy of foliar applied insecticides on flea beetles led growers to look for better alternatives. Due to their biology, predictive models have been difficult to develop for flea beetles, and, therefore, targeted application of seed treatments, as part of an IPM program, has contributed to grower profitability and overall pollinator success for canola production in Western Canada. Early evidence suggests that the recent restriction on NNI may negatively impact grower profitability and does not appear to be having positive impact on pollinator health. Further investigation on impact of NNI on individual bee vs. hive health need to be conducted. Predictive models for flea beetle emergence/feeding activity in canola/oilseed rape need to be developed, as broad acre deployment of NNI seed treatments may not be sustainable due to concerns about resistance/tolerance in flea beetles and other pest species.

Keywords: canola; oilseed rape; honeybee; pollinator; integrated pest management; IPM; ecosystem; seed treatment; neonicotinoid

1. INTRODUCTION

This workshop was held as part of the 14th International Rapeseed Congress (IRC) held at TCU Place (Saskatoons's Art and Convention Centre) in Saskatoon, Saskatchewan, Canada, 5–9 July 2015. The conference was co-organized by Ag-West Bio Inc. (www.agwest.sk.ca) and the Canola Council of Canada (www.canolacouncil.org). The Congress program comprised five Congress keynote presentations, 30 thematic keynote presentations, 31 concurrent oral sessions (four to five speakers in each) with a total of 143 speakers selected from the approximately 500 abstracts received. Approximately 300 posters were displayed and nine workshops on special topics were organized (see http://gcirc.org/event/single-view/article/14th-international-rapeseed-congress-july-2015.html). The workshops provided opportunity for researchers to present their findings, discuss and debate their work, and exchange viewpoints and information.

This workshop examined the use patterns of insecticide seed treatments, with focus on neonicotinoid insecticides (hereafter referred to as NNI or NNI ST), in canola/oilseed rape production in Western Canada and the EU. Since nearly all of the planted canola acres in Western Canada use seed treatments, it is worth examining whether treatment of 99.5% of acres with an insecticidal seed treatment (IST) is compatible with the principles of integrated pest management (IPM). Canola growers are focused on optimizing crop yield and quality at maximum profitability, both for their farm and the value chain, and this can include health of natural and managed pollinators and other arthropods present in and around the crop. Furthermore, the workshop provided discussion and perspective on two questions that are germane to successful (IPM) practices in the future. These are:

(1) Have seed treatments, as part of an IPM programme, led to more profitable canola crops, healthy pollinator populations and advanced arthropod biodiversity in canola/oilseed rape?

(2) Will seed treatments continue to be an important component of IPM in the future?

The workshop opened with a discussion on general principles of IPM and the importance of IPM for canola/OSR production globally. The consensus among workshop participants—consistent with the weight of evidence from the scientific literature—is that IPM in canola/OSR cropping systems has:

a. Achieved effective management of pests by using all tools available in the safest manner possible.

b. Utilized cultural, mechanical, biological and chemical control measures in a harmonized fashion to strive for balance of profitability, sustainability, and efficacy.

c. Enhanced the economic viability of farms.

Workshop Topics and Presenters

- Gregory Sekulic—Canola Council of Canada (CCC), Grande Prairie, AB Canada: "Evaluating the role of seed treatments in an Integrated Pest Management regime".
- Dr. Michael Schade—Syngenta Crop Protection, Basel, Switzerland: "Impact of neonicotinoid suspension in EU28".
- Caroline Nicholls, Agriculture and Horticulture Development Board (AHDB) Cereals and Oilseeds, Kenilworth, Warwickshire, UK: "Growing winter oilseed rape without neonicotinoid seed treatments—the UK perspective".
- Dr. Udo Heimbach (U. Heimbach, M. Stähler, D. Schenke, A. Dietzsch, N. Kunz, I.P. Wurz, J. Pistorius), Julius Kühn-Istitut, Federal Research Centre for Cultivated Plants, Braunschweig, Germany: "Impacts of neonicotinoid use in oilseed rape and their mitigation."
- Dr. Nicolas Cerrutti (N. Cerrutti, C. Rüger, M. Gayrard, J.-F. Odoux, P. Aupinel, A. Decourtye. M. Henry, V. Bretagnolle), Terres Inovia, ETOILE sur RHONE, France: "Impact of neonicotinoid insecticide

thiamethoxam on the lifespan of the honeybee *Apis mellifera* L. in a large scale study conducted under natural conditions".

- Sreten Terzić (Željko Milovac, Filip Franeta, Ljubiša Stanisavljević, Sreten Terzić, Ana Marjanović Jeromela, Velimir Radić), Institute of Field and Vegetable Crops, University of Belgrade, Novi Sad, Serbia: "Using Mason bees for rapeseed yield improvement".

NOTE: Every researcher submitting an abstract following public request for papers by the IRC organizing committee presented at the workshop. This workshop report contains the research findings and observations that were presented by the speakers, the literature that was highlighted in the presentations and discussions, as well as the contents of discussions following the presentations and round-table forum.

2. DEFINITION AND PRINCIPLES OF INTEGRATED PEST MANAGEMENT (IPM)

Authors Note: Due to time limitation for the workshop, and the presenters and participant's backgrounds in Integrated Pest Management, there was unanimous consensus not to spend time on definitions or principles of IPM. To guide the reader, we will briefly highlight some of the principles of IPM that are germane to this workshop.

IPM is the careful consideration of all available pest control techniques and subsequent integration of appropriate measures that discourage the development of pest populations and keep pesticides and other interventions to levels that are economically justified and reduce or minimize risks to human health and the environment [1]. IPM programs for canola combine cultural controls, such as planting dates, increased seeding rates, as well as seed treatments to ensure adequate populations of vigorous plants [2]. There are two important considerations for growers in managing early season pests: Preserve the targeted plant population and protect the overall health of the surviving plants.

A description and tools for implementation of IPM in canola for western Canada is found at http://www.canolacouncil.org/crop-production/canola-grower's-manual-contents/chapter-10-integrated-pest-management/chapter-10#canolaintegratedpestmanagement. A central tenet of IPM is the establishment of action thresholds to determine when a pest population warrants control; these thresholds do not exist for many of the pests that are controlled by seed treatments and even when they do, the availability of grower-viable testing is extremely challenging. Research efforts are needed to develop and test pest predictor models that will allow growers to reliably monitor their fields and surrounding habitats in order to make informed decisions about the need to use insecticide and/or fungicide treated seed. However, in absence of these models, it is an infringement of IPM principles to limit growers' access to the only tool that can protect the crop. This forces growers to use more costly and less efficacious foliar insecticides that have higher toxicity to non-target species.

3. HISTORY AND EVALUATION OF SEED TREATMENT INSECTICIDE USE IN CANOLA IN CANADA

The neonicotinoid insecticide seed (NNI) treatment use pattern has primarily evolved to prevent losses due to crucifer flea beetles and striped flea beetle (Phyllotreta cruciferae, Phyllotreta striolata) during emergence and establishment in the weeks following planting of the crop [3]. The crop is particularly vulnerable to yield loss at this stage. Flea beetles are the most economically damaging insect pests of canola on the Canadian prairies [3]. Yield losses due to flea beetles were in the range of 8%–10%/annum and resulted in $130–300 million/annum in lost farm income [4,5,6]. Severe economic damage typically occurs in the spring through defoliation of canola seedlings by adult flea beetles emerging from overwintering. Foliar feeding often leads to yield reductions of 10%–50% when flea beetles are present at medium to high infestation levels [7]. For the 2014 crop year 8,407,200 ha were planted to canola with production of 16,410,000 tonnes. The price of canola on 2 September 2014, Vancouver was $464/tonne. Thus, failure to control damage by flea beetles resulted in a potential loss of $760,960,000 to the Canadian rural economy [8].

The neonicotinoid class of insecticides—which includes imidacloprid, clothianidin, and thiamethoxam—was introduced in the mid-1990s. They have been well-received by farmers, regulators and ecotoxicologists due to their novel mode of action, efficacy against many serious insect pests and selectivity for insects over vertebrates. Due to their systemic activity, the neonicotinoids can be applied to seeds or soils at low rates and provide extended protection to crops during their vulnerable early growth stages. This reduces the number of foliar insecticide applications required, which are applied at much greater application rates and generally pose more hazards to non-target organisms. Neonicotinoid insecticides specifically target the insect nicotinyl receptor, which is structurally different from the mammalian receptor. Consequently, their mammalian toxicity is much lower than the products that they have replaced [9]. Prior to the development of NNI, lindane was used as a seed treatment, or in-furrow or foliar applied insecticides were used. These generally had high mammalian toxicity and significant non-target effects on pollinators, ground beetles, spiders, and other beneficial insects.

While occurrence of striped flea beetles and other flea beetle species found in prairie-wide surveys is increasing, the crucifer flea beetle still remains the predominant species in fields and is found in all areas of Western Canada where canola is produced and in high numbers, numbers which typically exceed an economic threshold [10].

The economic or damage threshold for the canola cotyledon or leaves is based on leaf area loss. The economic threshold for canola is 25% loss of leaf area of individual plants, at which control measures should be initiated [11]. While young seedlings often cannot compensate for extensive damage, once the crop is past the four-leaf stage, historic data indicate that, for spring-sown canola in Western Canada, significant losses do not occur as a result of flea beetle feeding [11]. Yield loss due to flea beetle injury is highest during the first one to three weeks after crop emergence and yield loss can be minimized if adult beetles are controlled during those weeks [7]. Most importantly, on warm days, flea beetle feeding and concomitant leaf damage can advance from 25% to 50% within hours [12]. Under high flea beetle pressure (linked to suitable environmental conditions) feeding can occur so rapidly that delaying treatment by 1–2 days can result in loss of the entire field [6]. Nearly every year on the Canadian prairies, farmers have to replant canola fields due to catastrophic loss from flea beetle feeding. The number of fields and geographic distribution is highly variable from year to year. Consensus among growers, extension agronomists, entomologists and scientists is that losses would be significantly greater and significantly higher volume and frequency of foliar insecticides would be applied if NNI ST were not used as a broad acre component of IPM. The widespread occurrence and rapidity that flea beetle feeding can go from little damage to catastrophic loss make the continued use of NNI ST necessary for a healthy, profitable canola crop. NNI ST provides protection against flea beetles during the critical period (0–21 days) of plant emergence and establishment.

NNI ST has been shown to protect or increase yields of canola when exposed to flea beetles. Research data has shown that research plots planted with insecticide seed treatment have 6%–43% more plants (av. 24%) with concomitant 0%–46% yield increase (av. 16%) over non-treated plots [13]. Nonetheless, and despite widespread use, insecticide seed treatments do not provide sufficient protection against significant economic loss when flea beetle populations are high and soil temperatures and ambient air temperature are cool. Under these conditions, the active ingredient is not taken up and translocated by the plant at a sufficient rate to inhibit flea beetle activity and prevent leaf damage. This situation occurs relatively frequently in parts of Western Canada with flea beetle outbreaks documented in both 2012 and 2014 necessitating the use of 2–3 separate foliar insecticide applications as rescue treatments on approximately 15% of canola acres. This can negatively impact profitability and arthropod biodiversity. Frequent and repeated use of foliar insecticides may negatively affect non-target insects such as pollinators and parasitoids [14]. When a seed treatment is used, the insecticide is applied prior to planting and consequently they are a viable option for reducing exposure to insecticides in non-target beneficial insect populations [15].

Applications of foliar insecticides are not necessarily harmful to bees early in the season. They do most damage to carabid and rove beetles that prey on root maggots. The Canola Council of Canada agronomists, provincial field entomologists, and disease specialists are observing significant late season root maggot damage and yield losses in canola fields that have used foliar insecticides for control of flea beetles. Research is currently underway (administered by CCC) to understand the impact on preserving biodiversity in canola crops and IPM practices that reduce risk to beneficial insects including pollinators.

Ideally, growers would like predictive models that could predict flea beetle emergence and activity prior to canola planting. Development of a flea beetle prediction model would help to guide growers to purchase/forego seed treated with IST or apply foliar insecticides only when appropriate. Despite significant historical investment in research, a functional predictive model has not been forthcoming [3,10,16,17,18,19,20]. Difficulty in monitoring overwintering populations, a wide window for emergence, rapid movement/high mobility, and aggressive feeding habits are reasons why predictive models have been difficult to develop.

To prevent significant annual yield losses from flea beetle outbreaks and to mitigate the high economic and environmental costs of multiple foliar insecticide applications, agronomists and entomologists in Canada have therefore agreed that an appropriate IPM program for canola includes the use of an IST on all planted canola acres [13].

4. NON-TARGET IMPACT OF SEED TREATMENT INSECTICIDE USE ON CANOLA

Seed treatment insecticides are systemic, which allows them to move from the treated seed into the young growing roots and leaves of the seedling and confer protection to the young plant for up to 40 days post emergence. Systemic insecticides move via the xylem to the leaves of the plant; transport back to other parts of the plant, including flowers, can only occur via the phloem and studies indicate that very little is transported this way [21,22]. As a result, residue levels in nectar and pollen are expected to be low. This is corroborated by existing data that confirm exposure via pollen and nectar of treated crops is negligible [23,24,25,26]. Nonetheless, numerous groups have questioned the potential for non-target insects, especially pollinators, to be exposed to seed treatment insecticides via the pollen and nectar of plants grown from treated seeds. This has led to numerous questions regarding the validity of using seed treatment insecticides as a preventative treatment and raised the profile of IPM among a wide range of stakeholder groups.

While seed treatment insecticides are used on a broad range of field crops across Canada—including canola, alfalfa, mustard, potato seed, cereals, lentils, peas, chickpeas, and soy—canola is perhaps the best model system to evaluate the potential impacts of seed treatment insecticides on beneficial and non-target species. This is for a number of reasons, some of which will be discussed briefly below.

Canola is extremely attractive to bees [27] and represents a primary forage crop for honeybees in Western Canada where it is grown on 18–21 million acres annually. Field surveys examining neonicotinoid insecticide concentrations in canola pollen in Alberta (pollen is the protein source for bees) indicate that residue was detected in 32% of samples at an average concentration of 1.06 μg/kg (1 part per billion). These data are consistent with those reported from a 2012 production scale study of five NNI treated and five untreated fields where researchers found NNI pollen residue from 0.5 to 1.1 ppb and observed found no evidence of increased mortality or altered behavior of bees during planting or associated with foraging on seed-treated canola [9]. At a macro-scale, number of managed honey bee colonies has continued to increase in Western Canada concurrent with an increase in canola acreage from 6.25 to 20.09 million in the period from 1990 to 2015. (STATSCAN CANSIM Table 001-0017—Figure 1).

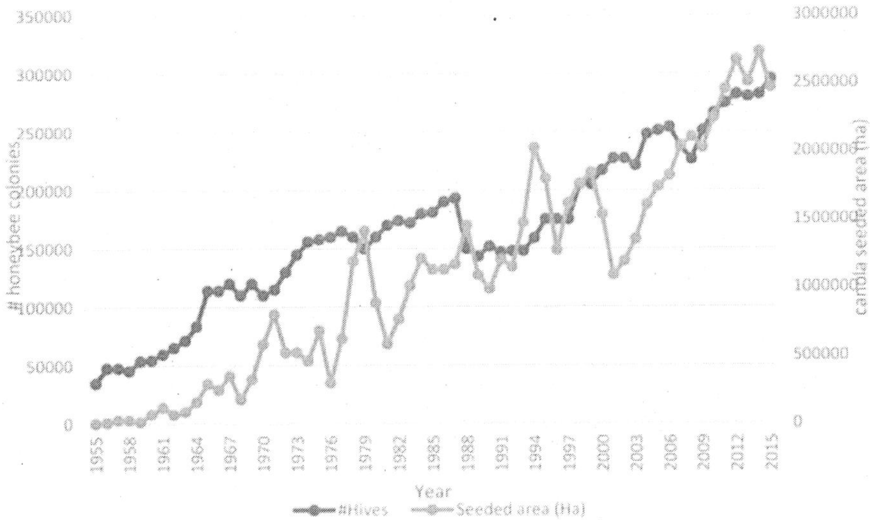

Figure 1. Historical canola planted acres and managed honeybee hives in Canada (1995–2015).

Pollen and nectar exposure studies were conducted in 5 field or semi-field locations planted to winter oilseed rape treated with clothianidin (10 g·a.i/ha) and beta-cyfluthrin (2 g·a.i/ha). Residues in nectar and pollen were measured and ranged from 0 to 2.9 ppb (mean 1.6 ± 0.43 nectar; 1.1 ± 0.37 pollen) [12]. No effects were found from exposure to nectar and pollen on honeybees, bumble bees, or *Osmia* sp., although it is worth noting sample sizes in this study were not always sufficient to permit residue detection. Residues from NNI in winter-sown oilseed rape pollen and nectar did not appear to affect bees but summer oilseed rape needs to be evaluated as well since seeding rates are higher for spring-sown oilseed rape and this may have an effect on exposure levels [28].

A French study indicated that ingestion of sublethal doses of thiamethoxam by honeybees could impact homing ability in semi-field tests [29]. Homing flight is a criterion used to assess sublethal effects of compounds on bees since it is a complex behavior that requires numerous biological processes including memory, motor function, orientation and flight [29]. A follow-up study conducted in 2013 and 2014 looked at free-ranging honeybees exposed to thiamethoxam (trade name Cruiser) as they foraged in an oilseed rape-dominant environment in western France [30]. In that study, researchers tagged 10,000 honeybees and studied the impact of thiamethoxam exposure on lifespan and homing behaviour on individual bees and the downstream impacts on colony viability. The researchers concluded that exposure to thiamethoxam under field-realistic conditions did result in reduced lifespan and increased homing failure in honeybees; however, these effects had no negative impact at the colony level.

Researchers at the Julius Kühn-Institut (JKI) in Germany have been investigating persistence and exposure of bees to NNI via guttation water, pollen, and nectar. Guttation is the excretion of droplets of xylem liquid at the tips or edges of leaves and it occurs more frequently in monocotyledon plants than in dicotyledons [28]. Guttation occurs when plants are sufficiently hydrated and transpiration is limited or inhibited (e.g., it can be triggered by relative humidity greater than 75%). While the risk of exposure of bees to neonicotinoids via guttation water was always assumed to be low, residues are detected in guttation at sufficiently high concentrations to warrant further research. Researchers at JKI conducted three field trials and 1 semi-field trial in winter oilseed rape looking at potential effects from exposures to guttation water and reported no observed impacts on colony development (including bee brood) or acute bee mortality [28]. The researchers observed that single bees collecting guttation drops in early development stages of winter oilseed rape could be killed or damaged if residue was present (LD_{50} oral toxicity = 3.7 ng/bee) but that risk is reduced significantly if sufficient water is available to the colony and bee hives are not placed directly in fields with emerging oilseed rape.

Biodiversity in canola fields is currently very high. Over 200 species of ground beetles, greater than 800 species of spiders, numerous *Staphylinids* (rove beetles), and at least 400 species of carabid beetles are commonly found in Canadian canola agroecosystems [31]. Canola fields also play host to a diverse range of beneficial insects including more than 1400 identified species of parasitoid wasps (comprising 1160 *Ichneumonidae* and 250 *Braconidae*) [31]; numerous species of native pollinators; and native predators such as lacewings (*Chrysoperla* spp.), which prey on *Lygus* nymphs (*Lygus* spp. are a frequent pest of canola). There is tremendous ecological and economic value in promoting and preserving these species in canola canopies; broad scale foliar application of foliar insecticides would dramatically reduce this beneficial insect biodiversity.

In summary, the Western Canadian canola experience indicates that ISTs, as part of an IPM program, have not led to a reduction in populations of managed pollinators and beneficial insects. Honeybee numbers in Western Canada have increased; arthropod biodiversity in IST canola is strong; and canola plays host to numerous beneficial species including predators, parasitoids, and pollinators that all have utility for canola production. The key for sustainable agricultural production will be to expand our monitoring of both pest and beneficial insect species, develop predictive models for their occurrence and density and re-examine biological control of key canola pests within context of better understanding of the role of beneficial insects such as predators and parasitoids.

Furthermore, the use of insecticide seed treatments has helped to increase crop yield and profitability by facilitating earlier planting and producing a vigorous crop that is better able to handle environmental stresses. This has led to improved canopy cover and more flowers per plant [13]. NNI ST have been reported to enhance plant vigor and stress tolerance (both biotic and abiotic) and better tolerate pest pressure, independent of their insecticidal role [32,33,34,35,36], in part through induction of salicylate-associated plant defense responses [37]. The opportunity to forego foliar insecticide applications due to application of NNI ST has also resulted in lower costs. NNI ST have contributed directly to increased canola yield and profitability and have reduced production risk and this is a key reason for expanded canola acreage in Western Canada over the past decade [28].

5. CONCERNS ABOUT SEED TREATMENT USE IN OTHER CROPS AND DISSIMILARITIES FOR CANOLA/OILSEED RAPE PRODUCTION

In Germany (like Eastern Canada), bee mortality has been observed following exposure to dust released from planters during the planting of neonicotinoid-treated corn (maize) [28,38]. Corn has a rough, waxy coating which tends to repel NNI seed treatment more than the oily seed coat of oilseed rape that results in a significant amount of material that can readily move from the seed when handled.

Corn seeds are also planted using precision, pneumatic planters that have air outlets and vent air externally into the surrounding environment. The precision planter ensures that there is optimal spacing between plants. Since corn seed is irregular in shape, they have a tendency to become stuck in the planting machinery so seed companies typically add a lubricant to the treatment or growers use a seedbox lubricant to prevent blockage. Until 2014, the lubricants of choice were either graphite or talc, both of which adsorbed traces of insecticide. When corn is sown during dry, windy conditions, the dust containing trace amounts of insecticide could be vented from the planter and, if bees are foraging in the vicinity, could result in inadvertent exposures. For example, Figure 2 shows the number of dead honeybees in traps in a semi-field trial following manual application of dust fractions (<160 μm) following differing rates of clothianidin (g·a.i./ha) [28].

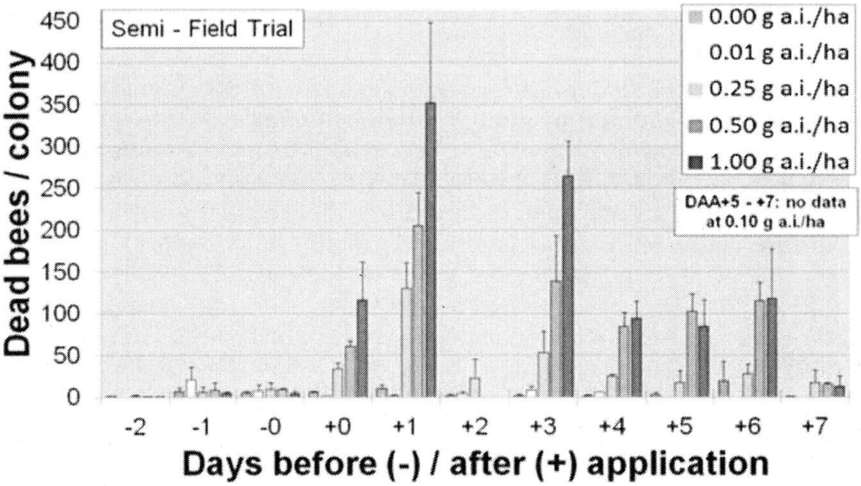

Figure 2. Honeybee mortality in a semi-field trial following manual application of dust fractions (<160 μm) length of time from application of different rates of clothianidin (g·a.i./ha) [28].

When considering the potential risk to non-target organisms from fugitive dust emitted during the planting of NNI-treated seeds, it is necessary to remember that risk is a function of hazard and exposure. As such, it is important to consider the following variables:

(1) Emission of active ingredient. This is a function of seeding rate per ha x amount of dust abrasion x amount of active ingredient present in dust PLUS particle size, planter drill type, soil humidity, wind speed, and direction.

(2) Exposure. This is influenced by emission of active ingredient (see above); distance of the non-target organism from the point of emission (for example, see Figure 2); wind speed; wind direction; stickiness of surfaces that the dust may land on (e.g., plants); plant leaf architecture; and the physical and chemical properties of the active ingredient.

(3) Organism. The sensitivity and biology of the organism (e.g., presence, place, type and intensity of activity) influence whether the non-target species is likely to be exposed and if so, the nature of their physiological response to that exposure [28,38].

The impact of distance from point of emission on potential exposure levels is illustrated in Figure 3, which shows clothianidin residues released during the planting of treated corn seed on oilseed rape plants and soil at various distances from the field edge. These data indicate that clothianidin residues on all media decrease with increasing distance from the site of emission.

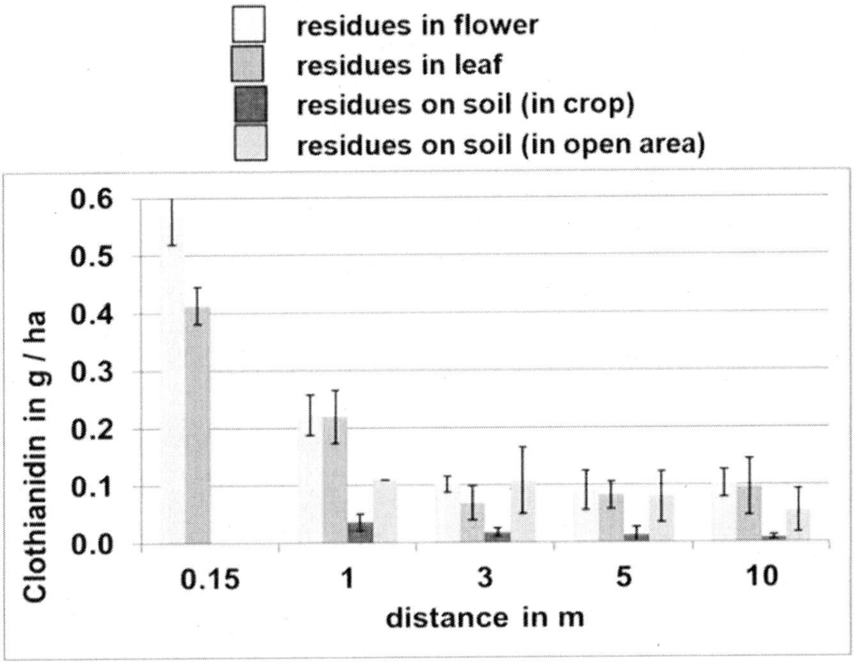

Figure 3. Clothianidin residue (g/ha) in oilseed rape crop or soil at various distances from the crop edge following sowing of corn (2010) [28].

Studies conducted with dust emitted from corn planters indicate that honeybees can be susceptible to fugitive dust containing NNI residues [39]. The levels of dust abraded from treated corn seeds are monitored using a Heubach Dustmeter; lower emissions, and therefore lower risks of exposure, are associated with Heubach values (expressed in g·a.i./ha). Exposure risks to non-target organisms in North America have been successfully reduced by limiting Heubach abrasion values; by using a wax-based fluency agent in place of talc or graphite; and by modifying planter design to minimize the amount of dust that is exhausted from the planter in the first place. It's important to note that the risks associated with the planting of neonicotinoid-treated seeds have been limited to corn and soybeans; there are no reports of any issues with the planting of canola or oilseed rape. Furthermore, researchers studying the potential for dust released during planting of corn and soybean to affect bees foraging on winter-seeded oilseed rape have concluded that there are no unacceptable effects on honeybees if the seed treatment in corn is re-formulated to reduce dust.

The planting equipment used for sowing canola in Western Canada is different from equipment used to sow corn. Differences in planting equipment and the fact that the canola seed coat is "oily" and uniform (does not have the ridges and dents of corn or soy seed) means that there is no "caking" and "dust" dispersion of the IST from the canola planting operation. This results in significant risk reduction for bee/pollinator exposure to the NNI ST.

6. INSECTICIDE SEED TREATMENT USE IN THE EU

In the European Union, NNI seed treatments (clothianidin, imidacloprid, and thiamethoxam (2–10 g·a.i./kg seed)) have been used to control major oilseed rape pests including cabbage flea beetle (*Phyllotreta* spp.), cabbage stem flea beetle (*Psylliodes chrysocephala*), peach-potato aphid (and other aphids as virus vectors—*Myzus persicae*), cabbage root fly (*Delia* brassicae) and turnip sawfly [16,18]. Similar to the advantages described above for Western Canada, since their introduction, NNI seed treatments in the EU have offered growers effective and extended pest control, coupled with low mammalian toxicity and low effects in the soil environment [28,40,41]. To date there are no reports of resistance to neonicotinoid insecticides for any of the target pests in oilseed rape (researchers in Canada have been monitoring for the past decade and tolerance to NNI is evident for striped flea beetle, *P. striolata*, in Western Canada [42]).

Crop establishment for a small seeded crop like canola or OSR is the foundation of crop management and yield security. Like Canadian canola growers, EU oilseed rape producers used insecticide seed treatments to protect the crop during its most vulnerable stage (cotyledon-4 leaf stage).

In recent years the use of insecticide seed treatment in the EU has been the subject of considerable debate. In April 2013, this culminated in the European Commission announcing a two-year moratorium on the use of imidacloprid, clothianidin, and thiamethoxam on all bee-attractive crops effective 1 December 2013 [32,43]. The effects of the ban on OSR cultivation were not expected to be evident until the 2015 growing season since much of the EU OSR crop is winter planted and therefore much of the 2014 harvested crop would have been grown from treated seed.

While opinions are mixed on the chain of events that led to this decision, it's noteworthy that Dr. M. Flüh (Head, Chemicals Unit, EU Health and Consumers Directorate, EC) stated "the recent suspension of certain uses of NNI insecticides was not imposed because they are the main threat to bee health but because they were the only factor that could be quickly regulated by the EC" [38].

6.1. Impact of Neonicotinoid Restrictions in the EU

The restrictions on NNI in the EU28 are beginning to have an economic impact and therefore provide an opportunity to investigate potential impacts on pollinators and farm economics.

6.2. Impact on Growers

Growers of affected crops have reported farm income losses of up to approximately 5% and the results of an independent analysis found that NNI protected €2.1 billion ($3.1 billion CDN) of oilseed rape, corn, sugar beet, wheat, barley and sunflower crops across the EU and improved production efficiency by €0.7 billion ($1.03 billion CDN) [38]. The NNI restrictions introduced at the end of 2013 have resulted in reduced competitiveness of crop production; a reduction in crop diversity; an increase in Europe's dependence on global imports for animal feed; and are putting 50,000 full-time jobs at risk [38].

Specific to oilseed rape production in some EU countries, the restrictions on NNI have resulted in a 6% decline in area planted to oilseed rape and farmers are reporting 90% of the planted crop is damaged by flea beetles, with 30% of this severely damaged [38,44]. In addition to substantially increased damage from flea beetle, farmers are facing higher input costs since they have to resort to foliar insecticide sprays to manage flea beetles. Across the EU, foliar applications have increased from two sprays up to 5.5 sprays/ha [38]. In the UK, foliar insecticide application has increased four-fold, reaching 100% treated ha in some regions [44]. This has resulted in costs that are now one quarter of the profit margin in some areas.

Oilseed rape producers in Germany report similar issues including no registered products for the control of cabbage root fly; limited registered options to control cabbage stem flea beetle; high risk of mortality for winter-seeded oilseed rape; reduced yield potential; and increased production risk [29]. Germany saw a five-fold increase of foliar pyrethroid insecticide applications in autumn 2014 compared to 2013 (prior to ban on NNI) and there are also reports of increased pyrethroid resistance in cabbage stem flea beetle. Germany saw a decrease in planted OSR acres in 2014 due to increased production risk (reduced yield security) and this trend is expected to continue [28]. Oilseed rape production has started to shift to Russia and the Ukraine, which now have a competitive advantage [32,34].

In addition to the yield loss, higher economic costs, and reduced profit margin from the adoption of foliar insecticide sprays, the restrictions on NNI use are also having a negative ecological impact. As farmers increase their reliance on pyrethroid foliar insecticides, agronomists and farm advisors anticipate insect resistance to pyrethroids will increase in the coming years [34]. This will mean that growers will run out of registered crop protection alternatives. As discussed earlier, foliar insecticide applications pose a higher risk to beneficial insects, and as a result it is reasonable to assume that secondary pests may also increase in significance. The use of targeted insecticide application via seed treatment is an important component in IPM because their use history clearly demonstrated a reduced likelihood of unintended negative consequences relative to the use of broadcast foliar insecticides.

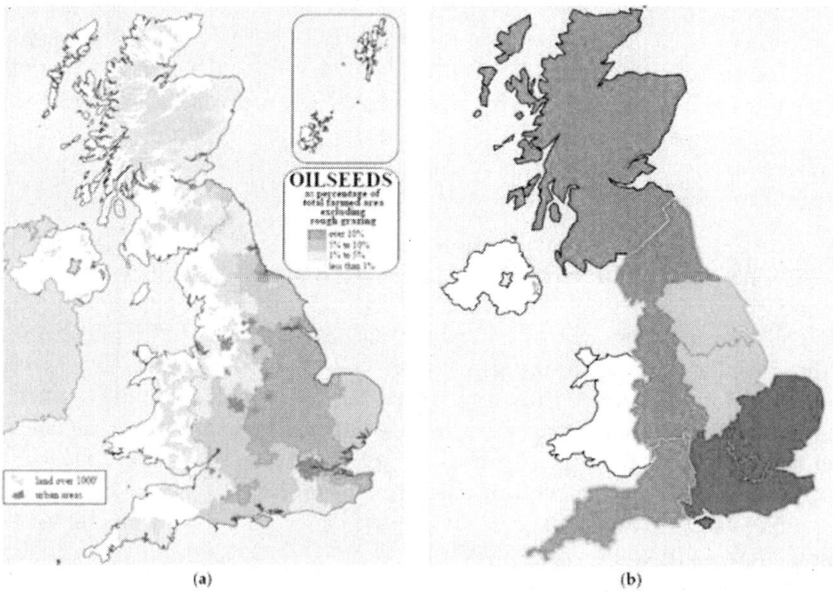

Figure 4. Incidence and severity of adult cabbage stem flea beetle in winter oilseed rape in the UK in 2015 [44] (**a**) Cropping Intensity (**b**) Insect severity. Red = Highest Loss and Risk.

In the UK, the area historically planted to winter oilseed rape treated with NNI was nearly identical to the area impacted by cabbage stem flea beetle, indicating growers were utilizing NNI as a key component in a pest management strategy. Figure 4 depicts the incidence and severity of adult cabbage stem flea beetle across the UK and indicates that they were highest in Eastern, South East, North East, and, Humber regions of the UK. Surveys across the UK confirmed the occurrence of pyrethroid resistant cabbage stem flea beetles in these regions.

In 2014, the greatest reduction in area not planted to winter oilseed rape due to NNI restrictions occurred in the eastern and southeastern regions and the highest losses due to cabbage stem flea beetle were reported in the eastern and southeastern regions. Areas in the UK with highest losses due to cabbage stem flea beetle infestation saw the largest reduction in oilseed rape planted area [34].

6.3. Impact on Pollinators

Impacts on pollinators (both native and managed as a result of the EU restrictions on neonicotinoids are harder to gauge since baseline data are limited and arguably lacking in scientific rigor. One important metric for measuring the health of honeybees is over-winter mortality. Historically, there has always been high spatial and temporal variability across European countries for over-winter bee mortality [45]. In winter 2012/2013 (i.e., the last year that NNI seed treatments were available and used), honeybee colony losses were particularly low; in contrast, in winter 2013/2014, reported losses were high [46]. In Sweden, in the first year of a field monitoring study (2014), they monitored 96 hives in total (48 were control) and found no difference in honey yield or winter survival between hives near crops grown from NNI-treated and control hives. Colony losses of up to 30% are historically common and can be explained by normal biological functions (Dr. Robin Moritz, Martin-Luther University, Halle-Wittenberg, IRC 2015). Researchers are not in agreement over the extent and pattern of colony losses across the EU.

Dr. Eike Genersch presented results of field studies during the Workshop on Bee Health at the Entomological Congress in Frankfurt, March 2015. The studies were designed to represent field-realistic exposure levels of bee colonies to crops grown from seeds treated with neonicotinoid seed treatments. These field-realistic studies are important to contextualize the data derived laboratory and greenhouse experiments that typically use artificially high exposures, which was the case with some of the studies described earlier in this paper. Dr. Genersch's study monitored a cohort of more than 1200 bee colonies from 2005 to 2014, evaluating a number of different variables, including exposures to over 400 pesticides. Hives were selected for inclusion based on proximity to NNI-treated oilseed rape, sunflower, and cornfields and bee losses were compared to control hives that had no NNI exposure. The field study results indicated that there was no correlation between the pesticides studied and colony survival. These data from 10 years of official monitoring did find correlations between bee mortality, *Varroa destructor*, viral infections, and age of the Queen. Dr. Genersch stated that the most important conclusion from the study was that beekeepers who want to successfully bring their colonies through the winter season need to monitor and apply effective treatment for control of *Varroa* mite.

Researchers evaluated the role of Mason bees (*Osmia* spp.) for pollinating oilseed rape in Serbia (reported by Sreten Terzić). Mason bees are solitary species and each female forms its own nest. While the results of the research did not show significant oilseed rape yield increase due to presence of Mason bees, further research should be conducted as these bees are immune to *Varroa* mite [47].

7. CONCLUSIONS

While there are data to indicate that the use of neonicotinoid seed treatments in oilseed rape could have impacts on individual honeybees, the weight of evidence suggests that NNI ST alone are not sufficient to be the only negative impact on the health, productivity, or viability of the colony.

Risks to bees from the use of insecticides in general can be mitigated by minimizing the likelihood of exposure. Insecticide seed treatments are designed to be used in such a way that non-target organisms, such as bees and other beneficial insects, should never be exposed in the first place. They also use substantially less active ingredient per unit area than a foliar application, which reduces the likelihood of resistance development and minimizes the potential for damage to non-target and beneficial insects.

The nature of flea beetle biology is such that, to date, the development of a predictive model that would allow growers to predict precisely where an insecticide will be needed has been impossible to develop. As a result, growers must either use a seed treatment insecticide or plan to make multiple foliar applications of an insecticide to protect the developing plants from flea beetle attack. In Western Canada, growers have overwhelmingly chosen the former option. To date, this choice has not resulted in decline in honeybee colony numbers in the prairie provinces (populations continue to increase). Furthermore, canola growers, production agronomists, entomologists and bee keepers who utilize canola crops are increasingly realizing a strong and diverse ecosystem in and surrounding canola fields containing many species of beneficial insects. It is uncertain whether discontinued use of NNI ST and resulting reliance on older ST chemistry or increase foliar insecticide application would maintain or increase populations of these beneficial insects. In western Canadian canola production, use of NNI ST does not have a negative impact on pollinators and other beneficial insects.

NNI seed treatments are an essential part of our current effective IPM program in Canada. The EU experience indicates that the consequences of NNI restrictions for Canadian growers would be significant, including reduced profitability, reduced yields, increased planting risk, increased use of broadcast foliar insecticide applications, and increased insect resistance to pyrethroids.

Workshop participants agreed that the EU NNI restrictions need to be revisited with better scrutiny of the scientific data. The weight of evidence indicates that seed treatment insecticide use in OSR, when label requirements are adhered to, does not have a negative impact on honeybees. Furthermore, while the data are still limited, the evidence indicates that the European NNI restrictions have not had a positive impact on honeybee health; therefore, further restrictions would continue to negatively impact oilseed rape production with little or no benefit to honeybees.

While we can, and should, strive to develop models that allow us to use insecticides only where (and when) they are needed, in the meantime seed treatment insecticides in canola represent the most environmentally sustainable approach to protecting the crop and the environment, As such they play, and will continue to play for the foreseeable future, an important role in an IPM approach for canola and oilseed rape.

ACKNOWLEDGMENTS

The authors wish to thank all speakers and participants at the workshop and the IRC 2015 Organizing Committee for facilitating this workshop.

AUTHOR CONTRIBUTIONS

All the authors organized and moderated the workshop and assembled this report from the notes of the workshop.

CONFLICTS OF INTEREST

The authors declare no conflict of interest.

REFERENCES

1. Williams, I.H. (Ed.) Biocontrol-Based Integrated Pest Management of Oilseed Rape Pests; Springer Netherlands: Dordrecht, The Netherlands, 2010.
2. Dosdall, L.M.; Stevenson, F.C. Managing flea beetles (Phylotretta.spp.) (Coleoptera: Chrysomelidae) in canola with seeding date, plant density, and seed treatment. Agron. J. 2005, 97, 1570–1578.
3. Lamb, R.J.; Turnock, W.J. Economics of insecticidal control of flea beetles (Coleoptera:Chrysomelidae) attacking rape in Canada. Can. Entomol. 1982, 116, 269–280.
4. Madder, D.J.; Stemeroff, M. The economics of insect control on wheat, corn, canola in Canada, 1980–1985. Bull. Entomol. Soc. Can. 1988, 20, 1–22.
5. Dosdall, L.M.; Mason, P.G. Key Pests and Parasitoids of Oilseed Rape or Canola in North America and the Importance of Parasitiods in Integrated Management. In Biocontrol-Based Integrated Management of Oilseed Rape Pests; Springer Netherlands: Dordrecht, The Netherlands, 2010.
6. Knodel, J.J.; Olson, D.L. Crucifer Flea Beetle: Biology and Integrated Pest Management in Canola; North Dakota State Univ. Coop. Ext. Serv. Publ. E1234; North Dakota State University: Fargo, ND, USA, 2002.
7. Thomas, P. Canola Growers Manual; Canola Council of Canada: Winnipeg, MB, Canada, 2003; Available online: http://www.canolacouncil.org /canola_growers _manual.aspx (accessed on 27 July 2016). [Google Scholar]
8. Statistics Canada CANSIM Table. Available online: http://www5.statcan.gc.ca/cansim/a26?lang=eng&retrLang=eng&id=0010017&tabMode=dataTable&srchLan=-1&p1=-1&p2=9 (accessed on 27 July 2016).
9. Cutler, G.C.; Scott-Dupree, C.D.; Sultan, M.; McFarlane, A.D.; Brewer, L. A large-scale field study examining effects of exposure to clothianidin seed-treated canola on honey bee colony health, development, and overwintering success. Peer J. 2014, 2, e652.
10. Soroka, J. Final Report—Canola Council of Canada CARP Project 2009–6. Potential Flea Beetle Species Composition Shift in Prairie Canola; AAFC Saskatoon Research Centre: Saskatoon, SK, Canada, 2012.
11. Gavloski, J.E.; Lamb, R.J. Compensation for herbivory in cruciferous plants: Specific responses to three defoliating insects. Environ. Entomol. 2000, 29, 1258–1267.
12. Soroka, J. Five things you need to know about flea beetles. Available online: http://www.canolawatch.org/2015/05/29/five-things-you-need-to-know-about-flea-beetles/ (accessed on 27 July 2016).

13. Sekulic, G. Evaluating the Role of Seed Treatments in an Integrated Pest Management Regime. In Proceedings of the IRC 2015, Saskatoon, SK, Canada, 5–9 July 2015.
14. Hassan, S.A.; Haves, B.O.; Degrande, P.E.; Herai, K. The side-effects of pesticides on the egg parasitoid Trichogramma. cacoeciae Marchal (Hym., Trichogrammatidae), acute dose-response and persistence tests. J. Appl. Entomol. 1998, 122, 569–573.
15. Cutler, G.C.; Scott-Dupree, C.D. Exposure to clothianidin seed-treated canola has no long-term impact on honey bees. J. Econ. Entomol. 2007, 100, 765–772.
16. Wylie, H.G. Observations on the distribution, seasonal life history and abundance of flea beetles (Coleoptera:Chrysomelidae) that infest rape crops in Manitoba. Can. Entomol. 1979, 113, 665–671.
17. Burgess, L. Crucifer-feeding flea beetles (Coleoptera:Chrysomelidae) occurring in the province of Saskatchewan, Canada. Coleopt. Bull. 1981, 35, 307–310.
18. Lamb, R.J. Phenology of flea beetle (Coleoptera:Chrysomelidae) flight in relation to their invasion of canola fields in Manitoba. Can. Entomol. 1983, 115, 1493–1502.
19. Turnock, W.J.; Lamb, R.J.; Bilodeau, R.J. Abundance, winter survival, and spring emergence of flea beetles (Coleoptera:Chrysomelidae) in a Manitoba grove. Can. Entomol. 1987, 119, 419–426.
20. Ulmer, B.J.; Dosdall, L.M. Emergence of overwintered and new generation adults of crucifer flea beetle, Phyllotreta. cruciferae (Goeze) (Coleoptera:Chrysomelidae). Crop Prot. 2006, 25, 23–30.
21. Alsayeda, H.; Pascal-Lorber, S.; Nallanthigal, C.; Debrauwer, L.; Laurent, F. Transfer of the insecticide [14 C] imidacloprid from soil to tomato plants. Environ. Chem. Lett. 2008, 6, 229–234.
22. Sur, R.; Stork, A. Uptake, translocation and metabolism of imidacloprid in plants. Bull. Insectology 2003, 56, 35–40.
23. Dively, G.P.; Kamel, A. Insecticide residues in pollen and nectar of a cucurbit crop and their potential exposure to pollinators. J. Agric. Food Chem. 2012, 60, 4449–4456.
24. Chauzat, M.P.; Faucon, J.-P.; Martel, A.-C.; Lachaize, J.; Cougoule, N.; Aubert, M. A survey of pesticide residues in pollen loads collected by honey bees in France. J. Econ. Entomol. 2006, 99, 253–262.
25. Halm, M.P.; Rortais, A.; Arnold, G.; Taséi, J.N.; Rault, S. New risk assessment approach for systemic insecticides: The case of honey bees and imidacloprid (Gaucho). Environ. Sci. Technol. 2006, 40, 2448–2454.
26. Maus, C.; Curé, G.; Schmuck, R. Safety of imidacloprid seed dressings to honey bees: A comprehensive overview and compilation of the current state of knowledge. Bull. Insectology 2003, 56, 51–58.
27. Nedicć, N.; Maccukanovicć-Jocicć, M.; Rancčicć, D.; Rørslett, B.; SŠosštaricć, I.; Stevanovicć, Z.D.; Mladenovicć, M. Melliferous potential of Brassica napus L. subsp. napus (Cruciferae). Arthropod-Plant Interact. 2013, 7, 323–333.
28. Heimbach, U. Impacts of Neonicotinoid Use in Oilseed Rape and Their Mitigation. In Proceedings of the IRC 2015, Saskatoon, SK, Canada, 5–9 July 2015.
29. Henry, M.; Beguin, M.; Requier, F.; Rollin, O.; Odoux, J.-F.; Aupinel, P.; Aptel, J.; Tchamitchian, S.; Decourtye, A. A common pesticide decreases foraging success and survival in honeybees. Science 2012, 336, 348–350.
30. Cerrutti, N. Impact of Neonicotinoid Insecticide Thiamethoxam on the Lifespan of the Honeybee Apis. Mellifera L. in a Large Scale Study Conducted under Natural Conditions. In Proceedings of the IRC 2015, Saskatoon, SK, Canada, 5–9 July 2015.
31. Giberson, D.J.; Cárcamo, H.A. (Eds.) Arthropods of Canadian Grasslands. Volume 4: Biodiversity and Systematics, Part II; Biological Survey of Canada: Ottawa, ON, Canada, 2014; ISBN: 978-0-9689321-7-9.
32. Afifi, M.; Lee, E.; Lukens, L.; Swanton, C.J. Maize (Zea mays) seeds can detect above ground weeds; thiamethoxam can alter the view. Pest Manag. Sci. 2015, 71, 1335–1345.

33. Afifi, M.; Lee, E.; Lukens, L.; Swanton, C.J. Thiamethoxam as a seed treatment alters the physiological response of maize (Zea mays) seedlings to neighbouring weeds. Pest Manag. Sci. 2015, 71, 505–514.
34. Thielert, W.; Metzlaff, M.; De BLock, M. Increase of Stress Tolerance by Application of Neonicotinoids on Plants Engineered to be Stress Tolerant. US Patent 20090270254 A1, 26 May 2006.
35. Thielert, W. A Unique Product: The Story of the Imidacloprid Stress Shield, in Science & Society. Pflanzenschutz-Nachr. Bayer 2006, 59, 73–86.
36. Gonias, E.D.; Oosterhuis, D.M.; Bibi, A.C. Physiologic response of cotton to the insecticide imidacloprid under high-temperature stress. J. Plant Growth Regul. 2008, 27, 77–82.
37. Ford, K.A.; Casida, J.E.; Chandran, D.; Gulevich, A.G.; Okrent, R.A.; Durkin, K.A.; Sarpong, R.; Bunnelle, E.M.; Wildermuth, M.C. Neonicotinoid insecticides induce salicylate-associated plant defense responses. Proc. Natl. Acad. Sci. USA 2010, 107, 17527–17532.
38. Schade, M. Impact of Neonicotinoid Suspension in EU28. In Proceedings of the IRC 2015, Saskatoon, SK, Canada, 5-9 July 2015.
39. Corn Dust Research Consortium (CDRC) Preliminary Report. 2015. Available online: http://www.pollinator.org/PDFs/July2015CDRCFINAL.pdf (accessed on 29 July 2016).
40. Jeschke, P.; Nauen, R. Neonicotinoids—From zero to hero in insecticide chemistry. Pest Manag. Sci. 2008, 64, 1084–1098.
41. Jeschke, P.; Nauen, R.; Schindler, M.; Elbert, A. Overview of the status and global strategy for neonicotinoids. J. Agric. Food Chem. 2010, 59, 2897–2908.
42. Tansey, J.A.; Dosdall, L.M.; Keddie, B.A.; Sarfraz, R.M. Differences in Phyllotreta. cruciferae and Phyllotreta. striolata (Coleoptera: Chrysomelidae) responses to neonicotinoid seed treatments. J. Econ. Entomol. 2008, 101, 159–167.
43. EUR-lex. Official Journal of the European Union. Available online: http://eur-lex.europa.eu/LexUriServ/LexUriServ.do?uri=OJ:L:2013:139:0012:0026:EN:PDF (accessed on 13 April 2016).
44. Nicholls, C. Growing Winter Oilseed Rape without Neonicotinoid Seed Treatments—The UK Perspective. In Proceedings of the IRC 2015, Saskatoon, SK, Canada, 5–9 July 2015.
45. COLOSS. Available online: http://www.coloss.org/announcements/losses-of-honey-bee-colonies-over-the-2014–15-winter-preliminary-results-from-an-international-study (accessed on 13 April 2016).
46. European Commission Health and Food Safety. Available on-line: http://ec.europa.eu/food/animals/live_animals/bees/study_on_mortality/index_en.htm (accessed on 13 April 2016).
47. Milovac, Ž.; Franeta, F.; Stanisavljević, L.; Terzić, S.; Jeromela, A.M.; Radić, V. Using Mason Bees for Rapeseed Yield Improvement. In Proceedings of IRC 2015, Saskatoon, SK, Canada, 5–9 July 2015.

CHAPTER 10

Global Transcriptomic Analysis Reveals the Mechanism of *Phelipanche aegyptiaca* Seed Germination

Zhaoqun Yao, Fang Tian, Xiaolei Cao, Ying Xu, Meixiu Chen, Benchun Xiang and Sifeng Zhao *

Key Laboratory at Universities of Xinjiang Uygur Autonomous Region for Oasis Agricultural Pest Management and Plant Protection Resource Utilization, Shihezi University, Shihezi 832003, China

ABSTRACT

Phelipanche aegyptiaca is one of the most destructive root parasitic plants of Orobanchaceae. This plant has significant impacts on crop yields worldwide. Conditioned and host root stimulants, in particular, strigolactones, are needed for unique seed germination. However, no extensive study on this phenomenon has been conducted because of insufficient genomic information. Deep RNA sequencing, including de novo assembly and functional annotation was performed on *P. aegyptiaca* germinating seeds. The assembled transcriptome was used to analyze transcriptional dynamics during seed germination. Key gene categories involved were identified. A total of 274,964 transcripts were determined, and 53,921 unigenes were annotated according to the NR, GO, COG, KOG, and KEGG databases. Overall, 5324 differentially expressed genes among dormant, conditioned, and GR24-treated seeds were identified. GO and KEGG enrichment analyses demonstrated numerous DEGs related to DNA, RNA, and protein repair and biosynthesis, as well as carbohydrate and energy metabolism. Moreover, ABA and ethylene were found to play important roles in this process. GR24 application resulted in dramatic changes in ABA and ethylene-associated genes. Fluridone, a carotenoid biosynthesis inhibitor, alone could induce *P. aegyptiaca* seed germination. In addition, conditioning was probably not the indispensable stage for *P. aegyptiaca*, because the transcript level variation of MAX2 and KAI2 genes (relate to strigolactone signaling) was not up-regulated by conditioning treatment.

Keywords: *Phelipanche aegyptiaca*; seed germination; ABA; strigolactones; GR24; transcriptome sequencing

1. INTRODUCTION

A total of 4100 plant species in approximately 19 families of flowering plants have evolved to parasitize other plants. These plants are widely distributed in Africa, Europe, Asia, America, and Australia. Haustoria are used by these plants to obtain water and nutrients from other living plants or hosts [1,2]. Parasitic weeds cause serious agricultural problems, resulting in remarkable crop losses in many regions worldwide. About $1.3 to 2.6 billion of annual food crop losses are caused by broomrape infestation [3]. *Phelipanche* spp. are holoparasites that lack chlorophyll. They parasitize more-temperate climate crops, such as sunflower, tomato, potato, tobacco, carrot, clovers, cucumber, rapeseed and legumes [4]. *Orobanche cumana* has damaged 20,000 ha of farmland in Greece and China, with estimated yield losses of 60% in Greece and 20%–50% in China. In 1994, *Phelipanche aegyptiaca* had extensive infestations of muskmelon and watermelon, leading to 20%–70% yield losses in Xinjiang Province, China [5]. The life cycle of *Phelipanche* spp. has a number of mechanisms that coordinate the life cycles of parasites to that of their host. The main steps in the life cycle are conditioning of seeds, germination under stimulants secreted by hosts, adhesion and formation of appressorium, penetration through host tissues, formation of haustorium to connect the host vascular tissues, development of a tubercle and apex, stem growth and emergence, and flowering and seed production [6,7,8]. The seeds of *Orobanche* spp. contain only little reserves. These seeds can survive for a few days only after germination unless they reach a host root to establish a xylem connection.

The *Orobanche* spp. parasitic strategy generally succeeds by coordinating early developmental stages with chemical signals from hosts. An important step in the life cycle of *Orobanche* spp. is their germination at the right place and time, enabling them to establish the connection they require to survive. *Orobanche* spp. usually use so-called germination stimulants secreted by roots of their hosts. To date, three different types of compounds, namely, dihydroquinones (dihydrosorgoleone), sesquiterpene lactones, and strigolactones (SLs), have been identified as chemical signals or germination stimulants for *Striga* spp. and *Orobanche* spp. Among these germination stimulants, SLs are the most active in inducing germination at 10^{-7} to 10^{-15} mol/L [9,10]. SLs are new plant hormones that control shoot branching, root architecture, cambial growth, and senescence [11,12]. SLs are synthesized from carlactone, which is derived from all-trans β-carotene via the action of an isomerase (D27) and two carotenoid cleavage dioxygenases (CCD7 and CCD8). Then, further ring closures and functionalizations involves in members of the CYP711 family (MAX1). Once synthesized, SLs may be transported by PhPDR1, a member of the ABC family within the plant and in the rhizosphere. Finally, MAX2 interacts with D14/KAI2 in an SLs-dependent manner, and this leads to SL ubiquitylation dependent degradation of D53 by the SCF^{MAX2}complex [10,11]. Further, these hormones also serve as extra organismal signals in soil that recruit symbioses with arbuscular mycorrhizal fungi and trigger Orobanchaceae plant seed germination [13].

Orobanche spp. exerts the greatest damage prior to their emergence, and the majority of field loss may occur before diagnosis of infection. Numerous physical, cultural, chemical, and biological approaches have been explored against root parasites. However, none of these techniques are effective and provide economical results [14,15,16,17,18]. SLs are regarded as potential new strategies to control *Orobanche* spp. [19]. The tomato SL-deficient mutant (*SL-ORT1*) displays resistance to *P. aegyptiaca* because of the inability of *SL-ORT1* roots to produce and secrete natural germination stimulants (SLs) to the rhizosphere. Silencing of the tomato *MAX3/CCD7* gene, which is the critical gene for SL production, reduces the number of *P. ramose* infection [20,21]. AM symbiosis in tomato also reduces SL production and *P. ramosa* infection [22]. In 2016, it was successful to reduce *O. ramosa* in tobacco fields by using SL analogues via suicidal germination approach [23]. However, genomic and molecular resources for *Phelipanche* are limited [24], and the mechanism of SLs inducing *Phelipanche* seed germination remains unclear.

Transcriptome and proteome technology can facilitate the understanding of the molecular basis of complex developmental processes [25]. De novo assembly and characterization of the transcriptome of *Cuscuta pentagona* and three parasites of Orobanchaceae have uncovered genes associated with plant parasitism [26,27,28]. Transcriptome sequencing successfully provided new insight into seed germination processes in *Arabidopsis*, wheat, and cotton [29,30,31]. In the present study, dormant, conditioned, and GR24 (a synthetic analog of SL)-treated *P. aegyptiaca* seeds were used to assemble and annotate a reference transcriptome. The transcriptome data were then used to analyze different expressions of genes during different seed germination stages. Finally, the role of plant hormones in *P. aegyptiaca* seed germination was investigated by physiological tests using differentially expressed genes (DEGs) and Kyoto Encyclopedia of Genes and Genomes (KEGG) pathway analysis.

2. RESULTS

2.1. De Novo Assembly of P. aegyptiaca Transcriptome

RNA-sequencing (RNA-seq) library was prepared from dormant, conditioned, and GR24-treated seeds and sequenced using the Hiseq 2500 platform. A total of 78,540,698 reads were obtained. The percentage of Q30 base in all samples was over 88.36%. Illumina adapters were trimmed, and low-quality bases were filtered. Trinity software package [32] was used to assemble clean data. A total of 274,964 transcripts (longer than 200 bp) were found. The average length of N50 of transcripts was 1066 bp, and the evaluated number of unigenes was 94,419 (Table 1 and Figure 1).

Table 1. Assembly statistics for *Phelipanche aegyptiaca* transcriptome.

Length Range	Contig	Transcript	Unigene
Total number	11,469,131	274,964	94,419
Total length (bp)	670,369,345	293,096,156	66,808,988
N50 (bp)	53	1558	836
Average (bp)	58.45	1065.94	707.58

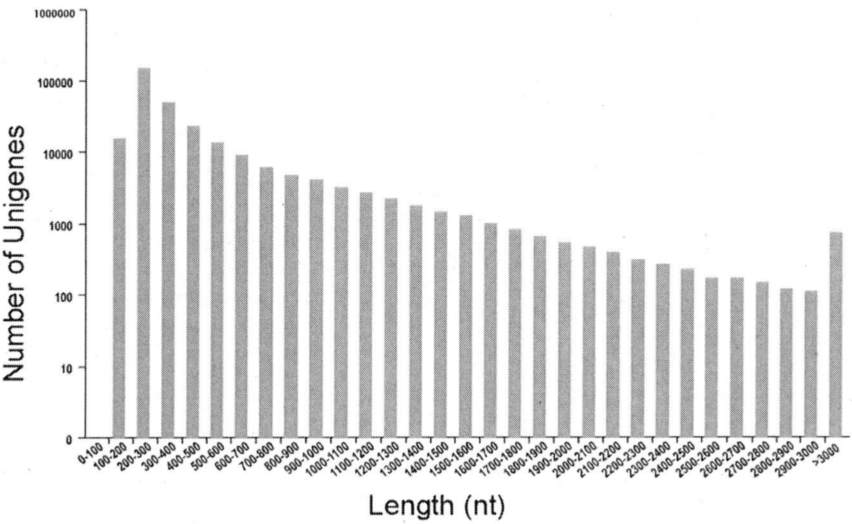

Figure 1. Length distribution of Trinity assembly for unigenes of *Phelipanche aegyptiaca*.

2.2. Transcriptome Functional Annotation

Various approaches were used to acquire comprehensive information on the assembled transcriptomes. First, unigenes were used for homology searches against the non-redundant (NR), Gene Ontology (GO), Cluster of Orthologous Groups of proteins (COG), KOG, and KEGG databases using the BLASTX algorithm, with an E-value $\leq 10^{-5}$. Subsequently, amino acid sequences from predicted unigenes were aligned against the Pfam database with an E-value $\leq 10^{-6}$ using HMMER software. Thus, 53,921 unigenes were annotated. However, numerous unigenes (40,498) remained unannotated. Annotation statistics of BLASTX hits and domain hits are summarized in Table S1.

BLASTX against the NR database provided insight into the taxonomic distribution of the transcripts. High mapping rates of contigs to the protein databases suggested that most of the contigs could be translated into proteins. BLASTX hits and top hits in terms of the total numbers of hits of all transcripts were mostly observed in *Sesamum indicum* (11,865 hits), *Mimulus guttatus* (3410 hits), and *Vitis vinifera* (2261 hits) (Figure 2).

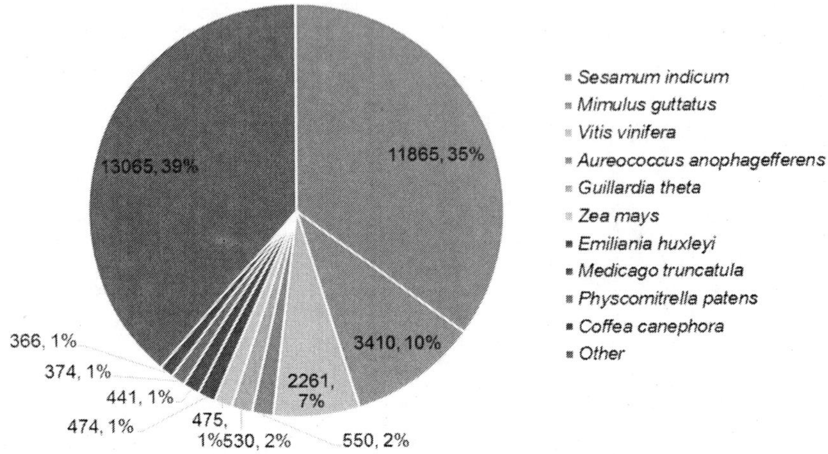

Figure 2. Hit species distribution of BLASTX matches of *P. aegyptiaca* transcriptome. Number and proportion of *P. aegyptiaca* transcripts with similarity to sequences from the NR database.

GO is an international classification system for standardized vocabulary, which is used in the comprehensive description of functions of uncharacterized genes. The three main independent GO categories are biological processes, molecular functions, and cellular components. Our annotated unigenes were assigned to GO terms for functional classification. A total of 21,851 (27.3%), 21,961 (27.5%), and 23,287 (29.2%) unigenes were annotated in the biological process, molecular function, and cellular component categories, respectively (Figure 3). The majority of unigenes under cellular component were involved in cells (23.42%), cell parts (23.68%), and organelles (17.05%). Approximately 43.14% and 41.33% of unigenes within the molecular function category were clustered in catalytic activity and binding, respectively. Sub-categories metabolic (28.19%), cellular (23.76%), and single-organism (17.93%) processes were the major sub-categories of biological process.

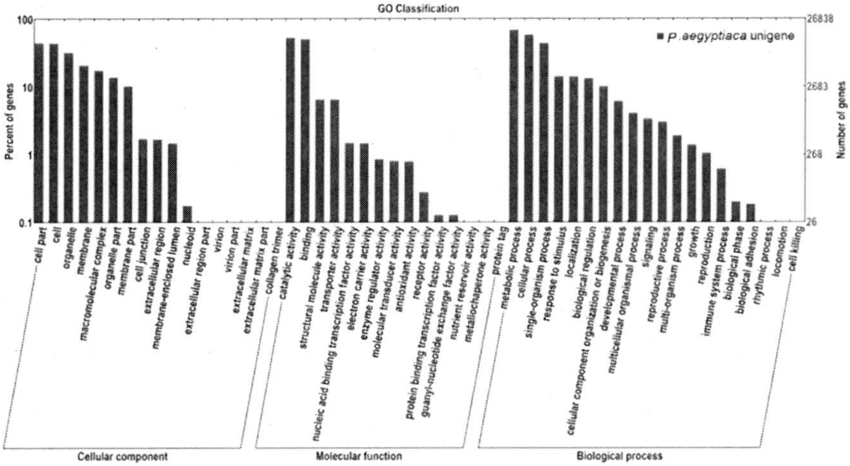

Figure 3. GO category distribution of *P. aegyptiaca* transcripts among level 1 GO categories: Biological process, molecular function, and cellular component.

2.3. qRT-PCR Validation

We also used RNA samples isolated for RNA sequencing to perform qRT-PCR analysis (Figures S1 and S2). The 12 randomly chosen genes and 8 genes relating to plant hormone (GA, ABA, ethylene and BR) biosynthesis were used to qRT-PCR validation. From the supplemental Figure S2, the 12 gene dynamic changes of qPCR and the RNA-seq results were consistent. Furthermore, expression levels of the 8 plant hormone genes are also supported by the changes of each hormone content change (Figure 4 and Figure 5). These results suggested the applicability of RNA-Seq to *Phelipanche* transcriptome analysis is an accurate and reliable way to find DEGs during *P. aegyptiaca* seed germination.

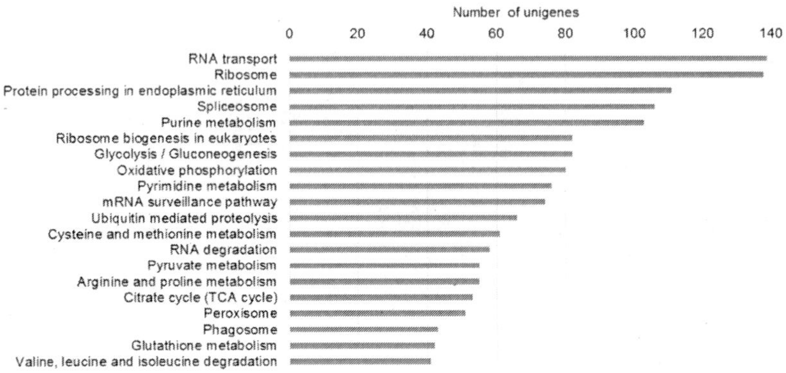

Figure 4. Top 20 KEGG pathways in *P. aegyptiaca* germinating seeds according to the numbers of DEGs. This Figure shows the KEGG metabolic pathways of plants that were well represented by unique sequences of *P. aegyptiaca*.

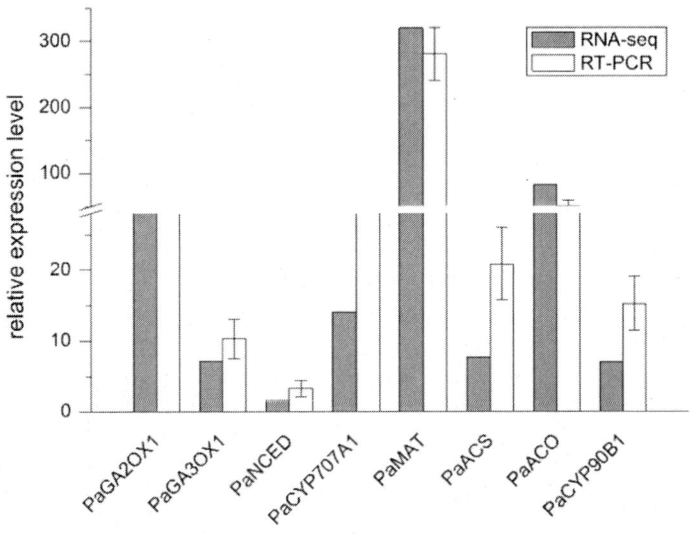

Figure 5. Validation of hormone-related genes at the GR24 treatment stage by qRT-PCR. Fold changes were originally estimated in the RNA-seq experiment (brown bar) and validated by qRT-PCR (white bar). The results are the average of three biological replicates. Abbreviations of genes: PaGA2OX1, gibberellin 2-beta-dioxygenase 1; PaGA3OX1, gibberellin 3-beta-dioxygenase 1; PaNCED, 9-*cis*-epoxycarotenoid dioxygenase; PaCYP707A1, ABA 8′-hydroxylase 1; PaMAT, methionine adenosyltransferase; PaACS, amino cyclopropanecarboxylate synthase; PaACO, aminocyclopropanecarboxylate oxidase; PaCYP90B1, cytochrome P450 90B1.

2.4. Comparative Analysis of Differential Expression during Germination

We used RPKM (reads aligned to gene per kilobase of exon per million mapped reads) to estimate gene expression values to better understand the dynamic performance among the three different transcriptomes during seed germination. False discovery rate (FDR) <0.01 and fold change ≥2 were used as cutoff values to identify differentially regulated transcripts. A total of 5324 DEGs among dormant, conditioned, and GR24-treatedseeds were identified. A total of 633 and 61 unigenes were up-regulated and down-regulated in dormant seeds, respectively, compared with conditioned seeds. A total of 3764 unigenes were up-regulated and 866 unigenes were down-regulated between conditioned and GR24-treated seeds (Figure S3).

Based on the DEGs among the three samples, clustering analysis was performed via MeV edition 4.90 using the algorithm of "hierarchical clustering" (Figure 6). Groups of unigenes with similar expression patterns under different stages were measured in nine clusters. Gradual changes in the gene expression patterns of clusters 1, 2, and 3 varied, and nearly half of DEGs belonged to these three clusters. DEGs of cluster 4 were not active until GR24 application. These genes were probably regulated by GR24, and play crucial roles in the last germination stage. Cluster 5 contained 1840 DEGs that were up-regulated during seed germination. These genes were evidently not only involved in plant hormones but also related to many metabolic pathways. However, the roles of these genes during *P. aegyptiaca* germination have not been elucidated. DEGs in cluster 6 were examined in dormant seeds, so these genes inhibited *P. aegyptiaca* seed germination, contrary to the expression patterns of genes of clusters 7 and 8. Thus, cluster 7 genes probably inhibited germination, in contrast to the promotion of germination by cluster 8 genes. The inhibition or promotion of the genes in clusters 7 and 8 was released or induced by conditioning, respectively. DEGs in the last group, cluster 9, were down-regulated gradually in chronological order. These genes played more important roles in dry seed than during seed germination.

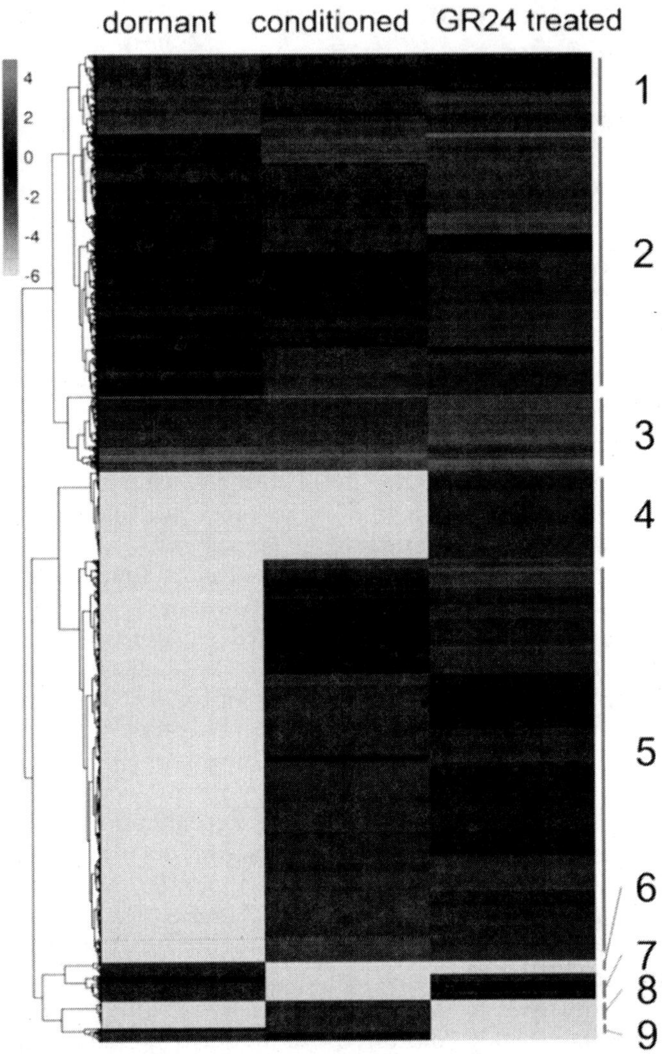

Figure 6. Heat map from hierarchical clustering of differentially expressed genes during *P. aegyptiaca* seed germination. Expression changes and cluster analysis of 5324 genes that were differentially expressed between any two of the three samples. Each row represents a differentially expressed gene (DEG), whereas each column represents a sample. Changes in expression levels are shown in color scales with saturation at >2.0-fold changes. Green and red color gradients indicate a decrease and increase in transcript abundance, respectively.

Furthermore, to determine the biological significance of the DEGs, GO terms enrichment analysis of the total DEGs was also carried out ($p \leq 0.05$). As shown in Figure S4, the biological processes of translation (GO:0006412), response to salt stress (GO:0009651), oxidation-reduction process (GO:0055114) were involved in 221, 44 and 43 DEGs, respectively, indicating mRNA translation, stress resistance and energy metabolism played a vital role in conditioned phase. Between the conditioned and GR24 treatment stage, 151, 119, 47 and 43 DEGs were enriched in the oxidation-reduction process (GO:0055114), translation (GO:0006412), metabolic process (GO:0008152), and protein phosphorylation (GO:0006468), respectively. This suggested energy metabolism and mRNA translation still play dominate roles, and protein and other metabolic process were carried out.

2.5. Gene Change during P. aegyptiaca Seed Germination

Removal of the seed coat permeability barrier, leaching of germination inhibitors, facilitating the access of stimulants to their putative cellular target, biosynthesis of plant hormones, and formation or activation of receptors for germination stimulants have been inferred to occur during conditioning [33,34]. DEGs between dormant and conditioned seeds showed the two biggest group of genes involved from DNA to protein repair and biosynthesis, as well as carbohydrate and energy metabolism (Figure 5). Tested *P. aegyptiaca* seeds and endogenous germination inhibitors have not been proven to inhibit the germination of this plant. We found that the gibberellin 20 oxidase 1 (GA20OX1) transcript was up-regulated and GA3 levels increased during conditioning. In addition, the 3-epi-6-deoxocathasterone 23-monooxygenase (CYP90C1) transcript was up-regulated. This cytochrome P450 enzyme plays an important role in early C-22 oxidation of brassinosteroid (BR) biosynthesis. The expression levels of the two genes were supported by quantitative real-time polymerase chain reaction (qRT-PCR) (Figure 6).

The second phase started after stimulation, during which sharp transition in *Phelipanche* seeds occurred [35]. Dominant DEGs were associated with DNA, RNA, protein, carbohydrate, and energy metabolism. More genes and higher expression levels were found during this phase. Additionally, genes involved in plant hormone biosynthesis and metabolism were dramatically up-regulated. Gibberellin 2-beta-dioxygenase 1 (GA2OX1) gene and gibberellin 3-beta-dioxygenase 1 (GA3OX1) transcript were up-regulated significantly, and GA3 levels increased. Cytochrome P450 90B1 (CYP90B1) gene, which encodes a BR biosynthesis enzyme, was activated dramatically, and the amount of BRs increased. The tendencies of the two phytohormones levels were apparently similar to those during the conditioning stage. Numerous studies have proven that ethylene can promote seed germination. Our data showed that in methionine adenosyltransferase (MAT), aminocyclopropanecarboxylate synthase (ACS), and aminocyclopropanecarboxylate oxidase (ACO), which are three key enzymes of ethylene biosynthesis, corresponding genes were up-regulated significantly after GR24 treatment (Figure 6). The 9-*cis*-epoxycarotenoid dioxygenase transcript, which plays a crucial role in abscisic acid (ABA) biosynthesis, was dramatically down-regulated. In addition, in ABA 8′-hydroxylase 1 (CYP707A1), which catalyzes hydroxylation at the C-8′ position of ABA, the transcript was up-regulated. These results coincided with the qRT-PCR results (Figure 6) and ABA level reduction.

As Kusumoto reported, receptors for germination stimulants are perhaps formed or activated during conditioning [34]. Transcriptome results showed that the F-box protein MAX2 expression level remained high in the whole germination process, which is associate with perception of SLs [36]. Furhter, two α/β-hydrolase KAI2-like genes were identified in our results, the expression of one was high, and the other expression level even down-regulated during the germination process (Figure S5). These results demonstrated that the SL receptor complex was not only active during conditioning or after GR24 application, but it was more likely active all the time.

2.6. Endogenous Hormone Levels and Germination Test

Changes in the endogenous plant hormone levels were tested to elucidate the roles of hormones. GA3, BRs, and ethylene levels continuously increased, similar to those in previous reports on seed germination. Interestingly, GA3 and BR levels gradually increased, contrary to the increase in ethylene level after GR24 application. An initial slight decrease in the ABA level was followed by a significant reduction after GR24 treatment (Figure 7).

A germination experiment was designed to explore the key changes in hormone levels for *P. aegyptiaca* seed germination. The results showed that individual application of the four hormones failed to induce *P. aegyptiaca* seed germination. However, fluridone, the inhibitor of carotenoid biosynthesis, promoted seed germination without GR24 application. GA3, BRs, and ethephon (it can be converted into ethylene at pH < 7) significantly help fluridone increase germination rates. GA3, fluridone, or GA3, as well as ethephon and fluridone, induced seed germination without GR24 application (Table 2 and Figure S6). These results were probably due to the enhanced GA/ABA ratio caused by exogenous fluridone, as well as the GA3 and ethylene-induced changes in the GA/ABA ratio value by adjusting the ABA content. Therefore, the change in ABA level may be the key factor in *P. aegyptiaca* seed germination, which is controlled by GR24.

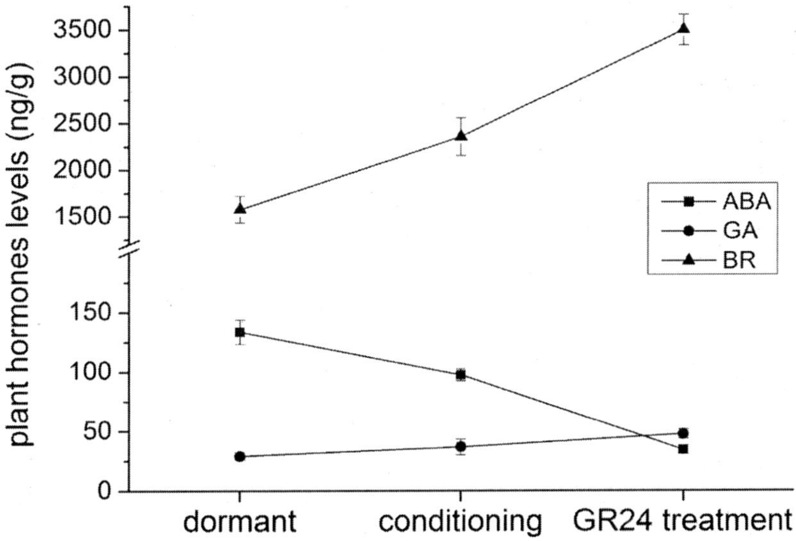

Figure 7. Time course of ABA, GA3, and brassinosteroid (BR) contents in *P. aegyptiaca* seed germination. Seeds were conditioned in water at 25 °C in the dark for 7 days and then conditioned in GR24 (10^{-6} mol/L) at 25 °C for 1 day.

3. DISCUSSION

Transcriptome sequencing was used to identify the role of GR24 in *P. aegyptiaca* seed germination. Three developmental stages from dormant to GR24-treated seeds were selected. Several key genes and pathways were determined. Protein biosynthesis and metabolism, as well as energy metabolism, were dominant in *P. aegyptiaca* seed germination. Additionally, many genes were involved in plant hormone biosynthesis and transduction.

Table 2. *P. aegyptiaca* seed germination rate under different conditions.

Treatment	Germination (%)		
	No Conditioned	Conditioned	Conditioned and GR24 Treatment
BR	0 a	0 a	83.3 ± 3.8 fg
GA3	0 a	2.2 ± 2.2 a	94.4 ± 1.1 h
ABA	0 a	0 a	0 a
Ethephon	0 a	0 a	90.0 ± 1.9 gh
Fluridone	26.5 ± 1.8 b	7.8 ± 2.9 a	86.7 ± 5.7 gh
GA3 + Fluridone	35.6 ± 1.1 c	10.0 ± 3.3 a	91.1 ± 2.2 gh
Ethephon + Fluridone	24.4 ± 2.2 b	7.8 ± 1.1 a	80.0 ± 5.0 f
GA3 + Ethephon + Fluridone	47.8 ± 2.9 e	20.0 ± 1.9 b	91.1 ± 4.0 gh
Water	0 a	0 a	75.6 ± 9.1 f

"No conditioned" means that the dormant seeds were treated for 9 days directly without conditioning. "Conditioned" represents seeds conditioned in water at 25 °C in the dark for 7 days and then treated with plant hormones or inhibitor for 2 days. "Conditioned and GR24 treatment" involved seeds conditioned in water at 25 °C in the dark for 7 days and then treated with GR24 and other plant hormones or inhibitor for 2 days. The concentrations of hormones or inhibitor were as follows: BR, 1 mg/L; GA3, 10 mg/L; ABA, 1 mg/L; fluridone, 100 mg/L; ethephon, 500 mg/L at pH 5; and GR24, 10^{-6} mol/L, all the hormones were dissolved in 0.1% ethanol (*v*/*v*), except GR24 and fluridone were dissolved in 0.1% acetone (*v*/*v*). Means within a column followed by the same letter do not differ significantly according to Fisher's protected LSD test ($p < 0.05$).

3.1. Transcriptome Assembly and Annotation

RNA-seq reads of dormant, conditioned, and GR24-treated seeds were used to assemble a reliable transcriptome. However, only 61% of the filtered transcripts were annotated. This result was similar to many other de novo-assembled plant transcriptomes, such as *Capsicum frutescens*, *Nicotiana benthamiana*, and *Cuscuta pentagona* [27,37,38]. We found that 35% hits belonged to *S. indicum*, the genome of which was sequenced in 2014 [39]. Although the taxonomic position of *Phelipanche* and *Striga* is closer compared with *Sesamum* and *Mimulus*, belonging to Lamiales, the two top hits results are *S. indicum* and *M. guttatus*. This resulted from a lack of *Striga* genome data.

3.2. DEGs Associated with Broomrape Seed Germination

3.2.1. DNA, RNA, and Protein

Numerous stored mRNAs, such as for seed reserve synthesis, late embryogenesis abundant proteins, protein synthesis, and degradation in mature dry seeds, are necessary for germination [40,41,42]. Degradation of stored mRNAs is a prerequisite to germination [43,44]. In our study, Trf4/5, CONT3, and PABP1 transcripts related to RNA degradation were up-regulated in the conditioned phase. By contrast, CNOT3 and DDX6 transcripts related to RNA degradation were down-regulated after GR24 application. Stored proteins were used to restart cellular activity, followed by mobilization of the stored mRNA pool [42].

Hunt and Macovei reported the DNA repair and replication in *Arabidopsis* and *Medicagotruncatula* seed germination [45,46]. We observed that xerodermapigmentosum complementation group C (XPC) and the HR23B genes of the DNA repair XPC complex were up-regulated. Replication protein A, which is a DNA replication protein, and ligase Lig I genes were up-regulated after GR24 treatment. Poly (ADP-ribose) polymerases (PARPs) were implicated in DNA repair in *Arabidopsis* [45]. We found that PARP transcripts were very high in dormant and conditioned seeds, but their expression was dramatically down-regulated after GR24 treatment. Thus, DNA repair may have been completed during the conditioning stage.

De novo RNA synthesis started after DNA repair and replication. α-amanitin, which is a highly potent and specific inhibitor of DNA-dependent RNA polymerase II, strongly affected the speed and uniformity of *Arabidopsis* seed germination [47]. Our data also showed that the DNA-dependent RNA polymeraseII transcripts were up-regulated during the entire germination period. Pentatricopeptide repeat proteins genes were up-regulated, which also participate in RNA splicing, cleavage, editing, stability, and translation [48].

De novo protein synthesis during early seed germination was indispensable. As Rajjou demonstrated, the germination of *Arabidopsis* seeds is entirely blocked in the presence of cycloheximide, a translation inhibitor [47]. However, free amino acids in dry seeds are not sufficient for protein synthesis during germination [49]. Therefore, storage protein degradation is an optimal way to solve this problem. The protein L-isoaspartyl-O-methyltransferase (PIMT) repair pathway facilitates the active elimination of deleterious protein products during germination [50]. Although we did not observe PIMT transcript activity, the two main protein degradation KEGG pathways, namely, proteasome- and ubiquitin-mediated proteolysis, were activated during the conditioning stage. Furthermore, majority of transcripts in both pathways were down-regulated in the presence of GR24.

3.2.2. Energy Metabolism

Energy is supplied from cellular respiration, including glycolysis of carbohydrates and β-oxidation of fatty acid, even amino acids, during seed germination. All biochemical reaction products feed into the tricarboxylic acid (TCA) cycle and provide power for mitochondrial electron transport chain to produce ATP [51].

The two main pathways for carbohydrate metabolism are the oxidative pentose phosphate pathway (OPPP) and glycolysis (Figure S7). Glucose-6-phosphate dehydrogenase (G6PDH) plays a key role in OPPP, and its expression level was stable during germination. By contrast, two key enzymes in the glycolysis pathway, namely, 6-phosphofructokinase (PFK) and pyruvate kinase (PK) transcripts, were up-regulated during the conditioning stage. After treatment with GR24, α-amylase and PFK gene expression levels was down-regulated. This phenomenon suggested that carbohydrates were depleted after conditioned treatment.

Thus, fatty acid or protein acted as an energy supplement in the next germination phase (Figure S7). Many lipoxygenases and fatty acyl-CoA synthetase, acyl-CoA dehydrogenase, enoyl-CoA hydratase, β-hydroxyacyl-CoA dehydrogenase, and acetyl-CoA acetyltransferase transcripts involved in fatty acid β-oxidation were active. Isocitratelyase (ICL), which is a key enzyme in the glyoxylate cycle, and its transcript was active. TAG provides energy to the embryo before the initiation of photosynthesis [52]. Although our results demonstrated that two crucial enzymes, namely, glycerol-3-phosphate dehydrogenase (G3PDHc) and triacylglycerol (TAG) lipase genes, were not activated, glycerol kinase expression increased after GR24 application. Therefore, fatty acid was the energy donor processed through both the glyoxylate cycle and glycerol shunt pathway (Figure S7).

Aspartate and alanine (AspAT) and aminotransferases (AlaAT) genes were activated during imbibition and inferred to participate in respiratory pathways [51]. Our data illustrated that the corresponding genes of both enzymes were not activated during the conditioning phase. However, only the AspAT transcript was activated after GR24 application. The activity and transcript levels of glutamate decarboxylase (GAD), which is the entry enzyme of the aminobutyrate (GABA) shunt, increase in germinating seeds [43,53]. GAD transcript expression level in *P. aegyptiaca* was high at the conditioning stage and upon GR24 application. Oxaloacetate is another product of aspartate. This product can feed into the TCA cycle either by transforming to PEP under PEPCK or by conversion to malate by cytosolic malate dehydrogenase (MDHc) (Figure S7). Our result showed that the PEPCK gene was activated during the whole developmental process, but MDHc transcripts were down-regulated after GR24 treatment. Thus, further studies are necessary to determine the manner by which the two pathways of amino acids supplying energy cooperate with each other during *P. aegyptiaca* seed germination at different stages.

Pyruvate, as the energy metabolism hub, feeds acetyl-CoA to the TCA cycle through pyruvate dehydrogenase complex (PDC) and alcohol dehydrogenase (ADH) or lactate dehydrogenase (LDH) under aerobic the condition, contributing to the transformation of pyruvate into ethanol or lactate, respectively. PDC-related genes were activated after imbibition and GR24 application. The genes of the entire coded TCA cycle enzymes, including malate dehydrogenase in mitochondria, citrate synthase, aconitatehydratase, isocitrate dehydrogenase, oxoglutarate dehydrogenase complex, succinate-CoA ligase, succinate dehydrogenase (SDH), and fumaratehydratase transcripts were up-regulated (Figure S7). The activity of these genes also highly increased during the early germination stage in many plants [51]. However, GR24 treatment resulted in the down-regulation of SDH and fumaratehydratase transcripts. This phenomenon needs further investigation. Additionally, the fermentation-related transcript, pyruvate dehydrogenase transcript, was up-regulated, contrary to LDH.

3.2.3. Phytohormone Activity

ABA inhibits both embryo growth and endosperm rupture to establish seed dormancy and inhibit germination. Nevertheless, GAs counteracts ABA responses and promotes seed germination [54]. Several studies have shown that ABA levels decrease during *P. aegyptiaca* seed conditioning [55]. ABA biosynthesis inhibitors, namely, fluridone and norflurazon, can promote dormancy release and germination [56,57]. Application of exogenous ABA inhibits *P. ramosa* seed germination [57]. In this study, changes in NCED and CYP707A1 transcripts coincided with the change in ABA level (Figure 7) and results from previous studies [57,58,59]. A recent study on *P. ramosa* seeds demonstrated that the seed response to SL is controlled by ABA-independent DNA methylation [60]. Thus, exogenous ABA biosynthesis inhibitor fluridone was examined and found to induce *P. aegyptiaca* seed germination without GR24. Although a large number of previous studies showed SLs act in a positive role with respect to seed germination, as ABA and SLs share a biosynthesis pathway, so this phenomenon resulting from the effect of ABA or SLs needs further discussion. In many studies on *Striga* and *Orobanche*, fluridone alone was not able to induce their seed germination [34,55,61,62,63], but our results showed that application of fluridone without GR24 could induce seed germination. We speculate the different results occurred because the *Phelipanche* seed used in studies was different, and this exciting phenomenon needs more investigation.

Several studies provided evidence that GA synthesis occurs during conditioning [64]. In our experiment, the GA20OX1 transcript was up-regulated after conditioned treatment and GR24 induced GA2OX1 and GA3OX1 transcript up-regulation. Some studies showed that application of GA biosynthesis inhibitors (e.g., uniconazole) during seed conditioning inhibits the subsequent germination of *Striga hermonthica*, *P. ramosa*, *P. aegyptiaca*, and *O. minor* in response to GR24 [57,62,65]. BR application after conditioning increased the rate of seed germination induced by stimulants [56]. CYP90C1and CYP90B1 transcripts were up-regulated during different germination stages. Although BR level increased during the whole process, BRs were inferred to alter the balance of other endogenous phytohormones, such as indole-3-acetic acid (IAA), ABA, GAs, and ethylene rather than acting directly [66]. However, we found that exogenous GA3 or BR promoted *P. aegyptiaca* seed response to GR24, but individual application of these two hormones failed to induce seed germination.

Ethylene, which is another plant hormone, also plays an important role in seed germination by promoting the weakening of the micropylar endosperm [67,68]. This hormone can neutralize many of the negative functions of ABA during germination [69,70,71]. Ethylene, ethephon, or aminocyclopropanecarboxylate (ACC) stimulates seed germination in numerous species, including several parasitic plants, such as *P. ramosa* [72] and some *Striga* species [73,74]. The results showed that methionine adenosyltransferase (MAT), ACS, and ACO expression levels were induced by GR24. This result was supported by Zehhar's study, in which ethylene synthesis was required for the induction of *P. ramosa* seed germination by GR24 [57]. However, we failed to quantify ethylene accurately; low endogenous ethylene was probably released into the medium. However, ethylene down-regulated ABA accumulation by both inhibiting its synthesis and promoting its inactivation, as well as negatively regulated ABA signaling [75]. Thus, indirect degradation by controlling ethylene requires further examination.

The ABA biosynthesis pathway was inhibited, and its degradation was promoted by GR24. This is consistent with the research on CYP707A in the seed germination of *P. ramosa* [58,60]. By contrast, the ethylene biosynthesis pathway was activated. Therefore, the bottleneck of *P. aegyptiaca* seed germination was endogenous ethylene and the ABA level, which were probably controlled by SLs.

SLs carry out their physiological functions via a specific reception system similar to other plant hormones. SLs bind to the D14/MAX2/SCF complex, and then lead to SMAX1, a D53 homolog, degradation, finally acting on seed germination, and this model is widely accepted [11]. However, recently, it was reported that the α/β-hydrolase KAI2 gene (D14-like, also known as HTL) plays a central role in the perception of SLs [36,76]; this gene was thought to be the key gene in reception of the strigolactone-analogous compound karrikin [11], originally found in forest-fire smoke. The transcriptome and qRT-PCR analysis of MAX2, D14, and two KAI2-like genes illustrated their expression levels are stable during seed germination. We only found two KAI2-like genes in our data; however, there are 11 KAI2 genes in *S. hermonthica* [76]. According to the results, we conservatively speculate the transcripts of partial components of the SL perception complex existed prior to conditioning.

4. MATERIALS AND METHODS

4.1. Sample Preparation

Mature seeds of *P. aegyptiaca* were collected from *Cucumis melo* L. in the Hami area of the Xinjiang Uygur Autonomous Region, China, in 2012. Dormant, conditioned, and GR24-treated seeds were used in this study. All seeds were surface-sterilized by 75% ethanol for 1 min and 1% sodium hypochlorite for 20 min followed by three rinses in sterile distilled water [77]. Subsequently, 10 g of seeds were frozen in liquid nitrogen as the dormant seed sample and then stored at −80 °C until RNA was extracted. The conditioned seed sample was treated as follows: 10 g of sterilized seeds were spread on sterilized glass fiber filter paper moistened with sterile distilled water. The seeds were placed into 10 cm Petri dishes and incubated in the dark at 25 °C/20 °C for 7 days [60]. The seeds were then frozen in liquid nitrogen and stored at −80 °C until RNA was extracted. The GR24-treated seed sample consisted of 10 g of conditioned seeds induced by applying GR24 at 10−6 mol/L (0.1% acetone, *v*/*v*) for 1 day, frozen in liquid nitrogen, and stored at −80 °C until RNA was extracted [78].

4.2. RNA-seq Library Preparation and Sequencing

Total RNA of each sample was isolated using Trizol Reagent (Invitrogen, Gaithersburg, MD, USA) according to the manufacturer's instructions. RNA quality was characterized by agarose gel electrophoresis and spectrophotometry. High-quality RNA, with 28S:18S of more than 1.5 and absorbance 260/280 ratio between 1.8 and 2.2, was used for library construction and sequencing. Illumina HiSeq2500 library was constructed according to the manufacturer's instructions (Illumina, San Diego, CA, USA). Magnetic beads with oligo (dT) (Thermo Scientific, Waltham, MA, USA) were used to enrich the mRNA from total RNA. mRNA was randomly cleaved by adding fragmentation buffer. The fragments were used to synthesize first-strand cDNA with random hexamers. After first-strand cDNA was synthesized, the buffer, dNTPs, RNase H, and DNA polymerase I were added to synthesize double-stranded cDNA (dscDNA), which was purified using AMPure XP beads. The purified dscDNA was end repaired, added to an A-tail, and linked with sequencing adapters [79]. AMPureXP beads (Beckman Coulter, CA, USA) were used to choose suitable fragments. The sequencing library was enriched with PCR amplification. After verification with Qubit 2.0 and quantification with Agilent 2100, the effective concentration was accurately quantified using qRT-PCR. The library was then sequenced with an Illumina HiSeq2500 platform (Biomarker Technology Co., Ltd., Beijing, China).

4.3. Preprocessing of Illumina Reads

The cDNA library of each sample was sequenced by Illumina HiSeq2500 based on sequencing by synthesis technology. This library produced amounts of high-quality reads (most bases reached or had excess Q30). Preprocessing of reads involved Q20-quality trimming (removal of low-quality reads with an average Phred quality score of 20, trimming of low-quality bases from the 3′ ends of the reads) and removal of adapter/primer contamination and duplicated reads using custom Perl scripts. Preprocessed reads were sorted into individual samples based on barcodes using fastx_barcode_splitter. Barcodes were trimmed using fastx_trimmer from Fastx_toolkit.

Prior to assembly, the raw reads were cleaned by removing adapter sequences using standard Illumina pipeline, namely, the CASSAVA program (http://support.illumina.com/sequencing/sequencing_software/casava.ilmn). Adaptor contamination was removed, and reads were screened from the 3′ to 5′ ends to trim the bases with a quality score of $Q < 20$ using 5 bp windows. Reads with final length < 25 bp were removed.

4.4. De Novo Transcriptome Assembly

The Trinity software package (http://trinityrnaseq.github.io/) was used for de novo assembly of *P. aegyptiaca* transcriptome from preprocessed RNA-seq reads [32]. A K-mer library was built using the sequencing reads, and K-mer, which probably contained errors, was deleted. The highest frequency of K-mer was used as seed to extend greedily to both ends. This process was continuously cycled until the K-mer ran out. The contigs were clustered to obtain components. A simplified De Bruijn diagram of each contig of each component was constructed. Finally, the graph was unlocked with true reads to obtain transcript sequences. Trinity combines reads of a certain length of overlap to form longer fragments and then processes them for sequencing. The resultant sequences were defined as unigenes for downstream analysis.

4.5. Functional Annotation and Analysis

The assembled contig sets were compared against various databases (NCBINR, Swissprot, GO, COG, KOG, and KEGG databases) using BLASTX (http://blast.ncbi.nlm.nih.gov/Blast.cgi) with an E-value threshold of 1×10^{-5} [80]. Predicted amino acid sequences of unigenes were searched against the Pfam [81] database using HMMER software (http://hmmer.org/) with an E-value $\leq 1 \times 10^{-10}$ to gain the unigene annotation information.

GO is a controlled vocabulary of terms split into three related ontologies consisting of molecular function, biological processes, and cellular components. It has a hierarchical structure that forms a directed acyclic graph in which each term has defined relationships to one or more other terms in the same domain. This characteristic can be described as a parent–child relationship. Each GO term is represented by a node in this graph, and the nodes are annotated with a set of genes. GO annotation was performed from gene expression data using TopGO [82].

The KEGG pathways were assigned to the assembled unigenes using the KAAS server automatic annotator to summarize the pathway information involved in *P. aegyptiaca* seed germination [83]. The transcripts were aligned to sequences in the COG database (http://www.ncbi.nlm.nih.gov/COG) and categorized under different functions accordingly.

4.6. Differential Expression Analysis

EBseq is a widely accepted and accurate analysis of RNA-seq data [84]. This process was performed to identify DEGs. RPKM values were used to measure gene expression levels. FDR was used to determine the threshold of the p-value in multiple tests [85]. FDR < 0.01 and fold change ≥ 2 were considered the cutoff threshold to determine the significance of expression. Cluster of significantly regulated genes was assessed with Cluster [86], MeV [87] and TreeView [88].

4.7. qRT-PCR Validation

Total RNA for RNA-seq samples was also used for qRT-PCR analysis. First-strand cDNA was synthesized using M-MLV reverse transcriptase (Roche, Switzerland) and oligo (dT) 18. We chose 8 plant hormone-related genes, 3 SL perception-related genes and randomly selected 12 genes for further study and validation of RNA-seq using qRT-PCR. Each transcript primer (Table S1) was designed using Primer Premier 5.0

(Premier Biosoft International, Palo Alto, CA, USA). OaTublin1 was used as an internal control [89]. qRT-PCR was conducted using SYBR GreenER™ qPCRSuperMix Universal (Invitrogen, Gaithersburg, MD, USA) according to the manufacturer's instructions. Thermal cycle conditions for PCR were as follows: 94 °C for 3 min, 40 cycles including 94 °C for 15 s and 60 °C for 30 s. Relative expression levels were calculated using the $2^{-\Delta\Delta Ct}$ method [90] and one-way ANOVA (IBM SPSS Statistics 19.0, Armonk, NY, USA).

4.8. Endogenous Hormone Level Analysis

Approximately 1 g (weighed after 5 min of air-drying) of seeds for each sample (for RNA-seq) was collected. Fresh material was frozen in liquid nitrogen and ground to a fine powder. ABA, IAA, and GAs were extracted according to the method of Durgbanshi with slight modification [91]. BRs were extracted using the method of Kim [92]. The four plant hormones were quantified using appropriate ELISA kits (Chengling, Beijing, China) according to the manufacturer's instructions [93]. The released ethylene level was quantified according to the method of Cristescu [94]. All quantifications were performed with triplicate samples of each treatment, and data were analyzed by one-way ANOVA (IBM SPSS Statistics 19.0, Armonk, NY, USA).

4.9. Seed Hormone Treatments

P. aegyptiaca seeds were prepared as previously described. IAA (0.1% ethanol, *v*/*v*), GA3 (0.1% ethanol, *v*/*v*), epi-brassinolide (0.1% ethanol, *v*/*v*), ethephon, ABA (0.1% ethanol, *v*/*v*), fluridone (0.1% acetone, *v*/*v*) (inhibitors of carotenoid biosynthesis), or distilled water (as a control) was added to each Petri dish.

Each hormone, inhibitor, or water was added before conditioning in the first set of treatments. Hormones, inhibitors, or water were applied after 7 days of conditioning as the second group of treatments. Germination percentage was examined under a binocular microscope (Zeiss, Oberkochen, Germany) 3 days after GR24 treatment. Each treatment was performed with three replicate Petri dishes. Data were analyzed by one-way ANOVA (IBM SPSS Statistics 19.0, Armonk, NY, USA).

4.10. Accession Numbers

The quality filtered and trimmed short read data set, which was used in this study, was deposited to the NCBI Short Read Archive (SRA) under accession numbers: SRR2477710, SRR2477711, and SRR2477712.

5. CONCLUSIONS

This study generated a well-annotated transcriptome for the root parasitic weed, *P. aegyptiaca*. The taxonomy of the transcripts of this plant was highly similar to that of *S. indicum*. Thus, as expected, the parasite belongs to the same family of *S. indicum*. We clarified DNA repair and replication, RNA biosynthesis, and energy metabolism, as well as the transcription, translation, and biosynthesis of proteins, in this plant. We also found several key genes involved in seed germination, and these genes were associated with changes in the plant hormones. These genes may be developed as new targets to control *P. aegyptiaca*. These genes were validated by q-PCR and hormones levels were verified in seed germination experiments. Fluridone alone induced *P. aegyptiaca* seed germination, whereas ABA and ethylene were inferred as the key phytohormones during *P. aegyptiaca* seed germination. Our results may provide a perspective for suicidal germination to control parasitic weeds. In addition, we found that the expression levels of MAX2 and a KAI2 gene, which were involved in SL signal transduction, were always stable during the whole germination process. Thus, we deduced that transcripts of partial components of the SL perception complex existed prior to conditioning.

ACKNOWLEDGMENTS

This work was supported by the National Natural Science Foundation of China (No. 31460467).

AUTHOR CONTRIBUTIONS

Zhaoqun Yao and Sifeng Zhao conceived and designed the experiments; Fang Tian, Xiaolei Cao and Meixiu Chen performed the experiments;

Zhaoqun Yao and Ying Xu analyzed the data; Benchun Xiang and Sifeng Zhao contributed reagents/materials/analysis tools; Zhaoqun Yao wrote the paper.

CONFLICTS OF INTEREST

The authors declare no conflict of interest.

REFERENCES

1. Parker, C. Parasitic weeds: A world challenge. Weed Sci. 2012, 60, 269–276.
2. Yoder, J.I.; Scholes, J.D. Host plant resistance to parasitic weeds; recent progress and bottlenecks. Curr. Opin. Plant Biol. 2010, 13, 478–484.
3. Ghannam, I.; Al-Masri, M.; Barakat, R. The effect of herbicides on the egyptian broomrape (Orobanche aegyptiaca) in tomato fields. Am. J. Plant Sci. 2012, 3, 346.
4. Westwood, J.H.; Yoder, J.I.; Timko, M.P. The evolution of parasitism in plants. Trends Plant Sci. 2010, 15, 227–235.
5. Parker, C. Observations on the current status of Orobanche and Striga problems worldwide. Pest. Manag. Sci. 2009, 65, 453–459.
6. Joel, D.; Hershenhorn, J.; Eizenberg, H.; Aly, R.; Ejeta, G.; Rich, P.; Ransom, J.; Sauerborn, J.; Rubiales, D. Biology and management of weedy root parasites. Horticult. Rev. Westport N. Y. 2007, 33, 267.
7. Parker, C.; Riches, C.R. Parasitic Weeds of the World: Biology and Control; CAB International: Wallingford, UK, 1993.
8. Press, M.C.; Graves, J.D. Parasitic Plants; Chapman & Hall Ltd: London, UK, 1995.
9. Joel, D.M. The long-term approach to parasitic weeds control: Manipulation of specific developmental mechanisms of the parasite. Crop Prot. 2000, 19, 753–758.
10. Xie, X.; Yoneyama, K.; Yoneyama, K. The strigolactone story. Annu. Rev. Phytopathol. 2010, 48, 93–117.
11. Koltai, H. Receptors, repressors, pins: A playground for strigolactone signaling. Trends Plant Sci. 2014, 19, 727–733.
12. Waldie, T.; Hayward, A.; Beveridge, C.A. Axillary bud outgrowth in herbaceous shoots: How do strigolactones fit into the picture? Plant Mol. Biol. 2010, 73, 27–36.
13. Ruyter-Spira, C.; Al-Babili, S.; van der Krol, S.; Bouwmeester, H. The biology of strigolactones. Trends Plant Sci. 2013, 18, 72–83.
14. Babaei, S.; Alizadeh, H.; Jahansouz, M.R.; Mashhadi, H.R.; Moeini, M.M. Management of Phelipanche aegyptiaca pomel: Using trap crops in rotation with tomato (Solanum lycopersicom L.). Aust. J. Crop Sci. 2010, 4, 437–442.
15. Grenz, J.; Iştoc, V.; Manschadi, A.; Sauerborn, J. Interactions of sunflower (Helianthus annuus) and sunflower broomrape (Orobanche cumana) as affected by sowing date, resource supply and infestation level. Field Crop. Res.2008, 107, 170–179.
16. Kannan, C.; Aditi, P.; Zwanenburg, B. Quenching the action of germination stimulants using borax and thiourea, a new method for controlling parasitic weeds: A proof of concept. Crop Prot. 2015, 70, 92–98.
17. MÜLLER-STÖVER, D.; Kohlschmid, E.; Sauerborn, J. A novel strain of Fusarium oxysporum from germany and its potential for biocontrol of Orobanche ramosa. Weed Res. 2009, 49, 175–182.
18. Sillero, J.C.; Rojas-Molina, M.M.; Ávila, C.M.; Rubiales, D. Induction of systemic acquired resistance against rust, ascochyta blight and broomrape in faba bean by

exogenous application of salicylic acid and benzothiadiazole. Crop Prot. 2012, 34, 65–69.

19. López-Ráez, J.A.; Matusova, R.; Cardoso, C.; Jamil, M.; Charnikhova, T.; Kohlen, W.; Ruyter-Spira, C.; Verstappen, F.; Bouwmeester, H. Strigolactones: Ecological significance and use as a target for parasitic plant control. Pest Manag. Sci. 2009, 65, 471–477.
20. Dor, E.; Yoneyama, K.; Wininger, S.; Kapulnik, Y.; Yoneyama, K.; Koltai, H.; Xie, X.; Hershenhorn, J. Strigolactone deficiency confers resistance in tomato line SL-ORT1 to the parasitic weeds Phelipanche and Orobanche spp. Phytopathology 2011, 101, 213–222.
21. Vogel, J.T.; Walter, M.H.; Giavalisco, P.; Lytovchenko, A.; Kohlen, W.; Charnikhova, T.; Simkin, A.J.; Goulet, C.; Strack, D.; Bouwmeester, H.J. SlCCD7 controls strigolactone biosynthesis, shoot branching and mycorrhiza-induced apocarotenoid formation in tomato. Plant J. 2010, 61, 300–311.
22. López-Ráez, J.A.; Charnikhova, T.; Fernández, I.; Bouwmeester, H.; Pozo, M.J. Arbuscular mycorrhizal symbiosis decreases strigolactone production in tomato. J. Plant Physiol. 2011, 168, 294–297.
23. Zwanenburg, B.; Mwakaboko, A.S.; Kannan, C. Perspective of suicidal germination for parasitic weeds control. Pest Manag. Sci. 2016.
24. Westwood, J.H.; Depamphilis, C.W.; Das, M.; Fernándezaparicio, M.; Honaas, L.A.; Timko, M.P.; Wafula, E.K.; Wickett, N.J.; Yoder, J.I. The parasitic plant genome project: New tools for understanding the biology of Orobanche and Striga. Weed Sci. 2012, 60, 295–306.
25. Ruiu, F.; Picarella, M.; Imanishi, S.; Mazzucato, A. A transcriptomic approach to identify regulatory genes involved in fruit set of wild-type and parthenocarpic tomato genotypes. Plant Mol. Biol. 2015, 89, 263–278.
26. Honaas, L.A.; Wafula, E.K.; Yang, Z.; Der, J.P.; Wickett, N.J.; Altman, N.S.; Taylor, C.G.; Yoder, J.I.; Timko, M.P.; Westwood, J.H. Functional genomics of a generalist parasitic plant: Laser microdissection of host-parasite interface reveals host-specific patterns of parasite gene expression. BMC Plant Biol. 2013, 13, 9.
27. Ranjan, A.; Ichihashi, Y.; Farhi, M.; Zumstein, K.; Townsley, B.; David-Schwartz, R.; Sinha, N.R. De novo assembly and characterization of the transcriptome of the parasitic weed dodder identifies genes associated with plant parasitism. Plant Physiol. 2014, 166, 1186–1199.
28. Wickett, N.J.; Honaas, L.A.; Wafula, E.K.; Das, M.; Huang, K.; Wu, B.; Landherr, L.; Timko, M.P.; Yoder, J.; Westwood, J.H. Transcriptomes of the parasitic plant family orobanchaceae reveal surprising conservation of chlorophyll synthesis. Curr. Biol. 2011, 21, 2098–2104.
29. Nambara, E. A comparative transcriptome analysis on growth regulation between seeds and buds of arabidopsis. Available online: https://pag.confex.com/pag/xxiii/webprogram/Paper15834.html (accessed on 13 July 2016).
30. Tao, T.; Zhao, L.; Lv, Y.; Chen, J.; Hu, Y.; Zhang, T.; Zhou, B. Transcriptome sequencing and differential gene expression analysis of delayed gland morphogenesis in Gossypium australe during seed germination. PLoS ONE 2013, 8, e75323.
31. Yu, Y.; Guo, G.; Lv, D.; Hu, Y.; Li, J.; Li, X.; Yan, Y. Transcriptome analysis during seed germination of elite chinese bread wheat cultivar Jimai 20. BMC Plant Biol. 2014, 14, 1–19.
32. Grabherr, M.G.; Haas, B.J.; Yassour, M.; Levin, J.Z.; Thompson, D.A.; Amit, I.; Adiconis, X.; Fan, L.; Raychowdhury, R.; Zeng, Q. Full-Length transcriptome assembly from RNA-Seq data without a reference genome. Nat. Biotechnol.2011, 29, 644–652.
33. Gonzalez-Verdejo, C.I.; Barandiaran, X.; Moreno, M.T.; Cubero, J.I.; di Pietro, A. An improved axenic system for studying pre-infection development of the parasitic plant Orobanche ramosa. Ann. Bot.-Lond. 2005, 96, 1121–1127.
34. Kusumoto, D.; Chae, S.H.; Mukaida, K.; Yoneyama, K.; Yoneyama, K.; Joel, D.M.; Takeuchi, Y. Effects of fluridone and norflurazon on conditioning and germination of Striga asiatica seeds. Plant Growth Regul. 2006, 48, 73–78.

35. Joel, D.; Steffens, J.; Matthews, D. Germination of weedy root parasites. Seed Dev. Germination 1995, 1, 567–597.
36. Tsuchiya, Y.; Yoshimura, M.; Sato, Y.; Kuwata, K.; Toh, S.; Holbrook-Smith, D.; Zhang, H.; McCourt, P.; Itami, K.; Kinoshita, T. Probing strigolactone receptors in Striga hermonthica with fluorescence. Science 2015, 349, 864–868.
37. Liu, S.; Li, W.; Wu, Y.; Chen, C.; Lei, J. De novo transcriptome assembly in chili pepper (Capsicum frutescens) to identify genes involved in the biosynthesis of capsaicinoids. PLoS ONE 2013, 8, e48156.
38. Nakasugi, K.; Crowhurst, R.N.; Bally, J.; Wood, C.C.; Hellens, R.P.; Waterhouse, P.M. De novo transcriptome sequence assembly and analysis of rna silencing genes of Nicotiana benthamiana. PLoS ONE 2013, 8, e59534.
39. Wang, L.; Yu, S.; Tong, C.; Zhao, Y.; Liu, Y.; Song, C.; Zhang, Y.; Zhang, X.; Wang, Y.; Hua, W. Genome sequencing of the high oil crop sesame provides insight into oil biosynthesis. Genome Biol. 2014, 15, 1–13.
40. Dal Degan, F.; Rocher, A.; Cameron-Mills, V.; Von Wettstein, D. The expression of serine carboxypeptidases during maturation and germination of the barley grain. Proc. Natl. Acad. Sci. USA 1994, 91, 8209–8213.
41. Dietrich, R.A.; Maslyar, D.J.; Heupel, R.C.; Harada, J.J. Spatial patterns of gene expression in Brassica napus seedlings: Identification of a cortex-specific gene and localization of mrnas encoding isocitrate lyase and a polypeptide homologous to proteinases. Plant Cell 1989, 1, 73–80.
42. Rajjou, L.; Duval, M.; Gallardo, K.; Catusse, J.; Bally, J.; Job, C.; Job, D. Seed germination and vigor. Annu. Rev. Plant Biol. 2012, 63, 507–533.
43. Howell, K.A.; Narsai, R.; Carroll, A.; Ivanova, A.; Lohse, M.; Usadel, B.; Millar, A.H.; Whelan, J. Mapping metabolic and transcript temporal switches during germination in rice highlights specific transcription factors and the role of rna instability in the germination process. Plant Physiol. 2009, 149, 961–980.
44. Xu, J.; Chua, N.-H. Arabidopsis decapping 5 is required for mrna decapping, p-body formation, and translational repression during postembryonic development. Plant Cell 2009, 21, 3270–3279.
45. Hunt, L.; Holdsworth, M.J.; Gray, J.E. Nicotinamidase activity is important for germination. Plant J. 2007, 51, 341–351.
46. Macovei, A.; Balestrazzi, A.; Confalonieri, M.; Faé, M.; Carbonera, D. New insights on the barrel medic MtOOG1 and MtFPQ functions in relation to oxidative stress response in planta and during seed imbibition. Plant Physiol. Biochem.2011, 49, 1040–1050.
47. Rajjou, L.; Gallardo, K.; Debeaujon, I.; Vandekerckhove, J.; Job, C.; Job, D. The effect of α-amanitin on the Arabidopsis seed proteome highlights the distinct roles of stored and neosynthesized mrnas during germination. Plant Physiol.2004, 134, 1598–1613.
48. Schmitz-Linneweber, C.; Small, I. Pentatricopeptide repeat proteins: A socket set for organelle gene expression. Trends Plant Sci. 2008, 13, 663–670.
49. Bewley, J.D.; Bradford, K.J.; Hilhorst, H.W.; Nonogaki, H. Development and maturation. In Seeds; Springer: Berlin, Germany, 2013; pp. 27–83.
50. Ogé, L.; Bourdais, G.; Bove, J.; Collet, B.; Godin, B.; Granier, F.; Boutin, J.-P.; Job, D.; Jullien, M.; Grappin, P. Protein repair l-isoaspartyl methyltransferase1 is involved in both seed longevity and germination vigor in Arabidopsis. Plant Cell 2008, 20, 3022–3037.
51. Weitbrecht, K.; Müller, K.; Leubner-Metzger, G. First off the mark: Early seed germination. J. Exp. Bot. 2011, 62, 3289–3309.
52. Wind, J.J.; Peviani, A.; Snel, B.; Hanson, J.; Smeekens, S.C. Abi4: Versatile activator and repressor. Trends Plant Sci.2013, 18, 125–132.
53. Yang, R.; Guo, Q.; Gu, Z. Gaba shunt and polyamine degradation pathway on γ-aminobutyric acid accumulation in germinating fava bean (Vicia faba L.) under hypoxia. Food Chem. 2013, 136, 152–159.

54. Linkies, A.; Leubner-Metzger, G. Beyond gibberellins and abscisic acid: How ethylene and jasmonates control seed germination. Plant Cell Rep. 2012, 31, 253–270.
55. Chae, S.H.; Yoneyama, K.; Takeuchi, Y.; Joel, D.M. Fluridone and norflurazon, carotenoid-biosynthesis inhibitors, promote seed conditioning and germination of the holoparasite Orobanche minor. Physiol. Plant. 2004, 120, 328–337.
56. Takeuchi, Y.; Omigawa, Y.; Ogasawara, M.; Yoneyama, K.; Konnai, M.; Worsham, A.D. Effects of brassinosteroids on conditioning and germination of clover broomrape (Orobanche minor) seeds. Plant Growth Regul. 1995, 16, 153–160.
57. Zehhar, N.; Ingouff, M.; Bouya, D.; Fer, A. Possible involvement of gibberellins and ethylene in Orobanche ramosa germination. Weed Res. 2002, 42, 464–469.
58. Lechat, M.-M.; Pouvreau, J.-B.; Péron, T.; Gauthier, M.; Montiel, G.; Véronési, C.; Todoroki, Y.; Le Bizec, B.; Monteau, F.; Macherel, D. PrCYP707A1, an aba catabolic gene, is a key component of Phelipanche ramosa seed germination in response to the strigolactone analogue GR24. J. Exp. Bot. 2012, 63, 5311–5322.
59. Lee, H.; Lee, K.; Seo, P. The Arabidopsis MYB96 transcription factor plays a role in seed dormancy. Plant Mol. Biol.2015, 87, 371–381.
60. Lechat, M.-M.; Brun, G.; Montiel, G.; Véronési, C.; Simier, P.; Thoiron, S.; Pouvreau, J.-B.; Delavault, P. Seed response to strigolactone is controlled by abscisic acid-independent DNA methylation in the obligate root parasitic plant, Phelipanche ramosa L. Pomel. J. Exp. Bot. 2015, 66, 1329–1340.
61. Matusova, R.; Rani, K.; Verstappen, F.W.; Franssen, M.C.; Beale, M.H.; Bouwmeester, H.J. The strigolactone germination stimulants of the plant-parasitic Striga and Orobanche spp. Are derived from the carotenoid pathway. Plant Physiol. 2005, 139, 920–934.
62. Song, W.; Zhou, W.; Jin, Z.; Cao, D.; Joel, D.; Takeuchi, Y.; Yoneyama, K. Germination response of Orobanche seeds subjected to conditioning temperature, water potential and growth regulator treatments. Weed Res. 2005, 45, 467–476.
63. Song, W.; Zhou, W.; Jin, Z.; Zhang, D.; Yoneyama, K.; Takeuchi, Y.; Joel, D. Growth regulators restore germination of Orobanche seeds that are conditioned under water stress and suboptimal temperature. Crop Pasture Sci. 2006, 57, 1195–1201.
64. Joel, D.; Back, A.; Kleifeld, Y.; Gepstein, S. Seed conditioning and its role in orobanche seed germination: Inhibition by paclobutrazol. In Progress in Orobanche Research, Proceedings of the International Workshop on Orobanche Research, Obermarchtal, Germany, 19–22 August 1989.
65. Uematsu, K.; Nakajima, M.; Yamaguchi, I.; Yoneyama, K.; Fukui, Y. Role of camp in gibberellin promotion of seed germination in Orobanche minor Smith. J. Plant Growth Regul. 2007, 26, 245–254.
66. Krizek, D.T.; Mandava, N.B. Influence of spectral quality on the growth response of intact bean plants to brassinosteroid, a growth-promoting steroidal lactone. I. Stem elongation and morphogenesis. Physiol. Plant. 1983, 57, 317–323.
67. Graeber, K.; Linkies, A.; Müller, K.; Wunchova, A.; Rott, A.; Leubner-Metzger, G. Cross-Species approaches to seed dormancy and germination: Conservation and biodiversity of aba-regulated mechanisms and the brassicaceae DOG1 genes. Plant Mol. Biol. 2010, 73, 67–87.
68. Morris, K.; Linkies, A.; Müller, K.; Oracz, K.; Wang, X.; Lynn, J.R.; Leubner-Metzger, G.; Finch-Savage, W.E. Regulation of seed germination in the close arabidopsis relative Lepidium sativum: A global tissue-specific transcript analysis. Plant Physiol. 2011, 155, 1851–1870.
69. Kucera, B.; Cohn, M.A.; Leubner-Metzger, G. Plant hormone interactions during seed dormancy release and germination. Seed Sci. Res. 2005, 15, 281–307.
70. Linkies, A.; Müller, K.; Morris, K.; Turečková, V.; Wenk, M.; Cadman, C.S.; Corbineau, F.; Strnad, M.; Lynn, J.R.; Finch-Savage, W.E. Ethylene interacts with abscisic acid to regulate endosperm rupture during germination: A comparative approach using Lepidium sativum and Arabidopsis thaliana. Plant Cell 2009, 21, 3803–3822.
71. Matilla, A.; Matilla-Vazquez, M. Involvement of ethylene in seed physiology. Plant Sci. 2008, 175, 87–97.

72. Chun, D.; Wilhelm, S.; Sagen, J.; Musselman, L.; Worsham, A.; Eplee, R. Components of record germination in vitro of branched broomrape, orobanche ramosa L. In Supplement to the Proceedings of the Second International Symposium on Parasitic Weeds, Raleigh, NC, USA, 16–19 July 1979.
73. Bebawi, F.F.; Eplee, R.E. Efficacy of ethylene as a germination stimulant of Striga hermonthica seed. Weed Sci. 1986, 34, 694–698.
74. Egley, G.; Dale, J. Ethylene, 2-chloroethylphosphonic acid, and witchweed germination. Weed Sci. 1970, 18, 586–589.
75. Arc, E.; Sechet, J.; Corbineau, F.; Rajjou, L.; Marion-Poll, A. ABA crosstalk with ethylene and nitric oxide in seed dormancy and germination. Front. Plant Sci. 2013, 4, 63.
76. Toh, S.; Holbrook-Smith, D.; Stogios, P.J.; Onopriyenko, O.; Lumba, S.; Tsuchiya, Y.; Savchenko, A.; McCourt, P. Structure-function analysis identifies highly sensitive strigolactone receptors in Striga. Science 2015, 350, 203–207.
77. Westwood, J.H. Characterization of the Orobanche–Arabidopsis system for studying parasite–host interactions. Weed Sci. 2009, 48, 742–748.
78. Fernández-Aparicio, M.; Rubiales, D.; Bandaranayake, P.C.; Yoder, J.I.; Westwood, J.H. Transformation and regeneration of the holoparasitic plant Phelipanche aegyptiaca. Plant Methods 2011, 7, 683–696.
79. Margulies, M.; Egholm, M.; Altman, W.E.; Attiya, S.; Bader, J.S.; Bemben, L.A.; Berka, J.; Braverman, M.S.; Chen, Y.-J.; Chen, Z. Genome sequencing in microfabricated high-density picolitre reactors. Nature 2005, 437, 376–380.
80. Altschul, S.F.; Madden, T.L.; Schäffer, A.A.; Zhang, J.; Zhang, Z.; Miller, W.; Lipman, D.J. Gapped blast and psi-blast: A new generation of protein database search programs. Nucleic Acids Res. 1997, 25, 3389–3402.
81. Finn, R.D.; Bateman, A.; Clements, J.; Coggill, P.; Eberhardt, R.Y.; Eddy, S.R.; Heger, A.; Hetherington, K.; Holm, L.; Mistry, J. Pfam: The protein families database. Nucleic Acids Res. 2013, 42, D222–D230.
82. Alexa, A.; Rahnenfuhrer, J. topGo: Enrichment analysis for gene ontology. Available online: http://bioconductor.uib.no/2.7/bioc/html/topGO.html (accessed on 13 July 2016).
83. Moriya, Y.; Itoh, M.; Okuda, S.; Yoshizawa, A.C.; Kanehisa, M. Kaas: An automatic genome annotation and pathway reconstruction server. Nucleic Acids Res. 2007, 35, W182–W185.
84. Leng, N.; Dawson, J.; Kendziorski, C. Ebseq: An R package for gene and isoform differential expression analysis of rna-seq data, R package version 1.5.3. Available online: https://www.biostat.wisc.edu/~kendzior/EBSEQ/ (accessed on 13 July 2016).
85. Benjamini, Y.; Hochberg, Y. Controlling the false discovery rate: A practical and powerful approach to multiple testing. J. R. Stat. Soc. Ser. B 1995, 57, 289–300.
86. Eisen, M.B.; Spellman, P.T.; Brown, P.O.; Botstein, D. Cluster analysis and display of genome-wide expression patterns. Proc. Natl. Acad. Sci. USA 1998, 95, 14863–14868.
87. Howe, E.A.; Sinha, R.; Schlauch, D.; Quackenbush, J. Rna-seq analysis in mev. Bioinformatics 2011, 27, 3209–3210.
88. Saldanha, A.J. Java treeview—extensible visualization of microarray data. Bioinformatics 2004, 20, 3246–3248.
89. Gonzalez-Verdejo, C.; Die, J.; Nadal, S.; Jimenez-Marin, A.; Moreno, M.; Roman, B. Selection of housekeeping genes for normalization by real-time RT-PCR: Analysis of Or-MYB1 gene expression in Orobanche ramosa development. Anal. Biochem. 2008, 379, 176–181.
90. Livak, K.J.; Schmittgen, T.D. Analysis of relative gene expression data using real-time quantitative PCR and the 2–ΔΔCt method. Methods 2001, 25, 402–408.
91. Durgbanshi, A.; Arbona, V.; Pozo, O.; Miersch, O.; Sancho, J.V.; Gómez-Cadenas, A. Simultaneous determination of multiple phytohormones in plant extracts by liquid

chromatography-electrospray tandem mass spectrometry. J. Agric. Food Chem. 2005, 53, 8437–8442.

92. Kim, T.-W.; Hwang, J.-Y.; Kim, Y.-S.; Joo, S.-H.; Chang, S.C.; Lee, J.S.; Takatsuto, S.; Kim, S.-K. Arabidopsis CYP85A2, a cytochrome P450, mediates the baeyer-villiger oxidation of castasterone to brassinolide in brassinosteroid biosynthesis. Plant Cell 2005, 17, 2397–2412.

93. Zhou, M.; Xu, M.; Wu, L.; Shen, C.; Ma, H.; Lin, J. CbCBF from Capsella bursa-pastoris enhances cold tolerance and restrains growth in Nicotiana tabacum by antagonizing with gibberellin and affecting cell cycle signaling. Plant Mol. Biol. 2014, 85, 259–275.

94. Cristescu, S.M.; Woltering, E.; Hermans, C.; Harren, F.J.; te Lintel Hekkert, S. Research tools: Ethylene detection. In Ethylene in Plants; Springer: Berlin, Germany, 2015; pp. 263–286.

CHPATER 11

Morphological and Molecular Characterization of Vicia faba and Phaseolus vulgaris Seed-born Fungal Endophytes

K.S. Akutse, N.K. Maniania, S. Ekesi, K.K.M. Fiaboe, J. Van den Berg, O.L. Ombura and F.M. Khamis

[1] *Institute of Applied Ecology and Research Center for Biodiversity and Eco-Safety, Fujian Agriculture and Forestry University, 350002 Fuzhou, China*

[2] *Plant Health Division, International Centre of Insect Physiology and Ecology (ICIPE), P.O. Box 30772-00100, Nairobi, Kenya*

[3] *Unit for Environmental Sciences and Management, North-West University, Private Bag X6001, 2520 Potchefstroom, South Africa*

ABSTRACT

Background: Fungal endophytes are heterotrophic microorganisms that occur inside plant tissues, with some showing adverse effects against insects, nematodes and plant pathogens. Initiatives are underway at the International Center of Insect Physiology and Ecology (ICIPE) to use these endophytes as a novel strategy for control of Liriomyza leafminer. Objective: The objective of this study was therefore to search for fungal endophytes from the Liriomyza major host plants that could be used in the management of this invasive insect pest. Materials and Methods: Bioprospecting was undertaken to isolate fungal endophytes from Phaseolus vulgaris and Vicia faba seeds collected from different local and super markets in Kenya. Fungal endophytes were isolated through seeds surface sterilization and characterized using morphological and molecular techniques. Fungal occurrence was analyzed using analysis of variance while Chi-square tests were performed to compare the various endophytes species occurring in the two host seeds. Results: Five fungal isolates were isolated from both P. vulgaris and V. faba seeds with no significant differences in their occurrence. However, there was significant difference between the endophyte species occurring in V. faba compared to P. vulgaris seeds ($p<0.0001$). Isolated fungi included Beauveria bassiana, Phialemonium sp., Phanerochaete chrysosporium and Metarhizium anisopliae. Beauveria bassiana ICIPE 693 occurred in all the V. faba seeds (100%) but not in P. vulgaris seeds. Similarly, P. chrysosporium had 66.7% occurrence in V. faba but absent in P. vulgaris seeds. The prevalence of Phialemonium sp. (55%) was only recorded in P. vulgaris, while that of M. anisopliae was recorded in both P. vulgaris and V. faba seeds with 55.4 and 70.8% occurrence, respectively. Conclusion: Fungal endophyte species occurrence in V. faba differed from P. vulgaris seeds. The characterization of these bean seed-born endophytes will not only create awareness but also facilitate studies on the role of these fungi in pest management strategies. The outcome of this study will stimulate further studies on the possible roles of these fungi in inhibiting the growth of artificially inoculated endophytes, assessing their pathogenicity and virulence effects on different sucking arthropod pests and evaluating the nutritional value of the seeds containing these endophytes.

INTRODUCTION

Endophytes are heterotrophic microorganisms[1] that live inside plants primarily for nutrition, protection and reproduction. Some of them are beneficial by showing adverse effects against insects, nematodes and plant pathogens[2,3]. Fungal endophytes have been detected in hundreds of plant species, including important agricultural crops such as wheat[4], bananas[5-7], soybeans[8] and tomatoes[9]. Endophytes occur in virtually all tissues of the host plant, however, Baker and Smith[10] and Schardl *et al.*[11] suggested that seeds may serve as good sources of endophytes or pathogens. Seeds are of particular interest as they may transmit endophytes vertically from generation to generation[12]. Indeed, plant seeds usually fall into the soil, a microbially rich habitat and lie dormant waiting for environmental cues to germinate, possibly recruiting surface microbes to help protect them against degradation or predation[13]. As seeds begin to germinate, endophytes may colonize the seedling as shown in rice[14,15], eucalyptus[16], maize[17] and beans.

However, most recent studies on seeds born-fungi were mainly focused on the fungal pathogens that transmit plant diseases to the host plants and consequently determined the quality of the seeds as well as their effects on the seeds germination[18-21]. In addition, until recently, most of the studies on these seed-born fungi were targeting fungi or microorganisms that were found to compromise the quality of the seed in terms of **NUTRITIONAL VALUE** and oil quality[22] but not necessarily in the prospect of exploring possible entomopathogenic fungi inside the seeds and that could be used not only as plant growth promoters but also as biocontrol candidates under the umbrella of pest management strategies. Gouda *et al.*[23] considered endophytes as a treasure of bioactive compounds and reported that they act as reservoirs of novel bioactive secondary metabolites, such as alkaloids, **PHENOLIC ACIDS**, quinones, steroids, saponins, tannins and terpenoids that serve as a potential candidate for antimicrobial, anti-insect, anticancer and many more properties. Hence, exploring possible seed endophytes could give a new research route to explore their potential for insect pest's management.

Horticultural crops are highly valued due to their importance in terms of nutrition, income generation and employment[24-26]. In Kenya where this study was conducted, *Phaseolus vulgaris* and ***VICIA FABA*** are among the high value horticultural crops, which are also attacked by the invasive Agromyzid leafminer species, *Liriomyza huidobrensis* (Blanchard) *L. sativae* Blanchard and *L. trifolii* (Burgess) (Diptera: Agromyzidae) the most damaging pests on these crops[27,28]. It is then important to assess the occurrence of fungal endophytes in *P. vulgaris* and *V. faba* seeds in the scope of developing a sustainable management strategy against these pests and indirectly, reduce the abusive use of insecticides.

Fungal endophytes have been recognized to play multiple roles such as protecting plants from pests and diseases and promoting plant growth[29,30] and are being considered as important component of IPM[7,31-34]. Some fungal endophytes protect host plants against plant pathogens[35,36] and herbivores, including insects[29,32,37-40]. For example, exposure of two aphid species, *Rhopalosiphum padi* and *Metopopophium dirhodum* (Hemiptera: Aphididae) and wheat stem sawfly, *Mayetiola destructor* (Diptera: Chloropidae) to wild barley infected with *Neotyphodium coenophialum* (Hypocreales: Clavicipitaceae) reduced their survival[41,42]. Wheat leaves colonized by either *B. bassiana* or *Aspergillus parasiticus* (Eurotiales: Trichocomaceae) reduced the growth rate of *Chortoicetes terminifera* (Orthoptera: Acrididae) nymphs[34]. Endophytic *B. bassiana* in banana significantly reduced larval survivorship of banana weevil, *Cosmopolites sordidus* (Coleoptera:

Curculionidae), resulting in 42-87% reduction in plant damage[43,44]. Reduction in feeding and reproduction by *Aphis gossypii* (Hemiptera: Aphididae) has also been reported on cotton endophytically colonized by either *B. bassiana* or *Lecanicillium lecanii* (Hypocreales: Clavicipitaceae)[34]. Another possible role of the fungal endophytes includes plant growth promotion as well as impact on tritrophic interaction[7,29,45,46]. Fungal entomopathogens that become established as endophytes can therefore play an important role in the regulation of insect populations. Most recently, Akutse *et al.*[31] demonstrated the effects of fungal endophytes on life-history parameters of *Liriomyza huidobrensis* (Diptera: Agromyzidae) leafminers and their associated parasitoids[47]. The researchers reported that endophytic fungal isolates of both *Hypocrea lixii* (F3ST1) and *Beauveria bassiana* (G1LU3, S4SU1 and ICIPE 279) could endophytically colonize ***VICIA FABA*** and *Phaseolus vulgaris* plants through seeds inoculation and cause detrimental effects on survival, fecundity, oviposition, emergence and longevity of *L. huidobrensis*. Thus, this study focuses on assessing the occurrence of fungal endophytes in *P. vulgaris* and *V. faba* seeds to develop a sustainable management strategy against *Liriomyza* leafminer pests. The aim was therefore to search for fungal endophytes that can be developed further for the control of the invasive *Liriomyza* leafminer specifically and other crop insect pests and diseases. Bioprospecting was subsequently carried out for isolation of fungal endophytes from *P. vulgaris* and *V. faba* seeds and we report here on their morphological and molecular characterization.

MATERIALS AND METHODS

Seeds: Seeds of open pollinated varieties of *Phaseolus vulgaris* (Rose coco beans) and ***VICIA FABA*** (Faba beans) were collected from local markets in Kenya (Nairobi, Kasarani, Eastleigh, Nyamakima, Gikomba and sometimes from Ethiopia to Kenya's markets) and sometimes procured from super markets (Naivas) since farmers often use these seeds for sowing.

Fungal isolation and culture: Seeds were surface-sterilized in 70% ethanol for 2 min followed by 1.5% sodium hypochlorite for 3 min after which they were rinsed three times with sterile distilled water[32]. Prior to incubation of seeds at the laboratory (25±1°C and 60% RH), half of the seeds were ground in a sterile motor and the other half used as a whole for detection of variations in the fungal endophyte species occurrence and diversity. Three to five seeds were plated onto 9 cm petri dishes containing either Potato Dextrose Agar (PDA), Sabouraud Dextrose Agar (SDA) or Yeast Extract Agar (YEA). A total of 320 seeds were used and incubated in 40 plates per seed species and replicated 5 times.

To determine the presence of fungal endophytes in both ground and whole seeds, the latter was placed on PDA, SDA and YEA plates (at a distant of 4 cm apart) containing 0.05% antibiotics (Streptomycin sulfate salt and chloramphenicol)[33,34]. Plates were incubated at 25±1°C for 14 days, after which the presence of endophytes was determined. Prior to incubation of the seeds, the last rinse water was also plated to assess the effectiveness of the surface sterilization procedure[32]. The presence or occurrence of fungal endophytes in the seeds was recorded by counting the number of seeds of the different varieties that showed fungal growth/mycelia according to Koch's postulates[48]. This assessment of the fungal endophytes occurrence was repeated daily from the 1st day of incubation until 14 days post-incubation.

Isolated fungi were sub-cultured 4-6 times to obtain pure cultures and preserved in the ICIPE'S Arthropod Germplasm Center.

Morphological identification: Morphological identification was carried out using the procedures described by Burgess *et al.*[49], Humber[50] and Goettel and Inglis[51]. Fungal colony features, such as appearance, texture (colonization pattern) and pigmentation on both the top and reverse plates were observed[52,53].

Slide culture preparations were used for observation of the following characteristics: (1) Presence or absence of microconidia and macroconidia, (2) Shapes and sizes of conidia, (3) Type of phialides bearing microconidia, (4) Presence or absence of chlamydospores, microconidial chains and sporodochia and (5) Type of fungal mycelia. The variability of colony appearance, pigmentation, growth rate, length of chains, production of bluish sclerotia and concentric ring aerial mycelium were also observed.

Morphological data analysis: Fungal occurrence was expressed as a percentage of the total number of seeds that were plated out. Percentages were square root transformed [$\sqrt{(x+1)}$] before applying analysis of variance (ANOVA) in R (2.13.1). Tukey HSD multiple comparisons of means was used to separate the means. The chi-square tests were used to compare the various endophytes species occurring in *V. faba* and *P. vulgaris* seeds. All the analyses were performed using R (2.13.1) statistical software packages[54], while relying heavily on the epicalc package[55].

Molecular characterization Preparation of fungal material: Conidia were harvested by scraping the surface of 2 week old cultures with a sterile spatula. Genomic DNA was extracted from the harvested conidia/mycelia using the Fast DNA Spin® kit for soil following the manufacturer's instructions (MP Biomedicals). Mycelia/conidia were freeze-dried for 24 h to facilitate easy grinding. Isolates of *B. bassiana* ICIPE 279 and *M. anisopliae* ICIPE 30 and S4ST7, previously used in the endophytic colonization study[31] were included as references in the phylogenetic trees for sibling or related isolate species characterization. These were obtained from the ICIPE'S Arthropod Germplasm Center.

Amplifications using ITS 5 and 4 and TW81 and AB28 primers: Amplifications were carried out for the rDNA region of the fungal isolates using[56] the ITS 5 and 4 and the TW81 and AB28[57] primers. The isolated DNA was amplified in 30 μL^{-1} PCR mix. The reaction mixture consisted of 3 μL^{-1} 10X PCR buffer (GenScript USA Inc), 1.5 μL^{-1} 25 mM $MgCl_2$, 0.6 μL^{-1} 10 mM dNTP mix, 1.5 μL^{-1} 10 μM of the primers (Table 1), 1 unit Taq polymerase (GenScript USA Inc) and 20 ng μL^{-1} genomic DNA. For the ITS region, the thermo-cycling conditions involved initial denaturation at 94°C for 3 min, followed by 35 cycles of denaturation at 94°C for 40 sec, annealing at 58°C for 40 sec and primer elongation at 72°C for 1 min followed by a final extension at 72°C for 10 min giving a product range of between 550-600 bp.

Table 1: Primer sequences used in the DNA amplification

Primer name	Sequence (5'-3')	Expected size of PCR product
ITS 5	[F]5' GGA AGT AAA AGT CGT AAC AAG G 3'	550-600 bp
ITS 4	[R]5' TCC TCC GCT TAT TGA TAT GC 3'	
TW81	[F]5' GTT TCC GTA GGT GAA CCT GC 3'	500-550 bp
AB28	[R]5' ATA TGC TTA AGT TCA GCG GGT 3'	

The same procedure was applied during amplification using the TW81 and AB28 primers with the exception that the reaction was set for 40 cycles with the annealing temperature of 57°C for 40 sec. This gave a target region ranging between 500-550 bp. Both reactions were done in a PTC 100 thermocycler (MJ Research, Gaithersburg).

DNA purification and sequencing Agarose gel electrophoresis and purification: The PCR products were resolved through 1% agarose gel for 1 h at 70 V (Bio-Rad model 200/2-0 power supply and wide mini-sub cell GT horizontal electrophoresis system, Bio-Rad laboratories, Inc., USA), followed by visualization of the DNA under UV-illumination. The gel was photographed, analyzed and documented using KODAK Gel Logic 200 Imaging System software (Raytest, GmbH, Straubenhardt).

The products were then gel purified using the QuickClean 5M Gel Extraction Kit II from GenScript (GenScript Corporation, Piscataway, NJ), following the manufacturer's instructions and subsequently sequenced in both directions using ABI 3700 genetic analyzers.

Sequencing and molecular data analysis: The sequences obtained were assembled and edited using Chromas version 2.13 (Technelysium Pty ltd, Queensland, Australia). Consensus sequences from both the forward and reverse strands were generated and were then queried through BLASTN in the GenBank database provided by the National Center of Biotechnology Information (NCBI) (http://www.ncbi.nlm. nih.gov/) for identification purposes and to check for similarity with organisms already identified. Any isolate exhibiting ≥95% sequence similarity to NCBI strains were considered as the correct species for that isolate[58].

Moreover, the consensus sequences were aligned using[59] ClustalX version 1.81. These alignments were used for phylogenetic and molecular evolutionary analyses that were conducted[60] using MEGA version 6. Neighbour-joining trees were constructed[61] with bootstrapping and using the Kimura 2 distance matrix[62,63] for both sets of sequences of the 2 gene regions. Tables of between species distances were also constructed using[60] MEGA version 6. The tables of distances were used to generate the principal component plots using the program[64] GenAlEx 6.41.

RESULTS

Morphological identification of seed-born fungal endophytes: No fungal growth was observed in any of the plated washing water. There were no significance differences (F = 0.97, df = 3, 20, p = 0.423) in the growth of fungal endophytes in ground and whole seeds of both *V. faba* and *P. vulgaris*. A total of 5 fungal isolates were obtained from the 320 seeds of *V. faba* and *P. vulgaris* using morphological features and they belonged to genera *Metarhizium, Beauveria, Phialemonium* and *Phanerochaete* (Figure 1). Among the key morphological features used to identify the isolates in the study that belong to the *M. anisopliae* included (1) The type of conidia formed that appeared either in short to long chains, (2) The short conidiogenous cells, with rounded to broadly conical apices (not having a distinctly narrowed and extended neck), (3) The conidiogenous cells borne at the apices of broadly branched and densely intertwined conidiophores that form a compact hymenium while the conidia are borne in parallel chains that are usually green in mass and (4) The cylindrical or ovoid conidia, forming chains that usually aggregated into a solid mass and appear as pale to bright green to yellow-green in colour (Figure 1). For the *B. bassiana* isolates, the conidia were produced singly on many separate denticles on each conidiogenous cell or, if they were growing in some sort of slime, they were either singly or in small groups in a slime droplet. Their conidiogenous cells had globose bases with extended denticulate raches/apex and each rachis normally bore a single conidium per denticle. The conidia were usually white in colour (Figure 1).

While the genus *Phialemonium* isolate was characterized by its abundance of adelophialides and few discrete phialides with no signs of collarettes. It also had distinct grayish white to brownish colonies pigmentation (Figure 1) and allantoid conidia which had a cylindrical to bean-shaped structure.

Lastly, the white rot fungus *Phanerochaete chrysosporium* isolate had colonies which were white in colour when cultured on Potato Dextrose Agar (Figure 1). The isolate also had hyaline mycelia which were well branched and lacked clamp-connections. Likewise, three types of conidia were observed, the aleuriospores, arthrospores and chlamydospores.

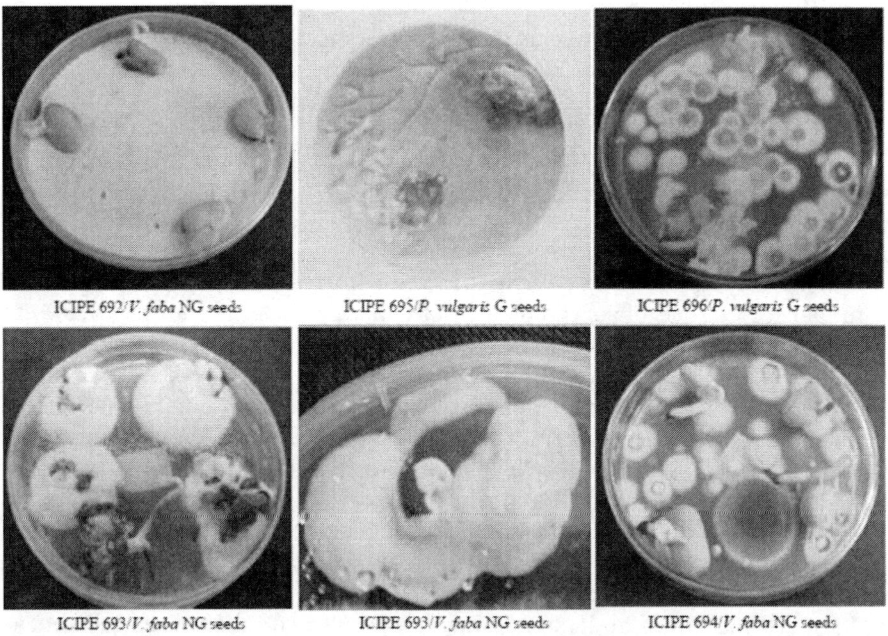

Figure 1: Fungal species isolated from Phaseolus vulgaris and Vicia faba seeds 2 weeks after incubation. ICIPE 692: Phanerochaete chrysosporium, ICIPE 695: Phialemonium sp., ICIPE 696: ICIPE 694: Metarhizium anisopliae and ICIPE 693: Beauveria bassiana, G: Grinded seeds, NG: Non grinded seeds

These isolates were given ICIPE'S Arthropod Germplasm Centre's accession numbers as follows: *Phialemonium* sp., isolate ICIPE 695, *Phanerochaete chrysosporium* isolate ICIPE 692, *Beauveria* sp., isolate ICIPE 693 and *Metarhizium* sp., isolates ICIPE 694 and ICIPE 696, in addition to the existing standards isolates (ICIPE 30, ICIPE 279 and S4ST7) (Table 2). GenBank accession numbers provided for the nucleotide sequences of the fungal isolates for both primers pairs are as follows: For ITS 5 and 4, S4ST7 = KM463106,ICIPE 279 = KM463107, ICIPE 694 = KM463108, ICIPE 696 = KM463109, ICIPE 692 = KM463110, ICIPE 695 = KM463111, ICIPE 693 = KM463112, ICIPE 30 = KM463113 and for TW81 and AB28, ICIPE 693 = KM463114, ICIPE 696 = KM463115, ICIPE 695 = KM463116, ICIPE 30 = KM463117, ICIPE 694 = KM463118, ICIPE 279 = KM463119, S4ST7 = KM463120, ICIPE 692 = KM463121 (www.ncbi.nlm.nih.gov/books/NBK51157/, GenBank Submissions Handbook).

There were no significant differences in the prevalence/occurrence of fungal endophytes between *V. faba* (F = 0.81, df = 3, 20, p = 0.502) and *P. vulgaris* seeds (F = 0.47, df = 3, 20, p = 0.707) (Figure 2). However, there was significant difference between the endophyte species, which occurred in *V. faba* compared to *P. vulgaris* seeds (χ^2 = 31.11, df = 1, p<0.0001). For instance, *B. bassiana* ICIPE 693 which occurred in all the *V. faba* seeds (100%) did not occur in *P. vulgaris* seeds (Figure 2). Similarly, the occurrence of *Phanerochaete chrysosporium* ICIPE 692 was 66.7% in *V. faba* whereas, the prevalence of *Phialemonium* sp. ICIPE 695 was of 55% in *P. vulgaris* only. However, the prevalence of *M. anisopliae* was recorded in both *P. vulgaris* (ICIPE 696, with 55.4% occurrence) and *V. faba* seeds (ICIPE 694, with 70.8% occurrence) (Figure 2).

Molecular characterization of seed-born fungal endophytes: The molecular identification of the endophyte isolate species after sequencing is consolidated in Table 2. All fungal isolates identified using the ITS 5 and 4 primer sets were also confirmed by the TW81 and AB28 primer sets. The two sets of primers depicted isolate species identities with 99-100% similarity and 0.0 E values (Table 2). Like the morphological identification, the molecular characterization also yielded four isolates species identities: *Metarhizium anisopliae* (isolates ICIPE 694, ICIPE 696, S4ST7 and ICIPE 30), ***BEAUVERIA BASSIANA*** (isolates ICIPE 693 and ICIPE 279), *Phanerochaete chrysosporium* (isolate ICIPE 692) and *Phialemonium* sp. (isolate ICIPE 695) (Table 2). Thus the molecular identification corroborates with the morphological identification for the endophyte isolates (Figure 1, Table 2).

Table 2: Identified fungal endophytes species using ITS 5 and 4 and TW81 and AB28

Sample codes	Length (base pair)	Accession No.	Identified fungal samples	E-value	Identities (%)
ITS 5 and 4					
ICIPE 30	541	FJ545302.1	*Metarhizium anisopliae* isolate CNXJ2 18S ribosomal RNA gene, partial sequence, internal transcribed spacer 1, 5.8S ribosomal RNA gene and internal transcribed spacer 2, complete sequence and 28S ribosomal RNA gene, partial sequence	0	99
S4ST7	541	FJ545302.1	*Metarhizium anisopliae* (same isolate CNXJ2 18S ribosomal RNA gene)	0	99
ICIPE 694	541	FJ545302.1	*Metarhizium anisopliae* (same isolate CNXJ2 18S ribosomal RNA gene)	0	99
ICIPE 696	541	FJ545302.1	*Metarhizium anisopliae* (same isolate CNXJ2 18S ribosomal RNA gene)	0	99
ICIPE 279	551	JQ266208.1	*Beauveria bassiana* strain MTCC_6286 18S ribosomal RNA gene, partial sequence, internal transcribed spacer 1, 5.8S ribosomal RNA gene and internal transcribed spacer 2, complete sequence and 28S ribosomal RNA gene, partial sequence	0	99
ICIPE 693	551	AJ560668.1	*Beauveria bassiana* ITS1, 5.8S rRNA gene and ITS2, isolate IMI 386701	0	100
ICIPE 692	621	GU966518.1	*Phanerochaete chrysosporium* strain TS03 18S ribosomal RNA gene, partial sequence, internal transcribed spacer 1, 5.8S ribosomal RNA gene and internal transcribed spacer 2, complete sequence and 28S ribosomal RNA gene, partial sequence	0	99
ICIPE 695	538	KF746155.1	*Phialemonium* sp., 1 AE-2013 strain FS070 18S ribosomal RNA gene, partial sequence, internal transcribed spacer 1, 5.8S ribosomal RNA gene and internal transcribed spacer 2, complete sequence and 28S ribosomal RNA gene, partial sequence	0	99
TW81 and AB28					
ICIPE 30	507	FJ545302.1	*Metarhizium anisopliae* isolate CNXJ2 18S ribosomal RNA gene, partial sequence, internal transcribed spacer 1, 5.8S ribosomal RNA gene and internal transcribed spacer 2, complete sequence and 28S ribosomal RNA gene, partial sequence	0	99
S4ST7	507	FJ609312.1	*Metarhizium anisopliae* strain M1311 18S ribosomal RNA gene, partial sequence, internal transcribed spacer 1, 5.8S ribosomal RNA gene, partial sequence	0	100
ICIPE 694	507	FJ609312.1	*Metarhizium anisopliae* strain M1311 18S ribosomal RNA gene, partial sequence, internal transcribed spacer 1, 5.8S ribosomal RNA gene and internal transcribed spacer 2, complete sequence and 28S ribosomal RNA gene, partial sequence	0	100
ICIPE 696	507	FJ609312.1	*Metarhizium anisopliae* strain M1311 18S ribosomal RNA gene, partial sequence, internal transcribed spacer 1, 5.8S ribosomal RNA gene and internal transcribed spacer 2, complete sequence and 28S ribosomal RNA gene, partial sequence	0	100
ICIPE 279	553	JQ999974.1	*Beauveria bassiana* strain YNSK1106 18S ribosomal RNA gene, partial sequence, internal transcribed spacer 1, 5.8S ribosomal RNA gene and internal transcribed spacer 2, complete sequence and 28S ribosomal RNA gene, partial sequence	0	99
ICIPE 693	517	AJ560668.1	*Beauveria bassiana* ITS1, 5.8S rRNA gene and ITS2, isolate IMI 386701	0	100
ICIPE 692	587	GU966518.1	*Phanerochaete chrysosporium* strain TS03 18S ribosomal RNA gene, partial sequence, internal transcribed spacer 1, 5.8S ribosomal RNA gene and internal transcribed spacer 2, complete sequence and 28S ribosomal RNA gene, partial sequence	0	99
ICIPE 695	504	KF746155.1	*Phialemonium* sp., 1 AE-2013 strain FS070 18S ribosomal RNA gene, partial sequence, internal transcribed spacer 1, 5.8S ribosomal RNA gene and internal transcribed spacer 2, complete sequence and 28S ribosomal RNA gene, partial sequence	0	99

Phylogenetic trees Phylogenetic tree using the ITS 5 and 4 gene region: The evolutionary history was inferred using the Neighbor-Joining method[61]. The optimal tree with the sum of branch length = 0.68264451 is shown in Figure 3. The percentage of replicate trees in which the associated taxa clustered together in the bootstrap test (500 replicates) is shown next to the branches[65]. The tree is drawn to scale, with branch lengths in the same units as those of the evolutionary distances used to infer the phylogenetic tree (Figure 3). The analysis involved 12 nucleotide sequences. Codon positions included were 1st+2nd+3rd+noncoding. All positions containing gaps and missing data were eliminated. There were a total of 506 positions in the final dataset.

Four groups resulted from this analysis (Figure 3). The first group consisted of the *M. anisopliae* isolates where each of the sample (ICIPE 696 and ICIPE 694) and the standards (ICIPE 30 and S4ST7) isolates branched separately. Furthermore, all the *Metarhizium* isolates linked to a *Metarhizium* sp., of accession number FJ545302.1 during blasting (Table

2). The second group consisted of *B. bassiana* isolates ICIPE 693 and ICIPE 279, likewise branching from the same node but occupying different branches, a clear indication that the ICIPE 693 from *V. faba* seeds is different from the standard ICIPE 279 isolate. The last two clusters of the phylogenetic tree consisted of *Phialemonium* sp., isolate ICIPE 695 and *P. chrysosporium* isolate ICIPE 692, respectively (Figure 3).

The **GENETIC DISTANCE**s between the isolates were also inferred using Kimura 2-parameter model. The estimates of evolutionary divergence between sequences of the various endophyte isolates ranged between 0.0 and 0.472 (Table 3).

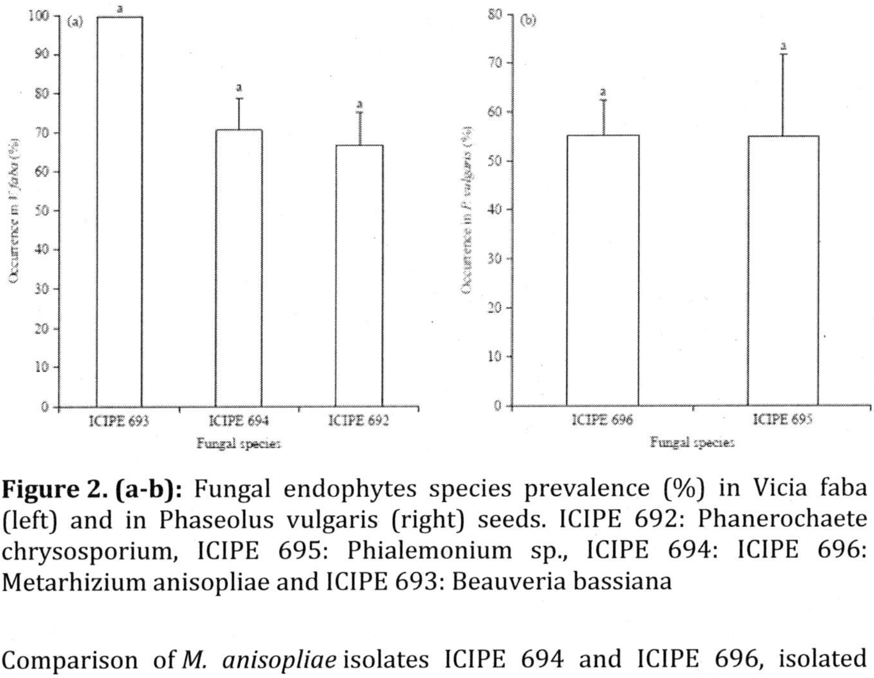

Figure 2. (a-b): Fungal endophytes species prevalence (%) in Vicia faba (left) and in Phaseolus vulgaris (right) seeds. ICIPE 692: Phanerochaete chrysosporium, ICIPE 695: Phialemonium sp., ICIPE 694: ICIPE 696: Metarhizium anisopliae and ICIPE 693: Beauveria bassiana

Comparison of *M. anisopliae* isolates ICIPE 694 and ICIPE 696, isolated from *V. faba* and *P. vulgaris* with the standards ICIPE 30 and S4ST7 gave a square distance of 0.0. Similarly, comparison of *B. bassiana* isolate ICIPE 693 isolated from *V. faba* with the standard ICIPE 279, gave a square distance of 0.0 (Table 3). Table 3 was used to generate principal component plots. In the Principal Coordinate Analysis (PCA) plot, the 1st two axes explained 76.9% of the variation (the 1st axis 49.8% and the second axis 27.1%) (Figure 4). The PCA separated the nine isolates into four distinct clusters. Each cluster was occupied by the isolates belonging to the different genera i.e., *Metarhizium*, *Beauveria*, *Phialemonium* and *Phanerochaete*, respectively (Figure 4).

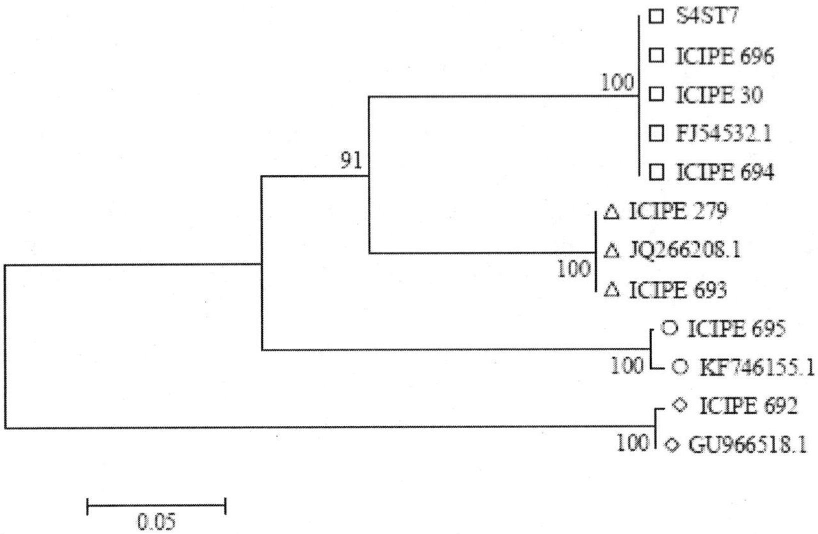

Figure 3: Phylogenetic tree using ITS 5 and 4 regions showing the evolutionary relationships of fungal endophyte isolates from Vicia faba and Phaseolus vulgaris

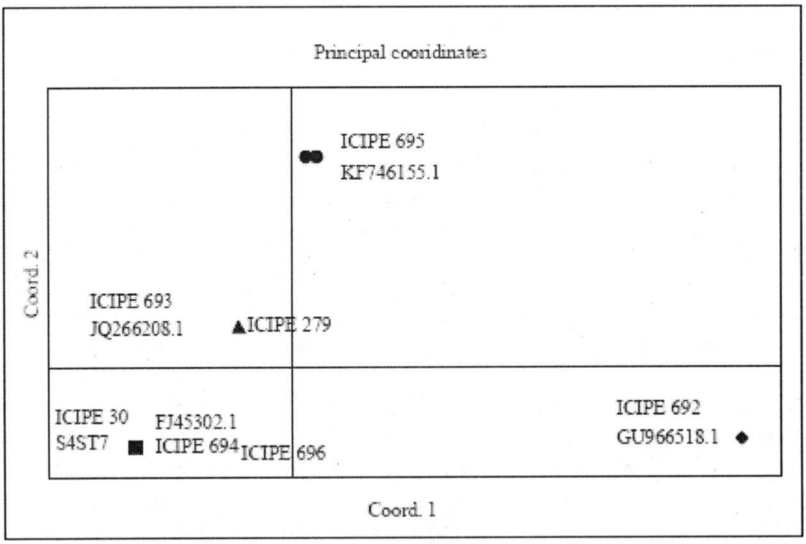

Figure 4: Plot of Principal Components Analysis (PCA) via the covariance matrix with data standardization calculated using GenAIEx for the various endophytes species isolated from Vicia faba and Phaseolus vulgaris using the ITS 5 and 4 regions (PC1 = 49.78 and PC2 = 27.11%)

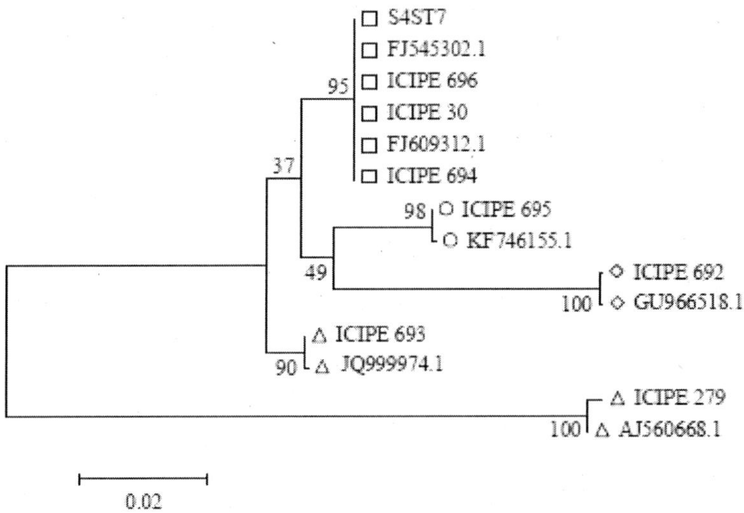

Figure 5: Phylogenetic tree using TW81 and AB28 regions showing the evolutionary relationships of fungal endophyte isolates from Vicia faba and Phaseolus vulgaris

Table 3: Estimates of evolutionary divergence between sequences using ITS 5 and 4

Fungal isolates	ICIPE 279	JQ266208.1	ICIPE 693	ICIPE 694	FJ545302.1	ICIPE 30	S4ST7	ICIPE 696	ICIPE 695	KF746155.1	ICIPE 692	GU966518.1
ICIPE 279												
JQ266208.1	0.000											
ICIPE 693	0.000	0.000										
ICIPE 694	0.176	0.176	0.176									
FJ545302.1	0.176	0.176	0.176	0.000								
ICIPE 30	0.176	0.176	0.176	0.000	0.000							
S4ST7	0.176	0.176	0.176	0.000	0.000	0.000						
ICIPE 696	0.176	0.176	0.176	0.000	0.000	0.000	0.000					
ICIPE 695	0.268	0.268	0.268	0.269	0.269	0.269	0.269	0.269				
KF746155.1	0.274	0.274	0.274	0.274	0.274	0.274	0.274	0.274	0.006			
ICIPE 692	0.431	0.431	0.431	0.463	0.463	0.463	0.463	0.463	0.472	0.461		
GU966518.1	0.427	0.427	0.427	0.459	0.459	0.459	0.459	0.459	0.468	0.457	0.002	

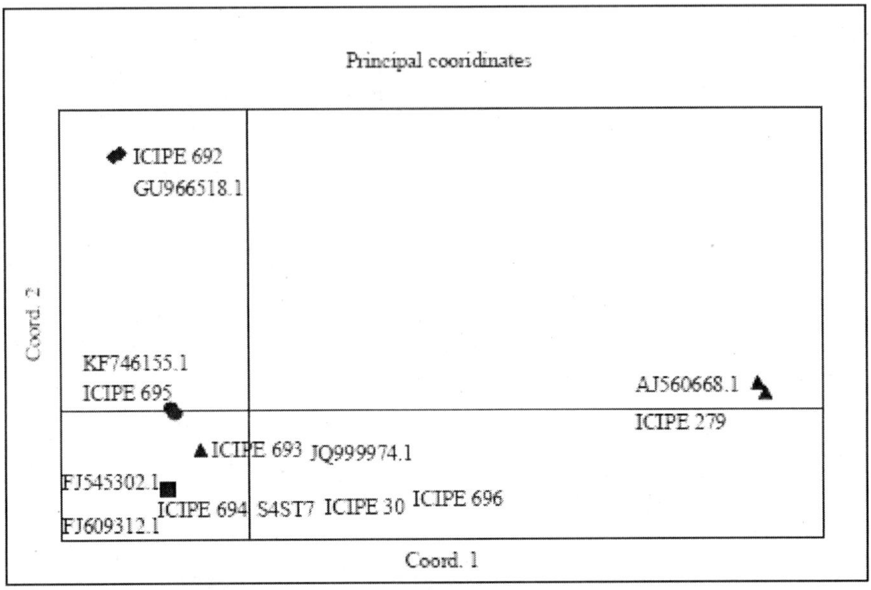

Figure 6: Plot of principal components analysis (PCA) via the covariance matrix with data standardization calculated using GenAIEx for the various endophytes species isolated from Vicia faba and Phaseolus vulgaris using the TW81 and AB28 regions (PC1 = 61.48% and PC2 = 20.99%)

Table 4: Estimates of evolutionary divergence between sequences using TW81 and AB28

Fungal isolates	ICIPE 695	KF746155.1	ICIPE 693	JQ999974.1	ICIPE 694	FJ609312.1	ICIPE 30	ICIPE 696	S4ST7	FJ545302.1	ICIPE 692	GU966518.1	ICIPE 279	AJ560668.1
ICIPE 695														
KF746155.1	0.007													
ICIPE 693	0.314	0.321												
JQ999974.1	0.321	0.328	0.005											
ICIPE 694	0.292	0.298	0.196	0.202										
FJ609312.1	0.292	0.298	0.196	0.202	0.000									
ICIPE 30	0.292	0.298	0.196	0.202	0.000	0.000								
ICIPE 696	0.292	0.298	0.196	0.202	0.000	0.000	0.000							
S4ST7	0.292	0.298	0.196	0.202	0.000	0.000	0.000	0.000						
FJ545302.1	0.292	0.298	0.196	0.202	0.000	0.000	0.000	0.000	0.000					
ICIPE 692	0.587	0.582	0.576	0.581	0.553	0.553	0.553	0.553	0.553	0.553				
GU966518.1	0.586	0.581	0.586	0.591	0.558	0.558	0.558	0.558	0.558	0.558	0.005			
ICIPE 279	1.565	1.593	1.409	1.417	1.476	1.476	1.476	1.476	1.476	1.476	1.952	1.923		
AJ560668.1	1.514	1.539	1.367	1.374	1.455	1.455	1.455	1.455	1.455	1.455	1.890	1.864	0.011	

Phylogenetic tree using the TW81 and AB28 gene region: The evolutionary history was inferred using the Neighbor-Joining method[61]. The optimal tree with the sum of branch length = 2.17571022 is shown in Figure 5. The percentage of replicate trees in which the associated taxa clustered together in the bootstrap test (500 replicates) is shown next to the branches[65]. The tree is drawn to scale, with branch lengths in the same units as those of the evolutionary distances used to infer the phylogenetic tree (Figure 5). The analysis involved 14 nucleotide sequences. Codon positions included were 1st+2nd+3rd+noncoding.

All positions containing gaps and missing data were eliminated. There were a total of 444 positions in the final dataset.

This analysis also clustered the *Metarhizium, Phanerochaete, Phialemonium* and *Beauveria* isolates into four distinct groups (Figure 5). The first cluster consisted of all the *M. anisopliae* isolates included in the study. The ICIPE 696 and ICIPE 694 isolates branched separately from the standards (ICIPE 30 and S4ST7). The second group consisted of *B. bassiana* isolates ICIPE 693 likewise branching separately from the standard ICIPE 279. This is a clear indication that the samples are different from the standards. The last two clusters consisted of the *Phialemonium* sp., ICIPE 695 and *P. chrysosporium* ICIPE 692, respectively (Figure 5).

The **GENETIC DISTANCE**s between the isolates were also inferred as for the other primer region and were used for **PRINCIPAL COMPONENT ANALYSIS**. The estimates of evolutionary divergence between sequences of the various endophyte isolates when using this primer region ranged between 0.0 and 1.952 (Table 4). Comparison of *M. anisopliae* isolates (ICIPE 694 and ICIPE 696) isolated from *V. faba* and *P. vulgaris* with the ICIPE standards, gave a square distance of 0.0, showing that the two are genetically closely related. However, comparison of *B. bassiana* isolate ICIPE 693 isolated from *V. faba* with the standard ICIPE 279, gave a square distance of 1.409 (Table 4), meaning that the 2 isolates are far apart. In the PCA plot, the first two axes explained 82.48% of the variation (the 1st axis 61.48% and the second axis 20.99%) (Figure 6). The PCA separated the 9 isolates into 4 discrete clusters as observed with the ITS 5 and 4 gene region. Each cluster was occupied by isolates belonging to the 4 different genera i.e., a cluster consisting of the *Metarhizium* isolates, a cluster consisting of *Beauveria, Phialemonium* and *Phanerochaete,* respectively (Figure 6).

DISCUSSION

Early approaches to fungal taxonomy or identification relied on morphological characterization at the macro and microscopic levels. The presence or absence of microconidia and macroconidia, shapes and sizes of conidia, type of phialides bearing microconidia, presence or absence of chlamydospores, microconidial chains and sporodochia, as well as the type of fungal mycelia as described by Burgess *et al.*[49], Humber[50] and Goettel and Inglis[51] have contributed to the morphological identification of especially the *M. anisopliae* and *B. bassiana* isolates. The variability of colony appearance, pigmentation, growth rate, length of chains, production of bluish sclerotia and concentric ring aerial mycelium[52,53] were other additional features which also assisted in the morphological identification of the isolated endophytes.

The key features used to identify *M. anisopliae* isolates (ICIPE 30, S4ST7, ICIPE 694 and ICIPE 696) are the conidia formed in short to long chains, short conidiogenous cells and with rounded to broad conical apices. Johnston[66] also proposed the morphological classification of *M. anisopliae* into long and short-spored forms. These features are also in line with the morphological idendification keys of Humber[50].

In *B. bassiana* isolates (ICIPE 279 and ICIPE 693), the conidia and conidiogenous shapes as well as the pigmentation (white in mass) corroborate with the features used by Humber[50] to describe *B. bassiana*. According to Thomas *et al.*[67], Viaud *et al.*[68], Alves *et al.*[69]and Kirkland *et al.*[70], the entomopathogen *B. bassiana* produces three distinct features ***IN VITRO***, single cell propagules: Aerial conidia, blastospores and submerged conidia. Different infectious *B. bassiana* propagules can be isolated and selected for host targeting. Thus, in addition to mycelial and hyphal growth, *B. bassiana* produces a number of mono-nucleated single cell types, including aerial conidia, blastospores and submerged conidia[71,72]. These cells display distinct morphological, biochemical and pathological characteristics and attempts are being made to exploit these properties in pest targeting and/or the enhancement of virulence.

Phialemonium sp., isolate ICIPE 695 was mainly characterized by its abundance of adelophialides and few discrete phialides with no signs of collarettes as well as its grayish white to brownish colonies pigmentation. These morphological features were also used by Gams and McGinnis[73] when describing *P. curvatum*. Perdomo *et al.*[74] when describing *Phialemonium* species, focused on key morphological features such as smooth or finely floccose, white colonies, becoming brown at maturity, short and cylindrical adelophialides and less commonly, long, tapering discrete phialides. Inconspicuous collarettes could also be observed on types of phialides, hyaline and cylindrical to curved conidia[74].

The white rot fungus *Phanerochaete chrysosporium* isolate ICIPE 692 was identified morphologically based on the white floccose colonies, hyaline mycelia and the three types of conidia (aleuriospores, arthrospores and chlamydospores). These features observed in *P. chrysosporium*, were also described by Burdsall and Eslyn[75] and their conidial productions were abundant[76].

However, identification using morphological (macro and micro) and physiological (growth rates, media) characters alone of fungi were not only frequently criticized among mycologists but also are time-consuming and sometimes hard to interpret due to ambiguous responses by some isolates in the media tested and some unclear expressed features[77]. Morphological identification was then combined with molecular characterization approach to confirm the identities of the seed-born fungal endophytes isolated in this study.

Results from the amplification of the rDNA region using the two sets of primers in this study confirmed the morphological identification of the isolates which belong to the genera *Metarhizium*, *Beauveria*, *Phialemonium* and *Phanerochaete*. They also linked the standards (ICIPE 30 and S4ST7) to the *M. anisopliae* species while isolate ICIPE 693 to *Beauveria* spp. Similar to the identity of *M. anisopliae* standard (S4ST7) using both primers, a study by Akello[78] reported the isolate S4ST7 as a *M. anisopliae* species using molecular characterization with the following primers: IGS [PNFo (CCCGCCTGGCTGCGTCCGACTC) and PN22 (CAAGCATATGACTACTGGC)] and ITS [ITS1 (TCCGTAGGTGA ACCTGCGG) and ITS4 (TCCTCCGCTTATTGATATGC)]. These results confirm the identity of our standard using different primers sets. In addition, Fernandes *et al.*[79] also used molecular-based techniques [AFLP and rDNA (ITS1, ITS2 and 5.8S) gene sequencing] to characterize morphologically identified *Metarhizium* spp., isolates from a wide range of sources.

BEAUVERIA BASSIANA ICIPE 279 was reported earlier to be endophytic in *V. faba* and *P. vulgaris* by Akutse *et al.*[31]. Since, *B. bassianna* ICIPE 693, isolated from *V. faba* seeds, belongs to the same cluster as ICIPE 279, it may also be endophytic, not just in seeds of *V. faba* but also in the plants through vertical transmission or when artificially inoculated. ***BEAUVERIA BASSIANA*** (Ascomycota: Hypocreales) has previously been reported as an endophyte in seeds and needles of *Pinus monticola* Dougl. ex. D. Don[80]. A number of studies have also reported on the presence of endophytes in seeds[81,82].

Rivero *et al.*[83] and Rao *et al.*[84] have also used ITS1 and ITS4 primers for molecular identification of *Phialemonium curvatum*. A species of *Phialemonium* (KF746155.1) has been reported as an endophyte[58] and having potential bioactivity as both anti-parasitic and anti-bacterial[85,86]. This is widely distributed in the environment, having been isolated from air, soil, industrial water and sewage[73]. Similar to the molecular characterization in this study, Lim *et al.*[76] have also identified 2 species of *Phanerochaete*, *P. chrysomonium* and *P. sordida* using ITS gene regions. *Phanerochaete chrysosporium* is known to be saprophytic capable of organic breakdown. Although, dying and dead plants serve as optimal substrate for *P. chrysosporium*, it has also been isolated from soil samples of petroleum refinery and was used for degradation of 5 polycyclic aromatic hydrocarbons (PAHs: Acenaphthene, anthracene, phenanthrene, fluoranthene and pyrene), simultaneously and individually in sterile and unsterile soil[87]. *Phanerochaete chrysosporium* has also been reported to be thermo-tolerant[88]. The white rot fungus *P. chrysosporium* is the only microbe capable of efficient depolymerization and mineralization of lignin. Furthermore, *P. chrysosporium* also has several features that might make it very useful. For instance, unlike some white rot fungi, it leaves the cellulose of the wood virtually untouched and it has a very high optimum temperature (about 40°C), which means it can grow on wood chips in compost piles, which attain very **HIGH TEMPERATURE**s. These characteristics point to some possible or potential roles of this fungus in biotechnology. The endophytic fungi isolated from *V. faba* and *P. vulgaris* seeds, may occur and propagate/accumulate naturally inside seeds under farming conditions or during seeds processing.

Although, plants are commonly colonized by a diverse array of endophytic organisms, it was previously reported that hardly any information exists on interactions between endophytic fungi and other ecological groups of fungi co-occurring in the same plant tissues/seeds[89]. It can therefore be expected that antagonistic interactions between these fungi are the rule rather than the exception. When entomopathogenic endophytic fungi are starting to grow systemically from the point of inoculation (from seeds for example) to other plant parts, they inevitably have to confront other fungi already established. Since, seeds are likely to already harbor endophytic fungi at the time of artificial inoculation, thus adding a component of interspecific interaction to the already complex set-up become important to clarify. In line with this hypothesis, a recent paper by Yan *et al.*[90] reported almost no systemic growth of endophytic fungi in *Silene dioica* (L.) Clairv., a non-mycorrhized forb, because most of the fungi were starting to exhibit antagonistic interactions when plated together in a kind of competitive setting on a growth medium. Vidal and Jaber[91] also underlined the possible interaction effects of the existing endophytes and the inoculated ones where they reported that these microorganisms may impact the effect of the endophytic entomopathogenic fungi on the herbivorous insects.

Therefore, the results of this study would create awareness on seed endophytes and facilitate/stimulate studies on the role of these fungi in inhibition/suppression of the growth of artificially inoculated endophytes used for pest control in general and for *Liriomyza* leafminer species in particular, explore the additive, symbiotic or synergic effects of these seeds endophytes in the management of *Liriomyza* leafminers and other pests. Regarding the exploration of possible inhibition of the growth of the artificially inoculated fungi, Kerr[92] has reported similar results where ***PSEUDOMONAS AERUGINOSA*** was found to significantly suppress or inhibit the growth of *Candida albicans* and other eleven strains of fungi ***IN***

VITRO. The presence of these endophytes in *V. faba* and *P. vulgaris* seeds shows a high chance that the seedlings of these seeds may be colonized by the same fungal endophytes either naturally or artificially through inoculation. ***BEAUVERIA BASSIANA*** and *M. anisopliae* are two key entomopathogenic fungi species isolated from these seeds, suggesting their potential use in biological control of arthropod pests. Entomopathogenic fungi are generally applied in inundative approach in the crops[93]. However, other strategies are currently being considered and include auto-dissemination[94] and endophytic colonization[30]. Fungal pathogens can endophytically colonize host plants and confer resistance against insect pests[30]. The pathogenicity and virulence effects of the isolated endophytes can also be assessed on different sucking arthropods ***IN VITRO*** and *in vivo* especially for the entomopathogenic fungi *M. anisopliae* and *B. bassiana* isolates. For instance, endophytic *B. bassiana* in banana significantly reduced larval survivorship of banana weevil, *Cosmopolites sordidus* (Coleoptera: Curculionidae), resulting in 42-87% reduction in plant damage[43,44]. Reduction in feeding and reproduction by *Aphis gossypii* (Hemiptera: Aphididae) has also been reported on cotton endophytically colonized by *B. bassiana*[34]. Vega *et al.*[29] reported similar effects of an endophytic *B. bassiana* SPCL 03047 on adult coffee berry borers, *Hypothenemus hampei* (Ferrari) (Coleoptera: Scolytidae) with an average survival period of 4.8±0.2 days compared to 15.0±0.6 days for the un-inoculated plants. The pathogenicity and virulence of fungal isolates against *Liriomyza* leafminer adults was also reported by Migiro *et al.*[95] using *M. anisopliae* through auto-dissemination approach. Akutse *et al.*[31] also reported the effects of fungal endophytes on life-history parameters of *Liriomyza huidobrensis* (Diptera: Agromyzidae) leafminers and their associated parasitoids[47]. The researchers reported that endophytic fungal isolates of *B. bassiana* (G1LU3, S4SU1 and ICIPE 279) could endophytically colonize *V. faba* and *P. vulgaris* plants through seeds inoculation and cause detrimental effects on survival, fecundity, oviposition, emergence and longevity of *L. huidobrensis*. However, no significant detrimental effects were observed on the development, survival and parasitism ability of the associated parasitoids *Diglyphus isaea* Walker (Hymenoptera: Eulophidae) and *Phaedrotoma scabriventris* Nixon (Hymenoptera: Braconidae)[47].

Assessment of the quality and the **NUTRITIONAL VALUE** of the seeds colonized by the endophytes can also be undertaken not only having reference to plant pathogens[18-22], but also when considering the possible presence of the entomopathogenic fungi in the seeds. For example, Van Du *et al.*[96] reported that some fungi species, such as *Fusarium graminum*, *Cephalosporium oryzae*, *Alternaria padwickii*, *Fusarium moniliforme*, *Fusarium pallidoroseum*, *Fusarium subglutinans*, *Phoma* sp. and *Sarocladium oryzae* found in rice seeds, affect the quality of the seeds through their effects on grain discoloration, milling recovery, cooking and palatability as compared to healthy seeds. In addition, McGee and Nyvall[97] reported that more than 30 fungi are seed-born on soybeans. Although, some cause quality problems[98] and reduce seed viability, most fungi that are associated with soyabean seeds are not known to cause diseases. Furthermore, the effects of seeds derived from endophytically inoculated plants on stored products insects can also be investigated.

CONCLUSION

BEAUVERIA BASSIANA isolate ICIPE 693, *M. anisopliae* isolates ICIPE 694, ICIPE 696, *Phanerochaete chrysosporium* isolate ICIPE 692 and *Phialemonium* sp., isolate ICIPE 695 were isolated as seed fungal endophytes. These isolates were clearly different from the standards *B. bassiana* ICIPE 279 and *M. anisopliae* ICIPE 30 and S4ST7 despite clustering into the same groups. Results of the study showed that bean seeds do not only bear in their inside plant pathogens but also entomopathogenic fungi. The outcome of this study will create awareness on seed endophytes and stimulate further studies on the role of these fungi in (1) Inhibiting or suppressing the growth of artificially inoculated endophytes used for pest control in general and *Liriomyza* leafminer species in particular, (2) Assessing the pathogenicity and virulence effects of these endophytes on different sucking arthropods, (3) Evaluating the **NUTRITIONAL VALUE** of the seeds containing these endophytes and finally (4) Assessing the effect on stored products insects when storing seeds derived from endophytically inoculated plants. Furthermore, since other fungi may block the systemic growth and colonization of the artificially inoculated fungi in the host plant parts distant to the point of inoculation, then there is need to know the already existing species of fungi in the seeds and study their interactions with the inoculated ones, not only for seedling growth promotion, but also for better biological pest management strategies development using endophytic fungi. And this study established some of the key entomopathogenic fungal endophytes that naturally exist in the bean seeds. Knowing that bean seeds contain already these fungi, the results of this study will also help to develop specific markers to trace the fate of the artificially inoculated endophytes when released in the ecosystem as biocontrol agents.

ACKNOWLEDGMENTS

The study was conducted with financial support from the Federal Ministry for Economic Cooperation and Development, Germany (BMZ) through the German Federal Enterprise for International Cooperation/Deutsche Gesellschaft für Internationale Zusammenarbeit (GIZ) (Grant No. 09.7860.1-001.00, Contract No. 81121261). The first author received a scholarship through a PhD fellowship provided by the German Academic Exchange Service/Deutcher Akademischer Austausch Dienst (DAAD) through the African Regional Postgraduate Programme in Insect Science (ARPPIS) at ICIPE. The authors are grateful to Dr. D. Masiga (ICIPE) for allowing use of the Molecular Biology and Bioinformatics laboratory facilities, to J. G. Kabii, E. Ouna and S. Wafula for their technical assistance.

REFERENCES

1. Akello, J., T. Dubois, D. Coyne and S. Kyamanywa, 2008. Endophytic Beauveria bassiana in banana (Musa spp.) reduces banana weevil (Cosmopolites sordidus) fitness and damage. Crop. Prot., 27: 1437-1441.
2. Akello, J., T. Dubois, D. Coyne and S. Kyamanywa, 2008. Effect of endophytic Beauveria bassiana on populations of the banana weevil, Cosmopolites sordidus and their damage in tissue-cultured banana plants. Entomologia Experimentalis Applicata, 129: 157-165.
3. Akello, J.T., 2012. Biodiversity of fungal endophytes associated with maize, sorghum and Napier grass and their influence of biopriming on the resistance of leaf mining, stem boring and sap sucking insect pests. Ph.D. Thesis, University of Bonn, Germany.
4. Akello, J.T., T. Dubois, C.S. Gold, D. Coyne, J. Nakavuma and P. Paparu, 2007. Beauveria bassiana (Balsamo) Vuillemin as an endophyte in tissue culture banana (Musa spp.). J. Invertebr. Pathol., 96: 34-42.
5. Akutse, K.S., K.K. Fiaboe, J. Van den Berg, S. Ekesi and N.K. Maniania, 2014. Effects of endophyte colonization of Vicia faba (fabaceae) plants on the life: History of leafminer parasitoids Phaedrotoma scabriventris (hymenoptera: braconidae) and Diglyphus isaea (hymenoptera: eulophidae). Plos One, Vol. 9. 10.1371/journal .pone.0109965
6. Akutse, K.S., N.K. Maniania, K.K.M. Fiaboe, J. Van den Berg and S. Ekesi, 2013. Endophytic colonization of Vicia faba and Phaseolus vulgaris (fabaceae) by fungal pathogens and their effects on the life-history parameters of Liriomyza huidobrensis (diptera: agromyzidae). Fungal Ecol., 6: 293-301.
7. Al-Amod, M.O., 2015. Seed-borne fungi of some peanut varieties from Hadhramout and Abyan Governorates in Yemen. J. Agric. Tech., 11: 1359-1370.
8. Aldrich-Markham, S., G. Pirelli, and A.M. Craig, 2007. Endophyte toxins in grass seed fields and straw effects on livestock. EM8598-E, Oregon State University Extention Service. https://ir.library.oregonstate.edu/xmlui/ bitstream/handle/1957 /20568/em8598-e.pdf.
9. Alves, S.B., L.S. Rossi, R.B. Lopes, M.A. Tamai and R.M. Pereira, 2002. Beauveria bassiana yeast phase on agar medium and its pathogenicity against Diatraea saccharalis (Lepidoptera: Crambidae) and Tetranychus urticae (Acari: Tetranychidae). J. Invertebrate Pathol., 81: 70-77.

10. Anjorin, S.T. and M. Mohammed, 2014. Effect of seed-borne fungi on germination and seedling vigour of watermelon (Citrullus lanatus thumb). Afr. J. Plant Sci., 8: 232-236.
11. Arnold, A.E. and F. Lutzoni, 2007. Diversity and host range of foliar fungal endophytes: Are tropical leaves biodiversity hotspots? Ecology, 83: 541-549.
12. Arnold, A.E. and L.C. Lewis, 2005. Ecology and Evolution of Fungal Endophytes and their Roles Against Insects. In: Insect-Fungal Associations: Ecology and Evolution, Vega, F.E. and M. Blackwell (Eds.). Oxford University Press, New York, pp: 74-96.
13. Arnold, A.E., L.C. Mejia, D. Kyllo, E.I. Rojas, Z. Maynard, N. Robbins and E.A. Herre, 2003. Fungal endophytes limit pathogen damage in a tropical tree. Proc. Natl. Acad. Sci. USA., 100: 15649-15654.
14. Azevedo, J.L., W. Maccheroni Jr., J.O. Pereira and W.L. de Araujo, 2000. Endophytic microorganisms: A review on insect control and recent advances on tropical plants. Electron. J. Biotechnol., 3: 40-65.
15. Backman, P.A. and R.A. Sikora, 2008. Endophytes: An emerging tool for biological control. Biol. Control, 46: 1-3.
16. Baker, K.F. and S.H. Smith, 1966. Dynamics of seed transmission of plant pathogens. Annu. Rev. Phytopathol., 14: 311-332.
17. Bishnoi, K., R. Kumar and N.R. Bishnoi, 2008. Biodegradation of polycyclic aromatic hydrocarbons by white rot fungi Phanerochaete chrysosporium in sterile and unsterile soil. J. Scient. Ind. Res., 67: 538-542.
18. Boucias, D.G., J.C. Pendland and J.P. Latge, 1988. Nonspecific factors involved in attachment of entomopathogenic Deuteromycetes to host insect cuticle. Applied Environ. Microbiol., 54: 1795-1805.
19. Burdsall, H.H. and W.E. Eslyn, 1974. A new Phanerochaete with a Chrysosporium imperfect state. Mycotaxon, 1: 123-133.
20. Burgess, L.W., B.A. Summerell, S. Bullock, K.P. Gott and D. Backhouse, 1994. Laboratory Manual for Fusarium Research. 3rd Edn., Fusarium Research Laboratory, University of Sydney and Royal Botanic Gardens, Sydney, Australia, ISBN-13: 9780867588491, Pages: 133.
21. Cao, L.X., J.L. You and S.N. Zhou, 2002. Endophytic fungi from Musa acuminata leaves and roots in South China. World J. Microbiol. Biotechnol., 18: 169-171.
22. Chabi-Olaye, A., N. Mujica, B. Lohr and J. Kroschel, 2008. Role of agroecosystems in the abundance and diversity of Liriomyza leafmining flies and their natural enemies. Proceedings of the Abstracts of the 23rd International congress of Entomology, July 6-12, 2008, Durban, South Africa -.
23. Chongsuvivatwong, V., 2012. Epicalc: Epidemiological calculator. R Package Version 2.14.1.6, Prince of Songkla University, Thailand.
24. Clement, S.L., L.R. Elberson, N.A. Bosque-Perez and D.J. Schotzko, 2005. Detrimental and neutral effects of wild barley-Neotyphodium fungal endophyte associations on insect survival. Entomol. Exp. Appl., 114: 119-125.
25. Clement, S.L., W.J. Kaiser and H. Eichenseer, 1994. Acremonium Endophytes in Germplasms of Major Grasses and Their Utilization for Insect Resistance. In: Biotechnology of Endophytic Fungi of Grasses, Bacon, C.W. and J.F. White Jr. (Eds.). CRC Press, Boca Raton, FL., USA., pp: 185-199.
26. Curran, J., F. Driver, J.W.O. Ballard and R.J. Milner, 1994. Phylogeny of Metarhizium: analysis of ribosomal DNA sequence data. Mycol. Res., 98: 547-552.
27. Dalling, J.W., A.S. Davis, B.J. Schutte and A.E. Arnold, 2011. Seed survival in soil: Interacting effects of predation, dormancy and the soil microbial community. J. Ecol., 99: 89-95.
28. El-Wakil, D.A., 2014. Seed-borne fungi of sunflower (Helianthus annuus L.) and their impact on oil quality. IOSR J. Agric. Vet. Sci., 6: 38-44.
29. FPEAK., 2012. An overview of the Kenya Horticulture Industry. Kenya Horticultural Council (KHC), Fresh Produce Exporters Association of Kenya, http://fpeak.org/index.php/kenya-horticulture-council-k-h-c/.

30. Felsenstein, J., 1985. Confidence limits on phylogenies: An approach using the bootstrap. Evolution, 39: 783-791.
31. Fernandes, E.K.K., C.A. Keyser, J.P. Chong, D.E.N. Rangel, M.P. Miller and D.W. Roberts, 2010. Characterization of Metarhizium species and varieties based on molecular analysis, heat tolerance and cold activity. J. Applied Microbiol., 108: 115-128.
32. Ferreira, A., M.C. Quecine, P.T. Lacava, S. Oda, J.L. Azevedo and W.L. Araujo, 2008. Diversity of endophytic bacteria from Eucalyptus species seeds and colonization of seedlings by Pantoea agglomerans. FEMS. Microbiol. Lett., 287: 8-14.
33. Foschino, R., S. Gallina, C. Andrighetto, L. Rossetti and A. Galli, 2004. Comparison of cultural methods for the identification and molecular investigation of yeasts from sourdoughs for Italian sweet baked products. FEMS Yeast Res., 4: 609-618.
34. Fraedrich, S.W. and M.M. Cram, 2011. Seed fungi. Conifer and Hardwood Diseases, Forest Nursery Pests, pp: 132-134. http://www.rngr.net/publications/forest-nursery-pests/conifer-and-hardwood-diseases/seed-fungi/at_download/file.
35. Gams, W. and M.R. McGinnis, 1983. Phialemonium, a new anamorph genus intermediate between Phialophora and Acremonium. Mycologia, 75: 977-987.
36. Ganley, R.J. and G. Newcombe, 2005. Fungal endophytes in seeds and needles of Pinus monticola. Mycol. Res., 110: 318-327.
37. Ghoneem, K.M., S.M. Ezzat and N.M. El-Dadamony, 2014. Seed-borne fungi of sunflower in Egypt with reference to pathogenic effects and their transmission. Plant Pathol. J., 13: 278-284.
38. Gitonga, Z.M., A. Chabi-Olaye, D. Mithofer, J.J. Okello and C.N. Ritho, 2010. Control of Invasive Liriomyza leafminer species and compliance with food safety standards by small scale snow pea farmers in Kenya. J. Crop Prot., 29: 1472-1477.
39. Goettel, M.S. and G.D. Inglis, 1997. Fungi: Hyphomycetes. In: Manual of Techniques in Insect Pathology, Lacey, L.A. (Ed.). Academic Press, New York, San Diego, USA., pp: 213-249.
40. Gouda, S., G. Das, S.K. Sen, H.S. Shin and J.K. Patra, 2016. Endophytes: A treasure house of bioactive compounds of medicinal importance. Front. Microbiol., Vol. 7. 10.3389/fmicb.2016.01538
41. Gurulingappa, P., G.A. Sword, G. Murdoch and P.A. McGee, 2010. Colonization of crop plants by fungal entomopathogens and their effects on two insect pests when in planta. Biol. Control, 55: 34-41.
42. Harish, S., M. Kavino, N. Kumar, D. Saravanakumar, K. Soorianasundaram and R. Samiyappan, 2008. Biohardening with plant growth promoting rhizosphere and endophytic bacteria induces systemic resistance against Banana bunchy top virus. Applied Soil Ecol., 39: 187-200.
43. Higginbotham, S., W.R. Wong, R.G. Linington, C. Spadafora, L. Iturrado and A.E. Arnold, 2014. Sloth hair as a novel source of fungi with potent anti-parasitic, anti-cancer and anti-bacterial bioactivity. PloS ONE., Vol. 9. 10.1371/journal.pone.0084549
44. Higginbotham, S.J., 2013. Direct Submission (22-OCT-2013). Plant Sciences, University of Arizona, 1140 E South Campus Drive, Forbes 303, Tucson, AZ 85721, USA.
45. Holder, D.J. and N.O. Keyhani, 2005. Adhesion of the entomopathogenic fungus Beauveria (Cordyceps) bassiana to substrata. Applied Environ. Microbiol., 71: 5260-5266.
46. Humber, R.A., 1997. Fungi: Identification. In: Manual of Techniques in Insect Pathology, Lacey, L.A. (Ed.). Academic Press, New York, pp: 153-185.
47. Ingram, D.S., 2002. The Diversity of Plant Pathogens and Conservation: Bacteria and Fungi Sensu Lato. In: Microorganisms in Plant Conservation and Biodiversity, Sivasithamparam, K., K.W. Dixon and R.L. Barret (Eds.). Chapter 9, Springer, Netherlands, ISBN: 978-1-4020-0780-4, pp: 241-267.

48. Istifadah, N. and P.A. McGee, 2006. Endophytic Chaetomium globosum reduces development of tan spot in wheat caused by Pyrenophora tritici-repentis. Aust. Plant Path., 35: 411-418.
49. Johnston, J.R., 1915. The entomogenous fungi of Porto Rico. Bulletin No. 10, Government of Porto Rico, Board of Commissioners of Agriculture, Rio Pedras.
50. Johnston-Monje, D. and M.N. Raizada, 2011. Conservation and diversity of seed associated endophytes in Zea across boundaries of evolution, ethnography and ecology. PLoS ONE, Vol. 6. 10.1371/journal.pone.0020396
51. Kaga, H., H. Mano, F. Tanaka, A. Watanabe, S. Kaneko and H. Morisaki, 2009. Rice seeds as sources of endophytic bacteria. Microbes Environ., 24: 154-162.
52. Kerr, J.R., 1994. Suppression of fungal growth exhibited by Pseudomonas aeruginosa. J. Clin. Microbiol., 32: 525-527.
53. Kimura, M., 1980. A simple method for estimating evolutionary rates of base substitutions through comparative studies of nucleotide sequences. J. Mol. Evol., 16: 111-120.
54. Kirkland, B.H., E.M. Cho and N.O. Keyhani, 2004. Differential susceptibility of Amblyomma maculatum and Amblyomma americanum (Acari:Ixodidea) to the entomopathogenic fungi Beauveria bassiana and Metarhizium anisopliae. Biol. Control, 31: 414-421.
55. Kornerup, A. and J.H. Wanscher, 1978. Methuen Hanbook of Colour. 3rd Edn., Eyre Methuen, London, UK., ISBN-13: 978-0413334008, Pages: 252.
56. Lacey, L.A. and M.S. Goettel, 1995. Current developments in microbial control of insect pests and prospects for the early 21st century. Entomophaga, 40: 3-27.
57. Larran, S., A. Perello, M.R. Simon and V. Moreno, 2002. Isolation and analysis of endophytic microorganisms in wheat (Triticum aestivum L.) leaves. World J. Microbiol. Biotechnol., 18: 683-686.
58. Larran, S., C. Monaco and H.E. Alippi, 2001. Endophytic fungi in leaves of Lycopersicon esculentum mill. World J. Microbiol. Biotechnol., 17: 181-184.
59. Larran, S., C. Rollan, H.B. Angeles, H.E. Alippi and M.I. Urrutia, 2002. Endophytic fungi in healthy soybean leaves. Invest. Agric. Prod. Prot. Veg., 17: 173-177.
60. Leslie, J.F. and B.A. Summerell, 2006. The Fusarium Laboratory Manual. Blackwell Publish Ltd., Iowa, USA., Pages: 388.
61. Liang, Y. and D.C. Li, 2010. Taxonomic studies and molecular phylogeny of thermotolerant fungi. Department of Plant Pathology, Shandong Agricultural University, Tai'an, China.
62. Lim, Y.W., K.S. Baik, J. Chun, K.H. Lee, W.J. Jung and K.S. Bae, 2007. Accurate delimitation of Phanerochaete chrysosporium and Phanerochaete sordida by specific PCR primers and cultural approach. J. Microbiol. Biotechnol., 17: 468-473.
63. Mano, H., F. Tanaka, A. Watanabe, H. Kaga, S. Okunishi and H. Morisaki, 2006. Culturable surface and endophytic bacterial flora of the maturing seeds of rice plants (Oryza sativa) cultivated in a paddy field. Microbes Environ., 21: 86-100.
64. McCulloch, N. and M. Ota, 2002. Export horticulture and poverty in Kenya. Working Paper 174, Institute of Development Studies, Sussex, UK. http://www.eldis.org/fulltext/mcculloch_neil_export_horticulture_2002.pdf.
65. McGee, D.C. and R.F. Nyvall, 1994. Soyabean seed health. Coorative Extension Service, Iowa State University of Science and Technology and the United States Department of Agriculture Cooparating. Distributed in Furtherance of the Acts of Congress of May 8 and June 30, 1914.
66. Medic-Pap, S., M. Milosevic and S. Jasnic, 2007. Soybean seed-borne fungi in the Vojvodina province. Phytopathol. Polonica, 45: 55-65.
67. Migiro, L.N., K.N. Maniania, A. Chabi-Olaye and J. Vandenberg, 2010. Pathogenicity of entomopathogenic fungi Metarhizium anisopliae and Beauveria bassiana (Hypocreales: Clavicipitaceae) isolates to the adult pea leafminer (Diptera: Agromyzidae) and prospects of an autoinoculation device for infection in the field. Environ. Entomol., 39: 468-475.

68. Nei, M. and S. Kumar, 2000. Molecular Evolution and Phylogenetics. Oxford University Press, Oxford, UK., ISBN-13: 9780195135855, Pages: 333.
69. Ownley, B.H., K.D. Gwinn and F.E. Vega, 2010. Endophytic fungal entomopathogens with activity against plant pathogens: Ecology and evolution. BioControl, 55: 113-128.
70. Ownley, B.H., R.M. Pereira, W.E. Klingeman III, N.B. Quigley and B.M. Leckie, 2004. Beauveria bassiana, A Dual Purpose Biological Control with Activity against Insect Pests and Plant Pathogens. In: Emerging Concepts in Plant Health Management, Lartey, R.T. and A.J. Cesar (Eds.). Research Signpost, India, ISBN-13: 9788177362275, pp: 255-269.
71. Paparu, P., T. Dubois, D. Coyne and A. Viljoen, 2009. Dual inoculation of Fusarium oxysporum endophytes in banana: Effect on plant colonization, growth and control of the root burrowing nematode and the banana weevil. Biocontrol Sci. Technol., 19: 639-655.
72. Peakall, R. and P.E. Smouse, 2006. genalex 6: Genetic analysis in Excel. Population genetic software for teaching and research. Mol. Ecol. Notes, 6: 288-295.
73. Perdomo, H., D.A. Sutton, D. Garcia, A.W. Fothergill and J. Gene et al., 2011. Molecular and phenotypic characterization of Phialemonium and Lecythophora isolates from clinical samples. J. Clin. Microbiol., 49: 1209-1216.
74. Petrini, O. and P.J. Fisher, 1986. Fungal endophytes in Salicornia perennis. Trans. Br. Mycol. Soc., 87: 647-651.
75. Pocasangre, L., R.A. Sikora, V. Vilich and R.P. Schuster, 2000. Survey of banana endophytic fungi from central america and screening for biological control of the burrowing nematode (Radopholus similis). InfoMusa, 9: 3-5.
76. R Development Core Team, 2011. R: A Language and Environment for Statistical Computing. R Foundation for Statistical Computing, Vienna, Austria.
77. Rao, C.Y., C. Pachucki, S. Cali, M. Santhiraj and K.L.K. Krankoski et al., 2009. Contaminated product water as the source of Phialemonium curvatum bloodstream infection among patients undergoing hemodialysis. Inform. Control Hosp. Epidemiol., 30: 840-847.
78. Rijavec, T., A. Lapanje, M. Dermastia and M. Rupnik, 2007. Isolation of bacterial endophytes from germinated maize kernels. Can. J. Microbiol., 53: 802-808.
79. Rivero, M., A. Hidalgo, A. Alastruey-Izquierdo, M. Cia, L. Torroba and J.L. Rodriguez-Tudela, 2009. Infections due to Phialemonium species: Case report and review. Med. Mycol., 47: 766-774.
80. Rudgers, J.A., J. Holah, S.P. Orr and K. Clay, 2007. Forest succession suppressed by an introduced plant-fungal symbiosis. Ecology, 88: 18-25.
81. Saitou, N. and M. Nei, 1987. The neighbor-joining method: A new method for reconstructing phylogenetic trees. Mol. Biol. Evol., 4: 406-425.
82. Schardl, C.L., A. Leuchtmann and M.J. Spiering, 2004. Symbioses of grasses with seedborne fungal endophytes. Annu. Rev. Plant Biol., 55: 315-340.
83. Schulz, B. and C. Boyle, 2005. The endophytic continuum. Mycol. Res., 109: 661-686.
84. Sinha, P. and A.K. Prakash, 2014. Seed-borne storage fungi of cowpea: Pathogenic to the seedlings. Proceedings of the 3rd World Conference on Applied Sciences, Engineering and Technology, September 27-29, 2014, Basha Research Centre, Kathmandu, Nepal -.
85. Suryanarayanan, T.S., 2013. Endophyte research: Going beyond isolation and metabolite documentation. Fungal Ecol., 6: 561-568.
86. Tamura, K., G. Stecher, D. Peterson, A. Filipski and S. Kumar, 2013. MEGA6: Molecular evolutionary genetics analysis version 6.0. Mol. Biol. Evol., 30: 2725-2729.
87. Thomas, K.C., G.G. Khachatourians and W.M. Ingledew, 1987. Production and properties of Beauveria bassiana conidia cultivated in submerged culture. Can. J. Microbiol., 33: 12-20.
88. Thompson, J.D., T.J. Gibson, F. Plewniak, F. Jeanmougin and D.G. Higgins, 1997. The CLUSTAL_X windows interface: Flexible strategies for multiple sequence alignment aided by quality analysis tools. Nucleic Acids Res., 25: 4876-4882.

89. Van Du, P., L.C. Loan, N.D. Cuong, H. Van Nghiep and N.D. Thach, 2001. Survey on seed borne fungi and its effects on grain quality of common rice cultivars in the Mekong Delta. Omonrice, 9: 107-113.

90. Vega, F.E., 2008. Insect pathology and fungal endophytes. J. Invertebr. Pathol., 98: 277-279.

91. Vega, F.E., F. Posada, M.C. Aime, M. Pava-Ripoll, F. Infante and S.A. Rehner, 2008. Entomopathogenic fungal endophytes. Biol. Control, 46: 72-82.

92. Vega, F.E., M.S. Goettel, M. Blackwell, D. Chandler and M.A. Jackson et al., 2009. Fungal entomopathogens: New insights on their ecology. Fungal Ecol., 2: 149-159.

93. Vega, F.E., P.F. Dowd, L.A. Lacey, J.K. Pell, D.M. Jackson and M.G. Klein, 2007. Dissemination of Beneficial Microbial Agents by Insects. In: Field Manual of Techniques in Invertebrate Pathology: Application and Evaluation of Pathogens for Control of Insects and Other Invertebrate Pests, Lacey, L.A. and H.K. Kaya (Eds.). Springer, Dordrecht, Netherlands, ISBN: 9789401715478, pp: 127-146.

94. Viaud, M., Y. Couteaudier and G. Riba, 1998. Molecular analysis of hypervirulent somatic hybrids of the entomopathogenic fungi Beauveria bassiana and Beauveria sulfurescens. Applied Environ. Microbiol., 64: 88-93.

95. Vidal, S. and L.R. Jaber, 2015. Entomopathogenic fungi as endophytes: Plant-endophyte-herbivore interactions and prospects for use in biological control. Curr. Sci., 109: 46-54.

96. Weinberger, K. and T.A. Lumpkin, 2007. Diversification into horticulture and poverty reduction: A research agenda. World Dev., 35: 1464-1480.

97. White, T.J., T.D. Bruns, S.B. Lee and J.W. Taylor, 1990. Amplification and Direct Sequencing of Fungal Ribosomal RNA Genes for Phylogenetics. In: PCR Protocols: A Guide to Methods and Applications, Innis, M.A., D.H. Gelfand, J.J. Sninsky and T.J. White (Eds.). Academic Press, San Diego, CA., USA., ISBN-13: 9780123721808, pp: 315-322.

98. Yan, J.F., S.J. Broughton, S.L. Yang and A.C. Gange, 2015. Do endophytic fungi grow through their hosts systemically?. Fungal Ecol., 13: 53-59.

CHAPTER 12

Effect of Flood Stress on Soybean Seed Germination in the Field

Chengjun Wu[1], Pengyin Chen[2*], Wade Hummer[1], Ailan Zeng[1], Mariola Klepadlo[3]

[1]*Department of Crop, Soil, and Environmental Sciences, University of Arkansas, Fayetteville, AR, USA*

[2]*Division of Plant sciences, University of Missouri, Delta Center, Portageville, MO, USA*

[3]*Department of Plant Sciences, University of Missouri, Columbia, MO, USA*

ABSTRACT

Flooding is an abiotic stress that impacts soybean [Glycine max (L.) Merr.] growth and reduces seed germination. Effect of flooding on soybean plant grown at different growth stage has been previously conducted and reported. However, soybean seed germination responses to flood stress are largely unknown. The objective of this study was to elucidate flooding influence on soybean seed germination after planting in the field. The research showed that seed germination rate (SGR) of each genotype, without flood stress, was significantly different and ranged between 64.7% to 84.0% and 69.0% to 90.7% while using untreated and fungicide-treated seed ($P < 0.0001$), respectively. Results indicated that fungicide treatment improved soybean seed survival and germination in the field. The average of SGR of high-yielding soybean group was significantly higher than those of non-high-yielding soybean (P < 0.0001). The results indicated that high-yielding trait of each genotype was correlated with seed germination and survival. Under flood stress in the field, SGR means of untreated and fungicide-treated seed significantly decreased over eight flooding treatment times (P < 0.0001). Flooding effect on germination between untreated and fungicide-treated seed was not significantly different (P = 0.1559). Furthermore, comparing the high-yielding and flood- tolerant soybean groups showed no difference in their SGR means over eight flooding treatment times (P = 0.7687 and P = 0.8490), indicating that soybean seed germination did not depend on genotype, yield, and flood tolerance trait, and seed treated by fungicide did not increase its germination in the field under the flood stress. Hence, it is necessary to develop new soybean seed pelleting to improve seed germination in the field under flooding conditions.

Keywords: Soybean, Flood Stress, Seed Germination Rate, Fungicide Apron Maxx RTA, Yield Trait Group, Flood Trait Group

1. INTRODUCTION

Flooding is one of the most hazardous natural occurrences caused by heavy rains, excessive irrigation, and low infiltration rate of soils, and its prolonged appearance severely reduces productivity of crops in major growing regions in the world. Flooding imposes a severe selection pressure on plants since excess water in the living surroundings can deprive plants of oxygen, carbon dioxide, and light [1] . Submerged plant shoots have a severely reduced photosynthesis level due to deficiency of external carbon dioxide causing progressive leaf chlorosis [2] . Root and shoot growth was also affected, and eventually the accumulation of dry matter and seed yield were reduced by flood [3] - [10] . Soybean is an important crop, which is widely used to provide protein, oil, carbohydrates, minerals, and other nutrients for humans and animals [11] . Most soybean cultivars are sensitive to flood stress causing chlorosis, necrosis, stunting, defoliation, reduction of nitrogen fixation, and plant death [12] [13] [14] [15] [16] . All of these symptoms occur at various vegetative (V) and reproductive (R) stages of the plant growth causing various level of yield deterioration [14] [16] [17] [18] [19] . Flooding as an abiotic stress causes approximately 16% reduction in soybean productivity worldwide and loss of billions of dollars for farmers [3] [20] [21] . Flooding regularly affected soybean growth and grain yield around the world including the United States [17] [22] . In the

Mississippi Delta region, flood stress can reduce overall soybean yield up to 25% [23] . Previous studies focused on understanding flooding influence on different soybean growth stages and yield cutback. Rhine et al. [24] concluded that soybean plants exposed to flood at R5 stage showed yield reduction of 20% - 39% in comparison to non-flooded control. Oosterhuis et al. [17] showed that flooding reduced soybean yields by 17% - 43% at the vegetative growth stage, and 50% - 56% at the reproductive stage. Scott et al. [18] demonstrated that only 2 days of flooding caused 18% of yield loss at late vegetative stage while it may exceed to 26% if flooding occurs at early reproductive stage of soybean and daily yield reductions are up to 1.6% at V4 and 3.6% at R2 stage. Similarly, Sullivan et al. [25] reported a 20% yield loss when soybean plots were flooded for three days at V2 and V3 growth stages. Shannon et al. [26] revealed 40% yield reduction in a soybean flood-tolerant group versus 80% reduction in a flood-susceptible group. These previous studies also demonstrated that soybean yield losses were the result of plant death due to occurrence of diseases, physiological stress mostly caused by hypoxia, reduced root and shoot growth, nodulation and nitrogen fixation, photosynthesis, biomass accumulation, and stomatal conductance [16] [17] [23] [27] . However, in contrary to soybean flood tolerance research with focus on plant injury and yield loss, the effect of flood stress on seed germination has not been determined. Hence, it is useful to not only investigate soybean plant growth, but also understand seed germination response to flood stress.

Seed germination is a critical developmental phase in plant life cycle and reproductive success [28] . In general, seed germination capacity is determined by genetic factors and environmental cues such as light, water, temperature, drought, and oxygen [29] [30] [31] [32] [33] . Under flooding, soybean seeds have poor survival and germination in the field due to quick loss of viability in hypoxic environment because oxygen supply is required for germination activation [34] [35] [36] . On the other hand, presence of soil-borne diseases caused by Phomopsis, Pythium, Phytophthora, Rhizoctonia, and Fusarium significantly impacted on soybean seed germination and seedling emergence [37] [38] [39] [40] . Using seeds covered by an appropriate fungicide increases seed germination of about 10% what results in a large plant emergence in the field [41] . Apron Maxx RTA (Syngenta Crop Protection Inc.) is a broad-spectrum fungicide widely used in the United States for seed treatment, and can control or suppress pathogens Phomopsis, Pythium, Phytophthora, Rhizoctonia, and Fusarium [42] . However, it is still unknown if this fungicide can maintain soybean seed survival and improve seed germination in the flood-affected field. Therefore, the objective of this study was to investigate the effect of flood stress on the soybean seed germination in the field and identify if broad-spectrum fungicide Apron Maxx RTA can protect seed survival and improve seed germination rate under flooding condition. Knowledge of seed germination interaction with flooding can help plant breeders develop lines and/or cultivars with increased seed germination and seedling growth at high water stress.

2. MATERIALS AND METHODS

2.1. Plant Materials

A total of 20 soybean genotypes including flood-tolerant (T), flood-moderately- tolerant (MT), and flood-sensitive (S) traits were used as experimental materials, of which, were selected from soybean breeding program at University of Arkansas (Table 1). Based on our previous yield trait data (not shown), 20 genotypes also were grouped into high-yielding (HY) and non-high-yielding traits (NHY) (Table 1).

2.2. Seed Treatment

Seeds of each genotype were separated two parts untreated and treated by fungicide Apron Maxx RTA with active ingredients Fludioxonil (0.73%) and Metalaxyl-M and S-isomer (1.10%) (Syngenta Crop Protection Inc.). Every 4.540 kg seeds were added 14.8 ml Apron Maxx RTA and mixed even in bucket. The untreated and treated seeds were packed 100 seeds per envelope for planting.

2.3. Seed Planting, Flooding Treatment, and Germination Data Collection

A total of sixteen blocks were planted in the field experiment in 2016, at the Rice Research Experiment Station at Stuttgart, Arkansas (34°30'N, 91°33'W). Levees were made to isolate each field block. In each block, all 20 genotypes were planted in a randomized complete block design (RCBD) with three replications. Soybean plants were grown on a crowley silt loam (fine, montmorillonitic, thermic TypicAlbaqualfs). A total of 100 seeds for each genotype were planted in a 3-m row with a 0.75-m row spacing. Three days after planting, flood stress was imposed to the field with 5 - 7 cm of water above the soil surface. Eight different durations which were 6, 12, 24, 36, 48, 72, 96, and 120 hours of flooding treatment, were applied in each type. After flooding treatment, water was drained for seed germination. Meanwhile, two control tests without flood stress including untreated and treated seed were performed in the field. Seed germination number of each genotype were recorded four weeks after removing the flooding water. The seed germination number of two control test without flood stress were also collected in same time. The seed germination rate (SGR) was calculated by dividing the number of emerged seedlings obtained at each counting and per 100 seeds.

Table 1. Twenty soybean genotypes with name, pedigree, flood trait, and yield trait. Foliar damage score (FDS) and plant survival rate (PSR%) of each genotype at R1 stage under flood stress in untreated and fungicide-treated seed tests.

Name	Pedigree	Flood Trait*	Yield Trait#	FDS (US)†	FDS (TS)‡	FDS Mean	PSR% (US)†	PSR% (TS)‡	PSR% Mean
Walters	Forrest × Narow	T	NHY	2.3	2.8	2.6	80.1	78.9	79.5
R10-4892	5002T × R01-3474F	T	NHY	2.2	2.6	2.4	81.2	79.5	80.4
R07-6669	Lonoke × R00-33	T	NHY	2.8	2.6	2.7	78.3	80.2	79.3
UA 5615C	5002T × R04-357	T	HY	1.9	2.2	2.1	86.9	84.3	85.6
R11-6870	5002T × R01-3474F	T	HY	2.6	2.4	2.5	79.6	82.2	80.9
R04-342	R97-1650 × 98601	T	NHY	3.1	2.8	3.0	72.1	75.6	73.9
R13-12552	5002T × 91210-350	T	NHY	3.2	3.1	3.2	71.5	72.4	72.0
Osage	Hartz 5545 × KS4895	MT	HY	4.6	5.0	4.8	58.6	50.5	54.6
UA 5612	R97-1650 × 98601	MT	HY	4.3	4.7	4.5	59.8	54.6	57.2
UA 5213C	R98-1523 × 98601	MT	HY	5.3	5.6	5.5	49.1	47.1	48.1
UA 5115C	BA 743303 × R00-684	MT	HY	4.8	4.5	4.7	52.6	56.9	54.8
UA 5414RR	R96-3427 × 98601	MT	HY	5.6	5.4	5.5	44.6	44.6	44.6
UA 5715GT	Lonoke × Hutcheson-RR	MT	HY	5.4	5.7	5.6	46.8	42.3	44.6
RM-22590	N/A	S	NHY	7.7	7.9	7.8	19.8	18.9	19.4
UA 5014C	Ozark × Anand	S	HY	8.4	8.5	8.5	8.7	8.4	8.6
R01-2731F	Caviness × PI 592947	S	NHY	7.6	8.0	7.8	18.7	15.6	17.2
R99-1613F	NKRA 452 × PI 290126B	S	NHY	7.9	7.7	7.8	19.8	19.5	19.7
R09-4095	S01-9265 × R00-1940	S	NHY	8.3	8.1	8.2	10.3	7.9	9.1
R06-4433	Lonoke × P9594	S	NHY	8.2	8.0	8.1	11.2	11.1	11.2
R10-2379	R01-52F × R02-6232F	S	NHY	8.1	7.9	8.0	12.5	12.8	12.7

*Flood trait, T = flood-tolerant, MT = flood-moderately-tolerant, and S = flood-sensitive; #Yield trait, HY = high-yielding, NHY = non-high-yielding; †FDS, foliar damage score; PSR, plant survival rate; US, untreated seed; ‡FDS, foliar damage score; PSR, plant survival rate; TS, fungicide-treated seed.

2.4. Plants Flooding Treatment and Response Data Collection

Once two control tests soybean plants reached R1 stage, flood stress was imposed (12 - 14 cm of water above the soil surface). After 8-day flooding treatment, water was drained. Foliar damage score (FDS) and plant survival rate (PSR) were recorded immediately after the termination of each flood treatment. FDS was based on a 1 - 9 scale, where 1 and 9 indicated less than 10 and over 85% of the plants showing foliar damage or death, respectively (1 = 0% - 10%; 2 = 11% - 20%; 3 = 21% - 30%; 4 = 31% - 40%; 5 = 41% - 50%; 6 = 51% - 60%; 7 = 61% - 70%; 8 = 71% - 85%; 9 = 86% - 100%) (Table 1). Genotypes were also grouped into three categories; flood tolerant when FSD < 4.0, moderately flood tolerant when FSD = 4.0 - 5.9, and flood sensitive when FSD = 6.0 - 9.0. To determine the plant survival rate (PSR), plants were counted in each row before and after flooding.

2.5. Data Analysis

Analysis of variance (ANOVA) was used for means of seed germination rating between untreated and treated seed test, and plant survival rate was performed using JMP Pro 12.1.0 (SAS Institute Inc., 2012). Significant differences among different treatments were calculated using the LSMeans Difference Student's test with a confidence level of $\alpha < 0.05$.

3. RESULTS

3.1. Germination of Untreated and Fungicide-Treated Seed without Flood Stress

In the untreated seed without flood stress test, the SGR of each genotype was variable and ranged from 64.7% to 84.0% (Table 2). LSMeans Difference Student's test showed the differences were significant ($P < 0.0001$). For the yield trait group comparing analysis, the SGR mean (79.3%) of high-yielding genotypes was significantly higher than that (67.4%) of non-high-yielding genotypes ($P < 0.0001$) (Figure 1). Comparing the flood trait group, the SGR mean (74.3%) of flood-tolerant genotypes was significantly different from the SGR mean (77.9%) of flood-moderately-tolerant genotypes and the SGR mean (68.4%) of flood- sensitive ($P = 0.0071$) (Figure 1).

In the fungicide-treated seed without flood stress test, seed germination rates (SGR) of 20 genotypes also largely varied in the field between 69.0% to 90.7% (Table 2). LSMeans Difference Student's test showed the differences were significant ($P < 0.0001$) (Table 2). For the yield trait group comparing analysis, the SGR mean (85.3%) of high-yielding genotypes was significantly higher than that (72.8%) of non-high-yielding genotypes ($P < 0.0001$) (Figure 1). Comparing the flood trait group, the SGR mean (78.0%) of flood-tolerant genotypes was significantly different from the SGR mean

(84.0%) of flood-moderately-tolerant genotypes and the SGR mean (74.5%) of flood-sensitive genotypes ($P < 0.0001$) (Figure 1).

Comparing SGR means between untreated and fungicide-treated seed without flood stress tests, the SGR mean off ungicide-treated seed test was 79.2% and dramatically higher than that 73.8% of untreated seed test (Figure 2). LSMeans Difference Student's test showed that the difference was significant. Likewise, treatment with fungicide Apron Maxx RTA also largely increased the SGR of each genotype in the field (Table 2 and Figure 3). In addition, our results showed that high-yielding genotypes had higher SGR in both untreated and fungicide-treated seed control tests without flood stress (Figure 1 and Figure 3).

Table 2. Seed germination rate (SGR) of twenty soybean genotypes in untreated and fungicide-treated seed without or with flood stress tests.

Name	SGR% (US/N)*	SGR% (TS/N)#	SGR% (US/F)†	SGR% (TS/F)‡
UA 5615C	84.0 a	90.7 a	25.0 a	25.1 a
UA 5612	83.3 ab	87.3 abc	24.3 a	24.7 a
Osage	81.7 abc	86.7 ab	23.7 a	25.4 a
UA 5414RR	78.7 abcd	84.0 bcd	25.1 a	24.0 a
UA 5715GT	78.7 abcd	84.0 bcd	24.5 a	24.6 a
UA 5014C	78.7 abcd	82.7 cd	25.8 a	23.3 a
R11-6870	78.3 abcd	86.7 abc	24.2 a	24.9 a
UA 5213C	78.3 abcd	88.0 abc	25.2 a	23.9 a
UA 5115C	76.3 bcde	83.3 bcd	25.4 a	23.7 a
R07-6669	75.0 cdef	76.3 ef	24.3 a	24.8 a
Walters	73.0 defg	74.3 efg	23.6 a	25.5 a
R10-4892	71.0 efgh	74.3 efg	24.7 a	24.5 a
R04-342	70.3 efgh	74.7 efg	23.2 a	25.9 a
R10-2379	68.7 fgh	72.3 fg	24.8 a	24.4 a
R13-12552	68.3 fgh	69.0 g	23.8 a	25.8 a
RM-22590	68.3 fgh	72.3 fg	25.3 a	23.8 a
R09-4095	67.3 gh	70.3 g	24.9 a	24.3 a
R01-2731F	66.0 gh	72.7 fg	25.0 a	24.1 a
R06-4433	65.3 h	78.7 de	24.2 a	25.0 a
R99-1613F	64.7 h	72.7 fg	25.1 a	24.0 a

*SGR, seed germination rate; US/N, untreated seed without flood stress; #SGR, seed germination rate; TS/N, fungicide-treated seed without flood stress; †SGR, seed germination rate; US/F, untreated seed with flood stress; ‡SGR, seed germination rate; TS/F, fungicide-treated seed with flood stress.

3.2. Effect of Flood Stress on Untreated Seed Germination

The effect of flood stress on untreated seed germination in the field was showed in Figure 4. The SGR mean without flood stress (flooding treatment time = 0 hour) was 73.8%. At eight different flooding treatment times (6, 12, 24, 36, 48, 72, 96, and 120 hours), the SGR mean of 20 genotypes in each flooding treatment time was 64.9%, 53.9%, 35.6%, 20.7%, 9.3%, 2.4%, 0.5%, and 0%, respectively (Figure 4(a)). LSMeans Difference Student's test showed that these SGR means reductions were significant between each other with flooding time extension ($P < 0.0001$) (Figure 4(a)). ANOVA analysis further confirmed that these SGR means were variable at different flooding treatment time ($P < 0.0001$) (Table 3). In order to analyze each genotype seed germination response to flood stress, the SGR mean of each genotype over eight flooding treatment times was showed in Figure 4(b). The SGR of each genotype was different and ranged between 19.7% to 26.5% after flooding treatment (Figure 4(b)). LSMeans Difference Student's test showed that these SGR variations among genotypes were not significant ($P = 0.9999$) (Table 2). In addition, the flood stress effect on untreated seed tests revealed that the SGR means of high-yielding and non-high- yielding groups were 25.7% and 21.9% (Figure 5(a)) and the SGR means of flood-tolerant, flood-moderately-tolerant, and flood-sensitive groups were 22.5%, 25.6%, and 22.4%, respectively (Figure 5(b)). LSMeans Difference Student's test showed that these group SGR means were not significant different after flooding ($P = 0.2237$ and $P = 0.9837$) (Figure 5).

Table 3. ANOVA analysis for seed germination rates (SGR) of untreated and fungicide-treated seed tests under different flood stress.

Source	DF	Sum of squares	F Ratio	Prob > F
Model†	8	406742.7	2478.4	
Flooding time	8	406742.7	2478.4	<0.0001
Error	531	10893.2		
Total	539	417636.0		<0.0001
Model‡	8	474223.6	3242.1	
Flooding time	8	474223.6	3242.1	<0.0001
Error	531	9708.7		
Total	539	483932.2		<0.0001

†Model, ANOVA analysis for untreated seed germination rates. ‡Model, ANOVA analysis for fungicide-treated seed germination rates.

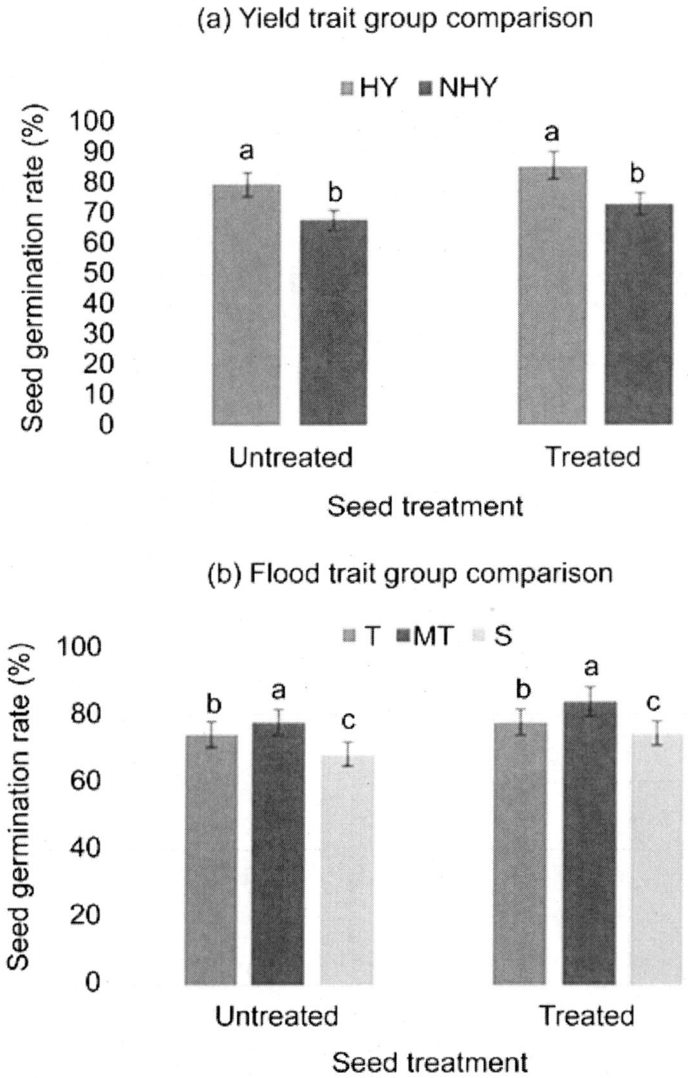

Figure 1. Seed germination rate (SGR) means of untreated and fungicide-treated seed tests without flood stress: (a) Yield trait groups with high-yielding (HY) and non-high-yielding (NHY) traits; (b) Flood trait groups with flood-tolerant (T), flood-moderately-tolerant (MT), and flood-sensitive (S) traits.

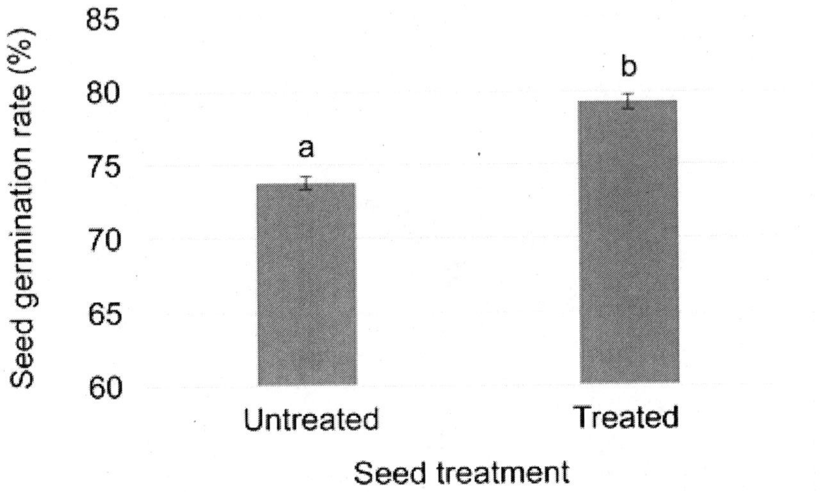

Figure 2. Comparison of seed germination rate (SGR) means between untreated and fungicide-treated seed test without flood stress.

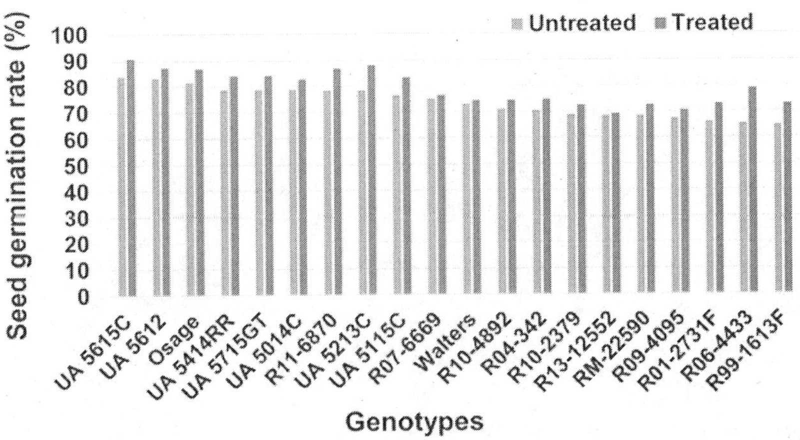

Figure 3. Summary of seed germination rate (SGR) of each genotype between untreated and fungicide-treated seed tests without flood stress.

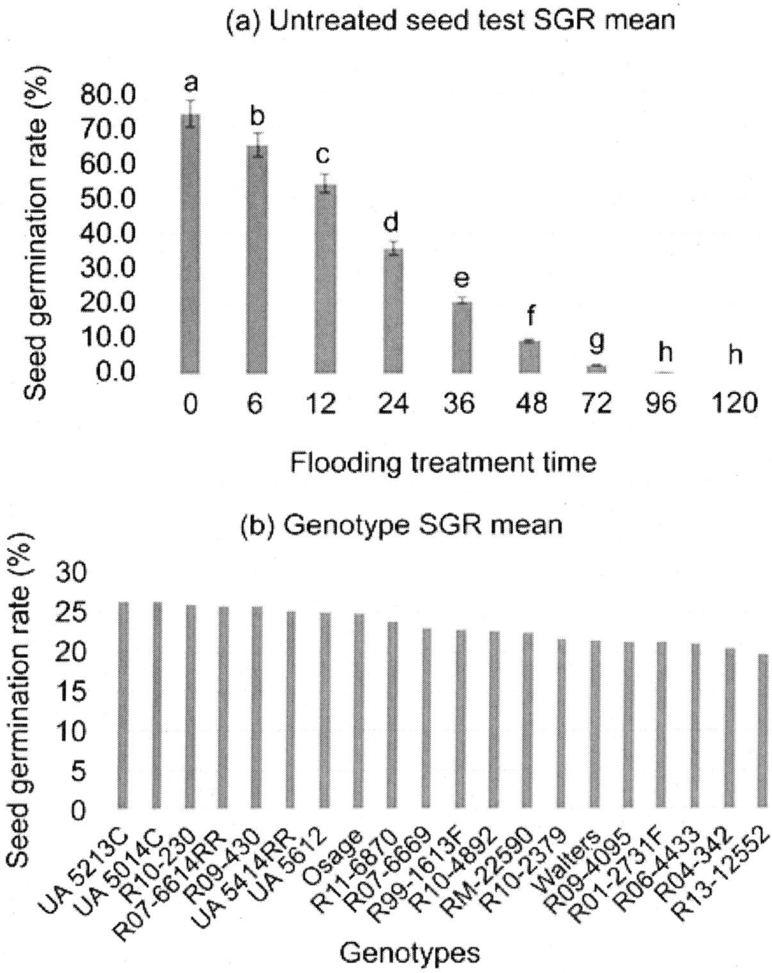

Figure 4. Comparing untreated seed germination rate (SGR) means at different time flood stress: (a) SGR means of untreated seed tests at eight flooding treatment times and control test without flood stress (0 hour flooding treatment time); (b) SGR means of 20 genotypes in untreated seed test over eight flooding treatment times.

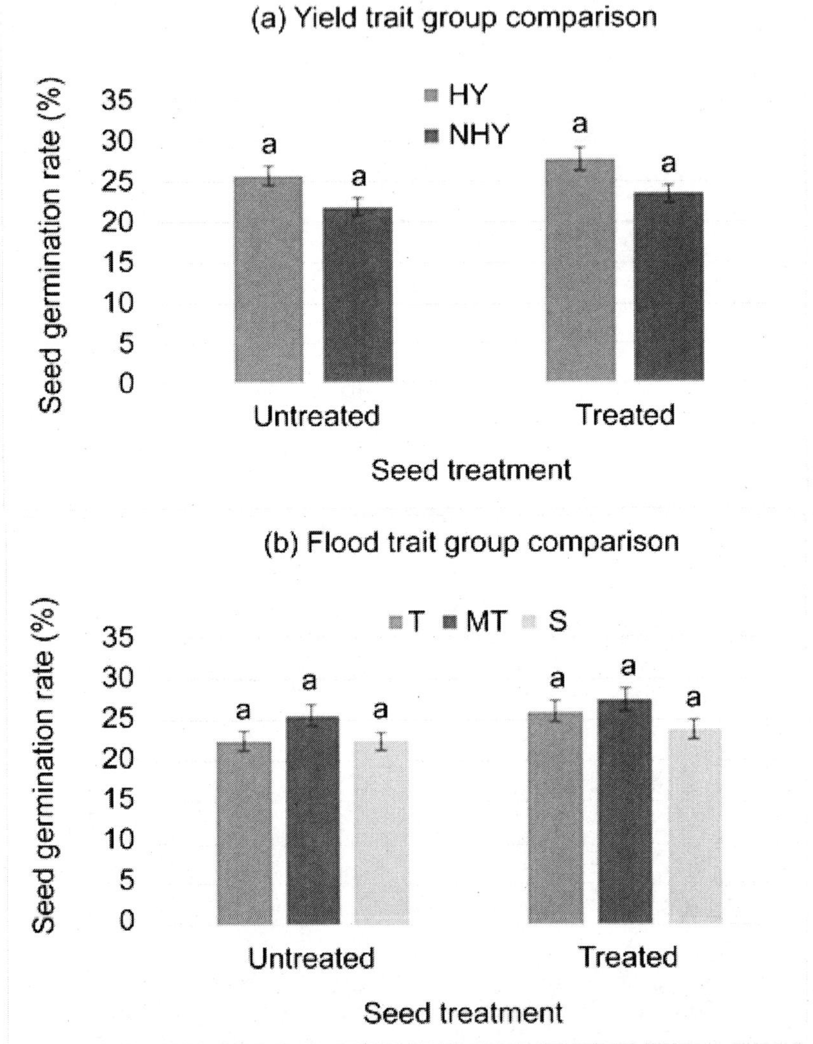

Figure 5. Comparing seed germination rate (SGR) means of different groups over eight flooding treatment times: (a) SGR means of yield trait groups between untreated and fungicide-treated seed tests (HY = high-yielding and NHY = non-high-yielding); (b) SGR means of flood trait groups between untreated and fungicide-treated seed tests (T = flood-tolerant, MT = flood-moderately-tolerant, and S = flood-sensitive).

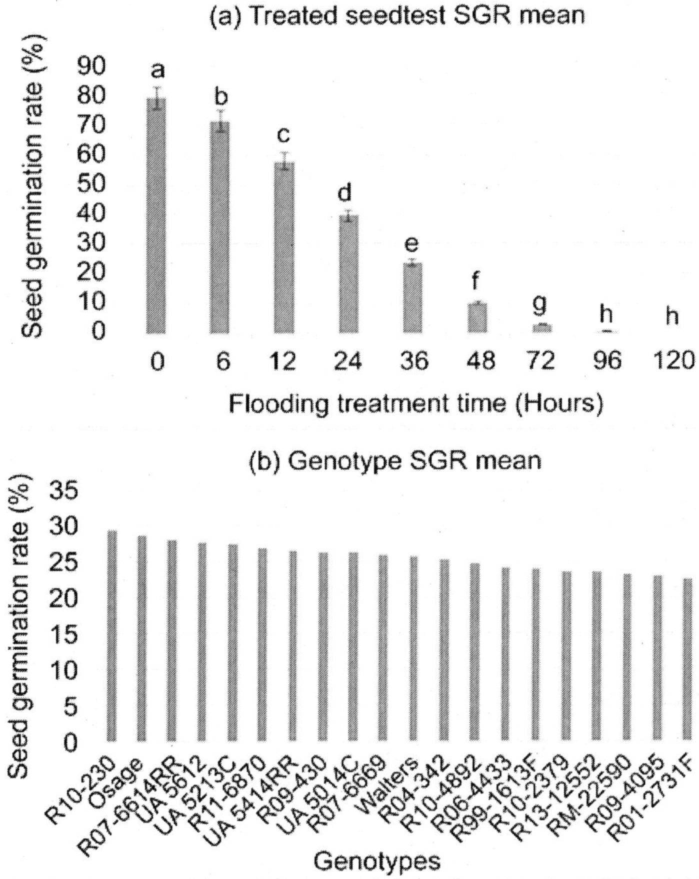

Figure 6. Comparing fungicide-treated seed germination rate (SGR) means at different time flood stress: (a) SGR means of fungicide-treated seed tests at eight flooding treatment times and control test without flood stress (0 hour flooding treatment time); (b) SGR means of 20 genotypes in fungicide-treated seed test over eight flooding treatment times.

3.3. Effect of Flood Stress on Fungicide-Treated Seed Germination

The effect of flood stress on fungicide-treated seed germination in the field was showed in Figure 6. The SGR mean without flood stress (flooding treatment time = 0 hour) was 79.2%. At eight different flooding treatment times (6, 12, 24, 36, 48, 72, 96, and 120 hours), the SGR mean of 20 genotypes in each flooding treatment time was 71.2%, 57.8%, 39.6%, 23.7%, 10.1%, 2.8%, 0.5%, and 0%, respectively (Figure 6(a)). LSMeans Difference Student's test showed that these SGR means were significantly decreased with flooding time extension (Figure 6(a)). ANOVA analysis further confirmed that these SGR means of fungicide-treated seed test were variations at different flooding treatment time (P < 0.0001) (Table 3). The results analysis of each genotype SGR mean over eight flooding treatment times showed that different SGR means between genotypes ranged from 22.7% to 29.6% (Figure 6(b)). LSMeans Difference Student's test showed that these SGR means' variations among genotypes were not significant (P = 0.9999) (Table 2). In addition, the flood stress effect on fungicide-treated seed tests revealed that the SGR means of high-yielding and non-high-yielding groups were 27.7% and 23.5% (Figure 5(a)) and the SGR means of flood-tole- rant, flood-moderately-tolerant, and flood-sensitive groups were 26.0%, 27.5%, and 23.9%, respectively (Figure 5(b)). LSMeans Difference Student's test showed that these group SGR means were not significant difference after flooding (P = 0.3928 and P = 0.8406) (Figure 5).

3.4. Germination Rate in Untreated and Fungicide-Treated Seed under Flood Stress

After flooding treatment, the SGR means of untreated and fungicide-treated seed tests over eight flooding treatment times were 23.4% and 25.7%, respectively. LSMeans Difference Student's test analysis showed that two SGR means were not significantly different (P = 0.1559) (Figure 7(a)). The SGR mean of each genotype over untreated and fungicide-treated seed tests were shown in Table 2. LSMeans Difference Student's test analysis showed that there was insignificant difference of SGR means between untreated and fungicide-treated seed (P = 0.9985) (Table 2). Groups comparing analysis showed that the SGR means of untreated and fungicide-treated seed tests were 24.8% and 24.4% in high-yield- ing group and 24.6% and 24.6% in non-high-yielding group, respectively (Figure 7(b)). The SGR means of untreated and fungicide-treated seed tests were 24.1% and 25.2% in flood-tolerant group, 24.7% and 24.4% in moderately tolerant group, and 24.9% and 24.3% in sensitive group, respectively (Figure 7(c)). LSMeans Difference Student's test showed that these groups SGR means were not significant (P = 0.7687 and P = 0.8490).

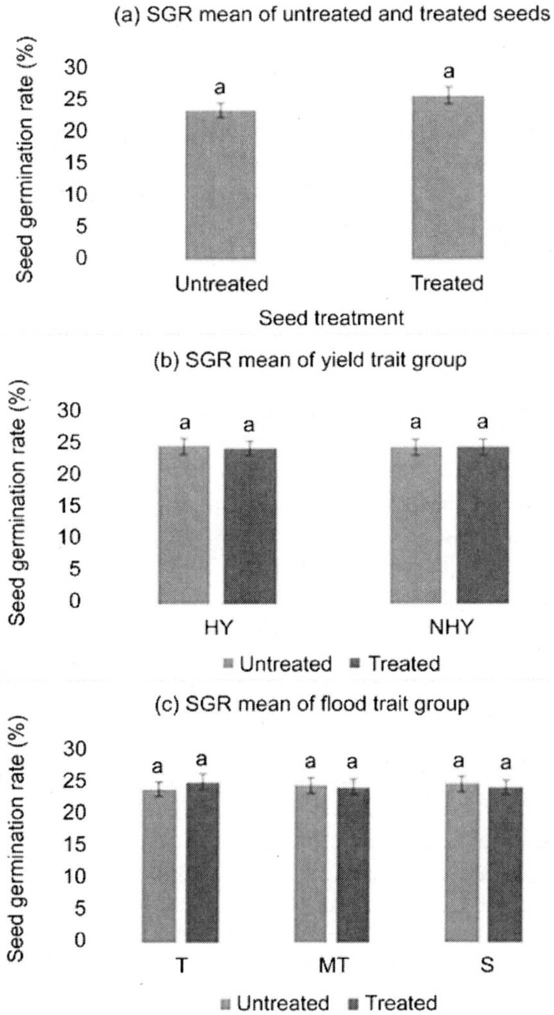

Figure 7. Comparison of seed germination rate (SGR) means between untreated and fungicide-treated seed tests over eight flooding treatment times: (a) SGR means between untreated and fungicide-treated seed tests; (b) SGR means of yield trait groups between untreated and fungicide-treated seed tests (HY = high-yielding and NHY = non-high- yielding); (c) SGR means of flood trait groups between untreated and fungicide-treated seed tests (T = flood-tolerant, MT = flood-moderately-tolerant, and S = flood-sensitive).

3.5. Evaluation of Plant Response to Flood Stress

In order to further validate each genotype plant response to flood stress, when plants both untreated and fungicide-treated seed without flood stress tests were grown until R1 stage and then flooded in the field, flood response of each genotype was scored and evaluated. Our results showed that flood responses of 20 genotypes were significantly different in both untreated and fungicide-treated tests (Table 1). For the flood-tolerant group, each genotype showed lower plant foliar damage score (FDS < 4.0) and higher plant survival rate (PSR > 70.0%). Likewise, each genotype showed middle plant foliar damage score (4.0 < FDS < 6.0) and plant survival rate (40% < PSR < 60.0%) in flood-moderately-tolerant group and high plant foliar damage score (FDS > 7.0) and low plant survival rate (PSR < 20.0%) in flood-sensitive group (Table 1). For the high-yielding group, two genotypes UA 5615C and R11-6870 showed flood tolerance, whereas six genotypes UA 5612, Osage, UA 5414RR, UA 5715GT, UA 5213C, and UA 5115C were moderately tolerant to flooding, and one genotype UA 5014C showed sensitivity to flood stress. For the non-high-yielding group, five genotypes including Waters, R10-4892, R07-6669, R04-342, and R13-12552 showed flood tolerance, and six genotypes including RM-22590, R01-2731F, R99-1613F, R09-4095, R06-4433, and R10-2379 were sensitive to flood stress (Table 1).

4. DISCUSSION

Based on previous studies, soybean seed germination rate (SGR) could be reduced by seed disease occurrence caused by soil-borne pathogens [37] - [42] . In our untreated and fungicide-treated seed tests without application of flooding, our results showed that fungicide treatment significantly increased soybean seed germination rate (SGR) in the field. The reasonable annotation is that fungicide provides protection to soybean seeds against damping-off and seed rots due to Pythium, Phtophthora, Fusarium, Rhizoctonia, seed-borne Sclerotinia, and Phomopsis and then improves seed survival, germination, emergence, and seedling stands in the field. In addition, our research also showed that SGR mean of high- yielding group was significantly higher than SGR mean of non-high-yielding group in both untreated and fungicide-treated seed tests without flood stress (Figure 1(a)). It indicates that high-yielding trait of each genotype is linkage to seed germination trait and normally high-yielding soybean cultivars have higher seed germination rate. The genes relevant to seed germination are always selected with high-yielding genes in soybean breeding and domestication. Furthermore, the comparison of SGR means between flood trait groups showed that moderately-flood-tolerant group had the highest SGR mean, the SGR mean of flood-tolerant group ranked second, and flood-sensitive group had minimum SGR mean (Figure 1(b)). This result addressed that soybean seed germination trait is not linkage to flood trait. On the other way, six genotypes of moderately-flood-tolerant group in this research all had high-yielding trait (Table 1). It was a reason that moderately-flood-tolerant group had highest SGR mean in our test and further confirmed that genes relevant to seed germination were linkage to high-yielding genes in soybean. Hence, soybean breeders should cross high- yielding varieties to flood-tolerant varieties and then select varieties with high- yielding and flood-tolerant traits adapting to flooding environment in soybean breeding process.

In this research, we evaluated different flood stress effect to soybean untreated seed germination in the field. When flooding treatment time was extended, the SGR dramatically reduced with longer flooding hours. After only 6 hours flooding, SGR significantly decreased 12.1% and continued to drop to 51.8% after 1-day flood stress. After 2-day flooding, SGR declined to less than 10% and almost no germination in the field (Figure 4(a)). Likewise, the different flood stress also largely impacted soybean fungicide-treated seed germination; the SGR also significantly reduced with extending flooding treatment time. After only 6 hours flooding, SGR significantly decreased 10.1% and continued to drop to 50.0% after 1-day flood stress. After 2-day flooding, SGR declined to less than 10% and almost no germination in the field (Figure 6(a)). Our results showed that flood stress significantly affected soybean seed germination in both untreated and fungicide-treated seed tests in the field. The longer flooding and the less soybean seed germination. One day flooding after planting will result to about 50% seed no germination and soybean yield dramatically reduce. Hence, heavy rain and excessive irrigation in planting season have a large affection to soybean germination and production.

Our research also further showed SGR of each genotype largely decreased with flooding prolong. Each genotype SGR means over eight flooding treatment times between untreated and fungicide-treated seed tests were not significantly different (Table 2, Figure 4(b) and Figure 6(b)). Comparing yield trait group, SGR means of high-yielding group were not significantly different from SGR means of non-high-yielding group in both untreated and fungicide-treated seed tests over eight flooding treatment times (Figure 5(a)). For the comparison of flood trait group, SGR means of flood-tolerant, moderately-flood-tolerant, and flood-sensitive were also not significantly different in both untreated and fungicide-treated seed tests over eight flooding treatment times (Figure 5(b)). These results indicated that soybean seed germination was not significantly affected by genotype, yield, and flood trait under the flood stress. In other words, soybean genotype with yield and flood traits cannot significantly increase seed germination in the field under flood stress.

Comparing flood stress effect between untreated and fungicide-treated seed tests over eight flooding treatment times, the SGR means were not significantly different (Figure 7(a)). For the yield trait group, the SGR means of high-yielding group between untreated and fungicide-treated seed tests were not significantly different and SGR means of non-high-yielding group between untreated and fungicide-treated seed tests were also not significantly different (Figure 7(b)). For the flood trait group, the SGR means of flood-tolerant, moderately-flood- tolerant, and flood-sensitive groups were not significantly different between untreated and fungicide-treated seed tests (Figure 7(c)). These results further indicated that soybean seed treated by fungicide Apron MaxxRTA did not increase SGR under flood stress in the field. The damage effect of flooding is the same to untreated and fungicide-treated soybean seeds in the field. The fungicide Apron MaxxRTA cannot effetely protect soybean seed survival and increase germination under flood stress. Therefore, it is necessary to develop new seed pelleting and improve seed germination on flooding environment.

ACKNOWLEDGEMENTS

We would like to thank the United Soybean Board and Arkansas Soybean Promotion Board for providing the financial support. We also thank Jonathan McCoy at the Rice Research and Extension Center in Stuttgart, AR for his support and assistance.

CITE THIS PAPER

Wu, C.J., Chen, P.Y., Hummer, W., Zeng, A. and Klepadlo, M. (2017) Effect of Flood Stress on Soybean Seed Germination in the Field. American Journal of Plant Sciences, 8, 53-68.

REFERENCES

1. Michael, B.J., Kimiharu, I. and Osamu, I. (2009) Evolution and Mechanisms of Plant Tolerance to Flooding Stress. Annals of Botany, 103, 137-142.
2. Jackson, M.B. and Colmer, T.D. (2005) Response and Adaptation by Plants to Flooding Stress. Annals of Botany, 96, 501-505.
3. Ahmed, F., Rafii, M.Y., Ismail, M.R., Juraimi, A.S., Rahim, H.A., Asfaliza, R. and Latif, M.A. (2013) Waterlogging Tolerance of Crops: Breeding, Mechanism of Tolerance,

Molecular Approaches, and Future Prospects. BioMed Research International, 2013, 1-10.

4. Buchanan Wollaston, V., Earl, S. and Harrison, E. (2003) The Molecular Analysis of Plant Senescence—Agenomics Approach. Plant Biotechnology Journal, 1, 3-22.
5. Drew, M.C. (1991) Oxygen Deficiency in the Root Environment and Plant Mineral Nutrition in Plant Life under Oxygen Deprivation. Academic Publishing, The Hague, The Netherlands, 301-316.
6. Drew, M.C. and Sisworo, E.J. (1977) Early Effects of Flooding on Nitrogen Deficiency and Leaf Chlorosis in Barley. New Phytologist, 79, 567-571.
7. Huang, B., Johnson, J.W., Nesmith, S. and Bridges, D.C. (1994) Growth, Physiological and Anatomical Responses of Two Wheat Genotypes to Waterlogging and Nutrient Supply. Journal of Experimental Botany, 45, 193-202.
8. Kozlowski, T.T. (1984) Extent, Causes, and Impact of Flooding in Flooding and Plant Growth. Academic Press, London, UK, 9-45.
9. Malik, A.I., Colmer, T.D., Lambers, H.T., Setter, L. and Schortemeyer, M. (2002) Short-Term Waterlogging Has Long-Term Effects on the Growth and Physiology of Wheat. New Phytologist, 153, 225-236.
10. Wang, S., He, L., Li, Z., Zeng, J., Cai, Y. and Hou, L. (1996) A Comparative Study of the Resistance of Barley and Wheat to Waterlogging. Acta Agronomica Sinica, 22, 228-232.
11. Hou, A., Chen, P., Alloatti, J., Li, D., Mozzoni, L., Zhang, B. and Shi, A. (2009) Genetic Variability of Seed Sugar Content in Worldwide Soybean Germplasm Collections. Crop Science, 49, 903-912.
12. Bacanamwo, M. and Purcell, L. (1999) Soybean Dry Matter and N Accumulation Responses to Flooding Stress, N Sources and Hypoxia. Journal of Experimental Botany, 50, 689-696.
13. Daugherty, C.J. and Musgrave, M.E. (1994) Characterization of Populations of Rapid-Cycling Brassica rapa L. Selected for Differential Waterlogging Tolerance. Journal of Experimental Botany, 45, 385-392.
14. Linkemer, G., Beard, J.E. and Musgrave, M.E. (1998) Waterlogging Effects on Growth and Yield Components in Late-Planted Soybean. Crop Science, 38, 1576-1584.
15. Reyna, N., Cornelious, B. Shannon, J.G. and Sneller, C.H. (2003) Evaluation of a QTL for Waterlogging Tolerance in Southern Soybean Germplasm. Crop Science, 43, 2077-2082.
16. Van Toai, T.T., Beuerlein, J.E., Schmitthenner, A.F. and St. Martin, S.K. (1994) Genetic Variability for Flooding Tolerance in Soybean. Crop Science, 34, 1112-1115.
17. Oosterhuis, D.M., Scott, H.D., Hampton, R.E. and Wullschleter, S.D. (1990) Physiological Response of Two Soybean [Glycine max (L.) Merr.]Cultivars to Short-Term Flooding. Environmental and Experimental Botany, 30, 85-92.
18. Scott, H.D., Deangulo, J. Daniels, M.B. and Wood, L.S. (1989) Flood Duration Effects on Soybean Growth and Yield. Agronomy Journal, 81, 631-636.
19. Scott, H.D., de Angulo, J., Wood, L.S. and Pitts, D.J. (1990) Influence of Temporary Flooding at Three Growth Stages on Soybean Growth on a Clayey Soil. Journal of Plant Nutrition, 13, 1045-1071.
20. Boyer, J.S. (1982) Plant Productivity and Environment. Science, 218, 443-448.
21. Rosenzweig, C., Tubiello, F.N., Goldberg, R., Mills, E. and Bloomfield, J. (2002) Increased Crop Damage in the US from Excess Precipitation under Climate Change. Global Environmental Change, 12, 197-202.
22. Stanley, C.D., Kaspar, T.C. and Taylor, H.M. (1980) Soybean Top and Root Response to Temporary Water Tables Imposed at Three Different Stages of Growth. Agronomy Journal, 72, 341-346.
23. Van Toai, T.T., Hoa, T.T.C., Hue, N.T.N., Nguyen, H.T. Shanner, J.G. and Rahman, M.A. (2010) Flooding Tolerance of Soybean [Glycine max (L.) Merr.] Germplasm from Southeast Asia under Field and Screen-House Environments. The Open Agriculture Journal, 4, 38-46.

24. Rhine, M.D., Stevens, G. Shannon, G. Wrather, A. and Sleper, D. (2010) Yield and Nutritional Responses to Waterlogging of Soybean Cultivars. Irrigation Science, 28, 135-142.
25. Sullivan, M., Vantoai, T., Fausey, N., Beuerlein, J., Parkinson, R. and Soboyejo, A. (2001) Evaluating On-Farm Flooding Impacts on Soybean. Crop Science, 41, 93-100.
26. Shannon, J.G., Stevens, W.E., Wiebold, W.J., McGraw, R.L., Sleper, D.A. and Nguyen, H.T. (2005) Breeding Soybeans for Improved Tolerance to Flooding. American Seed Trade Association, Chicago.
27. Armstrong, W. and Drew, M.C. (2002) Root Growth and Metabolism under Oxygen Deficiency. Plant Roots: The Hidden Half. 3rd Edition, Marcel Dekker, New York, 729-761.
28. Donohue, K., Rubio de Casas, R., Burghardt, L., Kovach, K. and Willis, C.G. (2010) Germination, Postgermination Adaptation, and Species Ecological Ranges. Annual Review of Ecology, Evolution, and Systematics, 41, 293-319.
29. Barnabás, B., Jager, K. and Fehér, A. (2008) The Effect of Drought and Heat Stress on Reproductive Processes in Cereals. Plant Cell Environ, 31, 11-38.
30. Bewley, J.D., Bradford, K.J., Hilhorst, H.W.M. and Nonogaki, H. (2013) Seeds: Physiology of Development, Germination and Dormancy. 3rd Edition, Springer, New York.
31. Holdsworth, M.J., Finch-Savage, W.E., Grappin, P. and Job, D. (2008) Post-Genomics Dissection of Seed Dormancy and Germination. Trends Plant Science, 13, 7-13.
32. Penfield, S. and King, J. (2009) Towards a Systems Biology Approach to Understanding Seed Dormancy and Germination. Proceedings of the Royal Society of London B, 276, 3561-3569.
33. Varshney, R.K., Bansal, K.C., Aggarwal, P.K., Datta, S.K. and Craufurd, P.Q. (2011) Agricultural Biotechnology for Crop Improvement in a Variable Climate: Hope or Hype? Trends Plant Science, 16, 363-371.
34. Hook, D.D. (1984) Adaptations to Flooding with Fresh Water. In: Kozlowski, T.T., Ed., Flooding and Plant Growth, Academic Press, New York, 265-294.
35. Kozlowski, T.T. (1997) Responses of Woody Plants to Flooding and Salinity. Tree Physiology, 17, 13-21.
36. Parolin, P. (2001) Seed Germination and Early Establishment of 12 Tree Species from Nutrient-Rich and Nutrient-Poor Central Amazonian Floodplains. Aquatic Botany, 70, 89-103.
37. Colyer, P.D. (1989) Soybean Disease Atlas. 2nd Edition, Associated Printing Professionals, Inc., Bossier City, 43.
38. Pratt, P., Bolin, P. and Godsey, C. (2011) Soybean Production Guide. OCES Circular E-967, 129.
39. Hartman, G.L., Sinclair, J.B. and Rupe, J.C. (1999) Compendium of Soybean Diseases. 4th Edition, APS Press, St. Paul, 100.
40. Heatherly, L.G. (2015) Soybean Seed Treatments and Inoculants. Mississippi Soybean Promotion Board. http://www.mssoy.org
41. Schulz, T.J. and Thelen, K.D. (2008) Soybean Seed Inoculant and Fungicidal Seed Treatment Effects on Soybean. Crop Science, 48, 1975-1983.
42. Swoboda, C. and Pedersen, P. (2009) Effect of Fungicide on Soybean Growth and Yield. Agronomy Journal, 101, 352-356.

CHAPTER 13

Use of image analysis techniques to determine the embryo size of Senna multijuga (Rich.) seeds and its relation to germination and vigor

José Luís de Marchi[1]*, Francisco Guilhien Gomes Junior[1]

[1]Departamento de Produção Vegetal, USP/ESALQ, Caixa Postal 9, 13418-900 - Piracicaba, SP, Brasil.

ABSTRACT

Semi-automatic evaluations of morphological attributes that are associated to seed germination performance such as the embryo size, have been of great interest when considering the study of image analysis techniques. The objective of this research was to evaluate the accuracy of the Tomato Analyzer software to quantify the variations on the embryo development degree of *Senna multijuga* seeds (Rich.) and its relationship with germination and tetrazolium tests. Seeds from four lots with different ripeness stages were X-rayed, evaluated according to the embryonic area and submitted to germination and tetrazolium tests. Results from the different parameters showed that the Tomato Analyzer[(r)] software is sensitive to assess the development degree of *Senna multijuga* seeds, when expressed by Pericarp Area parameter, but only the Area parameter was directly correlated with the physiological potential of the seed lots.

Index terms: silviculture; Tomato Analyser(r); quality; physiological potential

INTRODUCTION

Senna multijuga (Rich.), known as false sicklepod, belongs to the Mimosaceae family and occurs in almost all the country, mainly in the rainforest of the Atlantic slope (Carvalho, 2008) and in the semidecidual cloud forests of the states from Ceará to São Paulo. Due to the adaptation to various kinds of soils, it is proper to be used for revegetation programs in degraded areas and ciliary forests, as well as for urban forestry. Wood can be used for lightweight boxes, toys, fuel wood and coal (Lorenzi, 2002). It annually produces a great quantity of dormant seeds, dispersed by autochory, creating a seed bank (Carvalho, 2008); they were classified as orthodox (Carvalho and Nakagawa, 2000), also confirmed by Scalon et al. (2006). This species still presents silvicultural and ecological characteristics with potential for the implantation of forests by direct sowing. Thus, in order to rationally explore the potential of native species in the recovery of environments, it is extremely important to study seed quality, as well as the best way to produce seedlings.

The physiological quality of seeds is evaluated by fundamental parameters, such as viability and vigor. Vigor tests have been used to identify differences in the physiological quality of lots presenting similar germination power and to complement the information provided by the germination test (Carvalho and Nakagawa, 1988). In order to determine seed vigor, the first germination count test, the seedling emergence test, the electrical conductivity test and the accelerated aging test may be efficiently used. However, the tetrazolium test is a feasible alternative, because it is quick França-Neto (1999). In this sense, studies about vigor are important to obtain information about the period when seeds may remain feasible and about the deterioration and death of seeds.

The development of methods to quickly evaluate the physiological quality of seeds may help deciding about the use or discard of lots meant for the conservation and/or production of seedlings. With the use of image analysis techniques, objective information may be obtained, in a relatively short period, with less human interference and, in many cases, without destroying seeds. Moreover, recorded images can be catalogued, stored and used to define non-subjective and standardized evaluation criteria.

One of the morphological variations probably associated to the seed performance is the size of the embryo, that is, its development degree in relation to the available space in the seed internal cavity, estimated in a proportion expressed in percentage. This characteristic has been evaluated through image analysis in x-rayed seeds (Machado and Cicero, 2003; Pupim et al., 2008; Otoni et al., 2005), trying to associate the results to germination and vigor. However, aiming at a higher analysis precision, it is important to use specific softwares. The *Tomato Analyzer*(r) (TA) software allows substituting evaluations of morphological attributes through manual or subjective estimation measures (Brewer et al., 2006; Brewer et al., 2007; Brewer et al., 2008). The system makes quantitative evaluations in an objective and semi-automatic way, accelerating the phenotypic analysis and obtaining highly consistent and repeatable information. Thus, the goal of this study was to evaluate the efficacy of using image analysis techniques, including the *Tomato Analyzer*(r) (TA) software, in order to determine the internal morphology and its relation with the vigor of *Senna multijuga* seeds.

MATERIALS AND METHODS

The research was developed in the Seed Testing and Image Analysis Laboratory of the Plant Production Department (PPD) of the Escola Superior de Agricultura "Luiz de Queiroz", Universidade de São Paulo, in Piracicaba, São Paulo state, with seeds coming from the Instituto de Pesquisas e Estudos Florestais (IPEF), in Piracicaba, São Paulo state, Brazil.

Fruits were harvested with the help of trimmers and plastic canvas from fifteen mother plants, in different ripeness stages, when they presented green to green-yellow-brownish colors; pods were still closed and they were placed in a shaded environment, for drying and later seed extraction.

Seeds were separated by color using Munsell color paper (1994), creating four Lots: L1 - initial population (control sample); L2 - medium green (2.5GY 5/8, equal or darker tone in relation to the standard color 2.5GY 6/6); L3 - light olive green (2.5GY 2/4, darker tone in relation to the standard color 2.5GY 6/6); L4 - mix of L1, L2 and L3 treatment. It was considered that Lot 1 was composed by ripe seeds. On the other hand, Lots 2, 3 and 4 presented different ripeness stages.

After obtaining the lots, seeds were submitted to a treatment to overcome dormancy, constituted by acid scarification, where they were immersed in sulphuric acid (H_2SO_4) at 90%, for 20 minutes.

After treatment, seeds were evaluated by the following tests:

Determination of seed moisture content: four replications of twenty seeds were used for each lot, through oven method at 105±3 ºC, for 24 hours, according to the prescriptions from the Rules for Seed Testing (Brasil, 2009); results were expressed in percentage (wet basis).

Germination test: conducted with four replications of 50 seeds for each lot, which were placed in paper towel, dampened with distilled water, equivalent to 2.5 times the mass of dry paper. Seeds were placed in a germinator adjusted at the temperature of 25 °C, with photoperiod of 12 hours (Brasil, 2009).

Tetrazolium test: seeds were scarified with the help of sandpaper (80) on the opposite side of the embryo, and immersed for 14 hours in distilled water, at 25 °C. After hydration, the tegument and endosperm were removed with the help of a blade and the embryos were immersed in tetrazolium solutions at 0.1% concentration, remaining incubated for 5 hours in oven at 35 °C, in the dark. After incubation, embryos were washed with distilled water and longitudinally sectioned with a blade. Evaluations were performed with the help of a stereo microscope; a seed classification standard was elaborated, classifying them as viable (fully colored embryo), intermediate (colored areas represent the minimum area of the embryo necessary for normal germination) and unviable (non-colored embryo). The test was performed with four replications of 25 seeds for each lot.

X-ray analysis: in order to obtain x-ray images of the seeds, the Faxitron X-ray digital equipment was used, MX-20 DC-12 model, connected to a Core 2 Due computer (3.16 Ghz, 2 GB of RAM memory, 160 GB Hard Disk) and a 17' MultiSync LCD1990SX monitor. Two hundred seeds were x-rayed (4 replications of 50 seeds) for each lot and x-rays were obtained with seeds placed at 28.6 cm from the emission source. The radiation intensity and exposure time of seeds were automatically determined by the x-ray device. In order to properly position seeds during the exposure to x-rays, a transparent double-face tape was used, bonded to a rear projection transparent film (29.7 x 21.0 x 0.1 cm). After x-rays, seeds were placed in individualized cells on a plastic tray (100 cells/tray) for later germination. X-ray images were stored in the computer hard disk for future analysis of seed physical parameters by TA software.

Software Tomato Analyzer (r): after obtaining x-rays, seeds from the respective lots were individually submitted to evaluation by the software. The methodology described by Marcos-Filho et al. (2010) was adapted, using it to determine the parameter of the Pericarp Area, since in this research other parameters generated by the software were considered.

Germination after x-rays: seeds coming from the x-ray test, previously numbered, were distributed over two paper towel sheets, on the upper third of the substrate, covered with one more sheet and rolled. Paper sheets were moistened with a quantity of water equivalent to 2.5 times their mass. For each lot, eight replications of twenty-five seeds were used. Rolls were kept at 25 °C in BOD chamber with 8 light hours and 16 dark hours. Evaluations were performed on day 11, 16 and 28 after test establishment and germination data were expressed in percentage of normal seedlings (Brasil, 2009).

After analyzing and evaluating each seed, the software automatically generated numeric values, referring to morphologic parameters, that is: Perimeter, Area and Pericarp Area. It is important to highlight that the adopted procedure to interpret the Pericarp Area parameter was the one detailed by Marcos-Filho et al. (2010). Considering the different parameters evaluated by the software, images of the x-rayed seeds were submitted to the statistic analysis, with the goal of establishing different classes and relations with the results of other vigor tests, usually conducted for the species seeds, according to the methodology proposed by Silva and Cícero (2014).

The experimental design was completely randomized with four treatments and replications, with averages compared by Tukey's test ($p \leq 0.05$).

RESULTS AND DISCUSSION

Results of water content (U%), germination (G%), tetrazolium (Tz%) and germination after the x-rays tests for the four lots of *S. multijuga* seeds can be found in Table 1. Seeds presented water content between 10.4 and 11.2% and great variations were not found among seed lots; this fact must be considered in the evaluation of the physiological potential, since variations above 2% among samples may interfere with the results (Marcos-Filho and Kikuti, 2006; Tekrony, 2003).

Still, compared to the physiological classification of seeds presented by Amorim et al. (2008), this species shows a similar behavior to the one of orthodox species, since even with a low water content, they remain feasible after five months of storage. This behavior is also observed in species forming a seed bank as an establishment strategy, as well as abundant production and high-longevity seeds (Ferreira et al., 2004).

It was possible to observe that evaluations on germination, tetrazolium and x-rayed seed germination identically classified all four lots as for physiological quality (Table 1); lot 1 had high physiological quality and lot 4 had low physiological quality. Moreover, the compliance of the criterion used to obtained the different lots was verified, according to which some Munsell color paper was used to separate seeds in different ripeness stage, allowing the obtainment of the four lots at issue. Similarly, Lopes et al. (2005), in their studies with *Tibouchina granulosa* seeds, observed that the size of fruits and color of seeds are indicators of the physiological ripeness stage. Moreover, Martins and Silva (1997) reported that the moisture degree and dry matter mass were the parameters that best characterized the physiological ripeness in *Dalbergia nigra* (Vell.) Fr. All.ex Benth. seeds.

According to the classification proposed by França-Neto et al. (1998), results obtained by tetrazolium test (Table 1) allowed verify that seed lot 1 had a very high vigor lot, whereas lot 2 had high vigor and lots 3 and 4 had medium vigor. It is important to highlight that the tetrazolium test presented the same lot classification as the germination test. This demonstrated that the tetrazolium test may be used to efficiently evaluate the physiological quality of false sicklepod seeds.

Table 1 Evaluation of the initial viability of four *Senna multijuga* seed lots, by the results obtained in tests on seed moisture content (M), germination (G), tetrazolium (TZ) and germination after x-ray analysis (G x-ray). NS - normal seedlings; AS - abnormal seedlings and DS - dead seeds.

Lot	M	G	TZ	G (x-rays)		
				NS	AS	DS
		(%)			(%)	
1	11.22 A*	85 A	86 A	87 A	4 A	9 A
2	11.10 B	76 B	73 B	77 B	9 B	14 B
3	10.65 C	73 C	63 C	73 C	11 B	16 B
4	10.43 D	64 D	52 D	63 D	14 C	13 B
CV (%)	2.00	16.4	20.1	18.6	7.1	5.2

*Means followed by the same letters, in each column, do not differ among themselves by Tukey's test at 5% significance.

These results support the ones obtained by Piveta et al. (2010) and Ferreira et al. (2004), who report that the tetrazolium test (0.075% concentration and 30 °C) with incubation for 5 hours, was effective in evaluating the vigor and viability of *S. multijuga* seeds, in laboratory or coming from the soil seed bank.

By examining the images of *S. multijuga* seeds, and by the analysis of germination data obtained after the x-ray test (Table 1), it is possible to observe that most seeds presenting a fully filled embryonic cavity, with well formed embryonic axis and endosperm (Figure 1A) or embryonic cavity with more than 50% filling, with well formed embryonic axis and endosperm originated normal seedlings, regardless of the lot (Figure 1A). It is possible that the wider the seed internal area occupied by the embryo and endosperm, the higher the germination potential, since with a more developed embryo, a lower energy expenditure will be necessary for its development, as well as the higher endosperm quantity constitutes more available energy for the process. According to Marchi et al. (2011), some seeds presenting normal characteristics at the x-ray test, may present germination problems, possibly because of invisible infections with physiologically damaged or dead microorganisms and seeds, due to deterioration. Moreover, according to Socolowski and Cícero (2008), seeds that emitted the primary root and did not create normal seedlings could be at the initial stage of viability loss. However, it is important to highlight that various aspects shall be considered in the study of seed morphology. As well as the internal cavity occupied by the embryo and endosperm, the presence of mechanical damages, stains indicating tissue deterioration or seed malformations are characters that may influence germination.

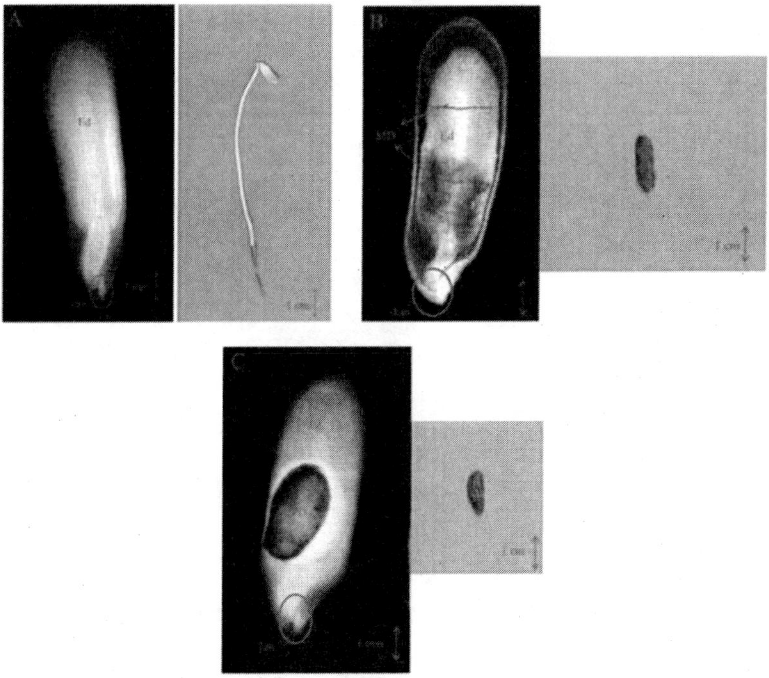

Figure 1 Seeds without damages from any species in the embryo (Em) and endosperm (Ed), originating normal seedlings (A); seeds with damages by tissue deterioration (TD) and mechanical damages (MD) originating dead seeds (B) and seeds with damages from insect predation (PD) originating dead seeds (C).

Generally speaking, results from the x-ray test classified lot 1 as the one with best quality and lot 4 as the one with worst quality. Results of malformed seed percentage, predated seeds, seeds with deteriorated tissues and seeds with mechanical damages are presented in Table 2. By data analysis, most seeds originated abnormal seedlings or dead seeds. However, these results must be compared with the ones observed in Table 1, that is, it is possible to observe that lot 4, with the highest percentage of malformed seeds and deteriorated tissues, presented lower percentage of normal seedlings and higher percentage of abnormal seedlings, in opposition to lot 1 (Table 1). However, lot 1, with the highest percentage of predated seeds, presented higher percentage of normal seedlings and lower percentage of abnormal seedlings (Table 1). Therefore, there is dependency of type and location of the damage on seeds and development of normal seedlings, abnormal seedlings or dead seeds. Similarly observed Marchi et al. (2011), who used the x-ray test as the procedure to evaluate the internal morphology and associate it with the physiological quality of peanut seeds. According to Socolowski and Cícero (2008), seed mass is an indicator of their physiological quality and x-ray images are important to observe the physical quality of seeds; this was similarly observed by Masetto et al. (2007), analyzing images of *Eugenia pleurantha*. Moreover, similarly to what was observed by Oliveira et al. (2004) in yellow ipe (*Tabebuia serratifolia* Vahl Nich.) and pink ipe (*Tabebuia impetiginosa* (Martius ex A. P. de Candolle) Standley) seeds, the x-ray test was effective in evaluating the internal morphology of pink trumpet tree (*Tabebuia heptaphylla*) seeds, and the internal damages detected in the x-rays affected germination, reducing the lot quality.

Table 2 Percentage (%) of *Senna multijuga* seeds from different damage classes characterized by the performance of x-ray tests. Seeds without damages (SD); malformed seeds (MS); predated seeds (PS); deteriorated seeds (DS); seeds with mechanical damages (MD).

Lot	SD	MS	PS	DS	MD
			(%)		
1	92 A*	1 A	1 A	4 A	1 A
2	86 B	3 AB	3 B	7 B	1 A
3	87 B	4 B	3 B	5 AB	1 A
4	79 C	8 C	3 B	8 B	2 A
CV (%)	9.6	10.1	1.1	4	1.2

*Means followed by the same letters, in each column, do not differ among themselves by Tukey's test at 5% significance

Parameters obtained by seed image analysis, using the *Tomato Analyzer*(r) software, can be found in Table 3. By the analysis of data, it is possible to observe that of all parameters evaluated by the software, the only one presenting significant differences among the lots the one evaluating seed area. Therefore, results are opposed to what was obtained by Gomes-Junior et al. (2013), who evaluated the effectiveness of the *Tomato Analyzer*(r)software in identifying variations in the development degree of the embryo and their possible relation with the physiological potential in cucumber seeds. The authors concluded that the semi-automated analysis, performed with the help of the *Tomato Analyzer*(r) software, allowed identifying close variations in the development degree of the embryos from different lots of cucumber seeds, observing that seeds whose embryonic area occupied less than 78.1% of the internal cavity were more likely to originate less developed seedlings. It is also worth mentioning that researches conducted by Machado and Cicero (2003), Pupim et al. (2008) and Otoni and McDonald (2005), which tried to associate the results of embryo development degree in relation to the available space in the seed internal cavity, estimated in a proportion expressed in percentage evaluated through image analysis techniques, to germination and vigor.

Studying *S. multijuga* seed morphology, Amorim et al. (2008) highlighted that the species present important characteristics that may help not only its identification, but that also affect the analysis of physiological quality, such as fruit characteristics, mainly dehiscence, shape, color and pericarp texture, common in the subfamily and seeds with the presence of closed pleurogram, areola and endosperm.

These characteristics may have caused the differences verified in the seed area (parameter A; Table 3), detected by the *Tomato Analyzer*(r) software, without significantly affect the relation between the existing free space area between embryo and tegument, influencing the physiological quality of different seed lots.

Thus, the results of this work demonstrated that for *S. multijuga* seeds, the measure of the seed Area parameter (A) expressed in a more evident way the differences in the physiological potential of lots compared to the Pericarp Area (PA) parameter, which evaluates the relation between the existing free space area between the embryo and the tegument.

Table 3 Mean values of perimeter (P), area (A), pericarp area (PA), generated from the analysis performed by the *Tomato Analyser*(r) software, of x-ray images obtained in the x-ray test of four *Senna multijuga* seed lots.

Lot	P (mm)	A (mm^2)	PA (mm^2)
1	22.94491 A*	81.654415 A	0.01203 A
2	22.6462 A	81.451482 A	0.01197 A
3	22.25488 A	79.669007 AB	0.01197 A
4	22.04137 A	76.915258 B	0.01197 A
CV (%)	2.36	2.37	0.36

*Means followed by the same letters, in each column for each period, did not differ among themselves by Tukey's test at 5% significance.

CONCLUSIONS

The x-ray image test is effective to evaluate the internal morphology of *Senna multijuga* seeds and their relation with vigor.

The *Tomato Analyzer*(r) software is sensitive to evaluate the development degree of *Senna multijuga* seeds.

REFERENCES

1. AMORIM, I.L.; DAVIDE, A.C.; FERREIRA, R.A.; CHAVES, M.M.F. Morfologia de frutos, sementes, plântulas e mudas de *Senna multijuga* var. *lindleyana* (Gardner) H. S. Irwin & Barneby - Leguminosae Caesalpinioideae. *Revista Brasileira de Botânica*, v.31, n.3, p.507-516, 2008.
2. BRASIL. Ministério da Agricultura, Pecuária e Abastecimento. *Regras para análise de sementes*. Ministério da Agricultura, Pecuária e Abastecimento. Secretaria de Defesa Agropecuária. Brasília: MAPA/ACS, 2009. 395p. http://www.agricultura.gov.br/arq_editor/file/Laborat%C3%B3rio/Sementes/Regras%20para%20Analise%20de%20Sementes.pdf
3. BREWER, M.T.; LANG, L.; FUJIMURA, K.; DUJMOVIC, N.; GRAY, M.; VAN DER KNAAP, E. Development of a controlled vocabulary and software application to analyze fruit shape variation in tomato and other plant species. *Plant Physiology*, v.141, n.1, p.15-25, 2006.
4. BREWER, M.T.; MOYSEENKO, J.B.; MUNFORTE, A.J.; VAN DER KNAAP, E. Morphological variation in tomato: a comprehensive study of quantitative trait loci

controlling fruit shape and development. *Journal of Experimental Botany*, v.58, n.6, p.1339-1349, 2007.

5. BREWER, M.T.; RODRIGUEZ, G.; GONZALO, M.J.; LANG, L.; SULLIVAN, D.; DUJMOVIC, N.; FUJIMURA, K.; GRAY, S.; VAN DER KNAAP, E. Tomato Analyzer (TA) user manual version 2.2.00. 18 p. 2008. http://www.oardc.ohio-state.edu/vander knap. Accessed on: March. 24th, 2016.
6. CARVALHO, N.M.; NAKAGAWA, J. *Sementes*: ciência, tecnologia e produção. Campinas: Fundação Cargill, 1988. 424p.
7. CARVALHO, N.M.; NAKAGAWA, J. *Sementes*: ciência, tecnologia e produção . 4.ed. Jaboticabal: FUNEP, 2000. 588p.
8. CARVALHO, P.E.R. *Espécies Arbóreas Brasileiras*. Brasília, DF: Embrapa Informação Tecnológica; Colombo: Embrapa Florestas, 2008, v.3, 593p.
9. FERREIRA, R.A.; DAVIDE, A.C.; MOTTA, M.S. Vigor e viabilidade de sementes de *Senna multijuga* (Rich.) Irwin et Barn. e *Senna macranthera* (Collad.) Irwin et Barn., num banco de sementes em solo de viveiro. *Revista Brasileira de Sementes*, v.26, n.1, p.24-31, 2004.
10. FRANÇA-NETO, J.B.; KRZYZANOWSKI, F.C.; COSTA, N.P. O teste de tetrazólio em sementes de soja. Londrina: EMBRAPA-CNPSo, 1998. 72p. (EMBRAPA-CNPSo. Documentos, 116).
11. FRANÇA-NETO, J.B. Teste de tetrazólio para determinação do vigor de sementes. In: KRYZANOWSKI, F.C.; VIEIRA, R.D.; FRANÇA-NETO, J.B. *Vigor de sementes*: conceitos e testes. Londrina. *ABRATES*, 1999. p.8.1-8.28.
12. GOMES-JUNIOR, F.G.; CHIQUITO, A.A.; MARCOS-FILHO, J. Semi-automated assessment of the embryonic area of cucumber seeds and its relationship to germination and seedling length. *Journal of Seed Science*, v.35, n.2, p.183-189, 2013.
13. LOPES, J.C.; DIAS, P.C.; PEREIRA, M.D. Maturação fisiológica de sementes de quaresmeira. *Pesquisa Agropecuária Brasileira*, v.40, n.8, p.811-816, 2005. http://www.scielo.br/pdf/pab/v40n8/a12v40n8.pdf
14. LORENZI, H. *Árvores Brasileiras*: manual de identificação e cultivo de plantas arbóreas do Brasil. 2.ed. São Paulo: Nova Odessa, 2002. v.2. 384p.
15. MACHADO, C.F.; CÍCERO, S.M. Aroeira-branca (*Lithraea molleoides* (Vell.) Engl. - Anacardiaceae) seed quality evaluation by the X-ray test. *Scientia Agricola*, v.60, n.2, p.393-397, 2003. http://www.scielo.br/pdf/sa/v60n2/15344.pdf
16. MARCHI, J.L.; CICERO, S.M.; GOMES JUNIOR, F.G. Utilização da análise computadorizada de plântulas na avaliação do potencial fisiológico de sementes de amendoim tratadas com fungicida e inseticida. *Revista Brasileira de Sementes*, v.33, n.4, p.652-662, 2011. http://www.scielo.br/pdf/rbs/v33n4/07.pdf
17. MARCOS-FILHO, J.; KIKUTI, A.L.P. Vigor de sementes de rabanete e desempenho de plantas em campo. *Revista Brasileira de Sementes*, v.28, n.3, p.44-51, 2006. http://www.scielo.br/pdf/rbs/v28n3/07.pdf
18. MARCOS-FILHO, J.; GOMES JUNIOR, F.G.; BENNETT, M.A.; WELLS, A.A.; STIEVE, S. Using tomato analyzer software to determine embryo size in x-rayed seeds. *Revista Brasileira de Sementes*, v.32, n.2, p.146-153, 2010. http://www.scielo.br /pdf/rbs/v32n2/v32n2a18.pdf
19. MARTINS, S.V.; SILVA, D.D. Maturação e época de colheita de sementes de *Dalbergia nigra* (Vell.) Fr. All. ex Benth. *Revista Brasileira de Sementes*, v.19, n.1, p.96-99, 1997. http://www.abrates.org.br/revista/artigos/1997/v19n1/artigo18.pdf
20. MASETTO, T.E.; DAVIDE, A.C.; SILVA, E.A.A.; FARIA, J.M.R. Avaliação da qualidade de sementes de *Eugenia pleurantha* (Myrtaceae) pelo teste de raios X. *Revista Brasileira de Sementes*, v.29, n.3, p.170-174, 2007. http://www.scielo.br /scielo.php?script=sci_arttext&pid=S0101-31222007000300020
21. MUNSELL COLOR COMPANY. Macbeth, Division of Kolmorgen Instruments Corporation. Munsel. *Soil Color Charts*, New Windsor, 1994. 10p.
22. OLIVEIRA, L.M .; CARVALHO, M.L.M .; GUIMARÃES, R.M .; MASETTO, T.E. Avaliação da qualidade de sementes de *Tabebuia serratifolia* Vahl Nich. e *T. impetiginosa* (Martius ex A. P. de Candolle) Standley - (Bignoniaceae) pelo teste de

raios X. *Revista Brasileira de Sementes*, v.26, n.2, p.138-143, 2004. http://www.scielo.br/pdf/rbs/v26n2/24500.pdf

23. OTONI, R.R.; McDONALD, M.B. Moisture and temperature effects on maize and soybean seedlings using the seed vigor imaging system. *Seed Technology*, v.27, n.2, p.243-247, 2005. http://www.seedtechnology.net/journal/st27no2.pdf
24. PIVETA, G.; MENEZES, V.O.; PEDROSO, D.C.; MUNIZ, M.F.B.; BLUME, E.; WIELEWICKI, A.P. Superação de dormência na qualidade de sementes e mudas: influência na produção de *Senna multijuga* (L. C. Rich.) Irwin & Barneby. *Acta Amazônica*, v.40, n.2, p.281-288, 2010. http://www.scielo.br/pdf/aa/v40n2/v40n2a06.pdf
25. PUPIM, T.L.; NOVEMBRE, A.D.L.C.;CARVALHO, M.L.M.; CICERO, S.M. Adequação do teste de raios-X para avaliação da qualidade de sementes de embaúba (*Cecropia pachystachya* Trec). *Revista Brasileira de Sementes*, v.30, n.2, p.28-32, 2008. http://www.scielo.br/pdf/rbs/v30n2/a04v30n2.pdf
26. SCALON, S.P.Q.; MUSSURY, R.M.; SCALON FILHO, H.; FRANCELINO, C.S.F.F.; FLORENCIO, D.K.A. Armazenamento e tratamentos pré-germinativos em sementes de jacarandá (*Jacaranda cuspidifolia* Mart.) *Revista Árvore*, v.30, n.2, p.179-185, 2006. http://www.scielo.br/pdf/rarv/v30n2/a04v30n2
27. SILVA, V.N.; CICERO S.M. Análise de imagens de plântulas para avaliação do potencial fisiológico de sementes de berinjela. *Horticultura Brasileira*, v.32, n.32, p.145-151, 2014. http://www.scielo.br/pdf/hb/v32n2/0102-0536-hb-32-02-00145.pdf
28. SOCOLOWSKI, F.; CÍCERO, S.M. Caracterização morfológica de embriões por imagens de raios X e relação com a massa e a qualidade fisiológica de sementes de *Tecoma stans* L. Juss ex Kunth (Bignoniaceae). *Revista Brasileira de Sementes*, v.30, n.2, p.200-208, 2008. http://www.scielo.br/pdf/rbs/v30n2/a25v30n2.pdf
29. TEKRONY, D.M. Precision is an essential component in seed vigour testing. *Seed Science and Technology*, v.31, n.2, p.435-447, 2003. https://doi.org/10.15258/sst.2003.31.2.20

CHAPTER 14

Seed Coating Increases Seed Moisture Uptake and Restricts Embryonic Oxygen Availability in Germinating Cereal Seeds

Linda Gorim [1] and Folkard Asch [2,*]

[1]*Department of Plant Sciences College of Agriculture and Bioresources, University of Saskatchewan, 51 Campus Drive, Saskatoon, SK S7N5A8, Canada*

[2]*Institute for Agricultural Sciences in the Tropics (Hans-Ruthenberg-Institute), University of Hohenheim, Garbenstr. 13, 70599 Stuttgart, Germany*

ABSTRACT

Seed coating is a technology to improve germination and homogenize stand establishment. Although coating often results in lower germination rates, seeds that do germinate grow more vigorously and show strongly reduced respiratory losses during reserve mobilization. We hypothesize that the higher mobilization efficiency is due to a shift in the enzymatic cleavage of sucrose from invertase to sucrose synthase in the embryonic tissue caused by a reduced oxygen availability induced by oversaturation with water caused by the coating during early germination. We investigated the effect of coating on barley, rye, and wheat seed imbibition during the first 30 h after seeds were placed in moisture. We profiled oxygen in the embryos and measured sucrose and acid invertase levels as imbibition progressed. We found that seeds within coatings absorbed significantly more moisture than uncoated seeds. Coating resulted in near anoxic oxygen concentrations in the developing embryonic tissues in all three species. In barley, sucrose was not cleaved via the invertase pathway, despite the fact that invertase activity in coated seeds was increased. In rye and wheat, invertase activities were significantly lower in embryos from coated seeds without significantly changing the sugar composition.

Keywords: barley; imbibition; invertase; oxygen profile; rye; sucrose metabolism; wheat

1. INTRODUCTION

Seed coating with hydro-absorbers as a technology to overcome water-related problems in drought-prone agricultural systems generally results in low or reduced germination rates [1,2,3,4]. Earlier studies attributed low germination rates to effects of the coating on water imbibition and available oxygen in the seed [5,6]. In contrast, in those seeds germinating, hydro-absorber coated barley, rye, and wheat seeds with coat shares greater than 75% of the average seed have been shown to promote better seedling growth compared to those seedlings growing from uncoated seeds [4] by increasing efficiency of grain reserve mobilization via a possible switch from invertase-based to sucrose synthase-based embryonic sucrose breakdown [7]. We have argued that this apparent switch in activity of sucrose metabolic enzymes could be due to a reduced oxygen supply to the embryo caused by a higher saturation of the seed with water supplied from the coating [7,8].

In general, germination progresses in three phases that are delineated by water uptake characteristics. Germination begins with rapid imbibition of water in phase I, followed by a plateau phase of seed moisture content in phase II, during which the enzymatic and energetic basis for reserve mobilization is laid [9,10,11]. Phase III begins with the radicle emergence and is characterized by a rapid increase in seedling water content and massive reserve mobilization from storage tissues. In an earlier study [7], we reported on early seedling growth and sucrose and glucose mobilization during phase III of germination. When growing from coated seeds, barley, rye, and wheat differed strongly in mobilization efficiency and thus, in the respiratory losses of reserves mobilized during germination as compared to seedlings growing from uncoated seed. This indicated a strong effect of the coating on the enzymatic breakdown of starch and sucrose during germination, suggesting a major influence on the seed metabolism during phase II of germination (the first 48 h of seed germination) when major enzymes involved in carbohydrate and protein metabolism are activated [11]. Carbohydrate metabolism has been shown to be active one hour after imbibition based on metabolite levels, suggesting an immediate increase in the activity of glycolysis and the Krebs cycle that facilitates early energy-demanding processes [8,12]. Transcripts and mRNAs encoding starch breakdown and several sucrose synthesizing and cleaving enzymes already present in dry seeds are activated during phase I and II of germination [12], with the level of activation (up- or downregulation) depending on environmental conditions and the physiological state of the seed [10]. The degradation of sucrose is a major pathway fueling glycolysis [13]. Sucrose can be cleaved by either invertase (into glucose and fructose) or by sucrose synthase into fructose and uridine diphosphoglucose, depending on tissue oxygen availability [13,14]. The recent development of oxygen-sensitive micro-sensors allows measuring oxygen profiles in living tissues. So far, research is focused mainly on developing and maturing seeds [15,16] and little is known about the oxygen distribution in tissues of germinating cereal seeds.

For the current study, we hypothesize that coating strongly affects water uptake to the seed, this in turn strongly influences the oxygen availability in tissues critical for successful mobilization of endosperm reserves which in turn has an effect on the activities of the sucrose cleaving enzymes, providing the developing embryo with the sugar required for growth. The focus of our study is, therefore, three pronged: (1) we investigated imbibition during the first 30 h in seeds of wheat, rye, and barley coated with the hydro-absorber, Stockosorb® (Evonik Nutrition & Care GmbH, Krefeld, Germany) compared to that of the uncoated seed employing the widely and successfully used (e.g., [17,18]) imbibition model developed by Peleg [19]. (2) We determined the oxygen concentration in the various tissues of the germinating seed with and without coating. (3) We investigated sucrose and glucose abundance in early embryonic tissue with the aim to determine which enzymatic activity may govern sucrose metabolisms in either coated or uncoated seeds.

2. MATERIALS AND METHODS

2.1. Plant Material and Treatments

Seeds of spring barley (*Hordeum vulgare* L., cv. Maltasia), winter rye (*Secale cereal* L., cv. Jobaro), and spring wheat (*Triticum aestivum* L., cv. Thasas) were obtained from Freudenberger Feldsaaten GmbH and used in all experiments. Seeds of all three cereal species were either uncoated or coated with a seed coating developed by Freudenberger Feldsaaten GmbH containing a specific amount of hydro-absorber (Stockosorb). The term "coated" refers to grains with a coat share greater than 75% of the total mass. The term "uncoated" refers to the original seed.

2.2. Determination of Grain Initial Moisture Content

The initial moisture content of the seed was determined following the standard procedure laid down by International Seed Testing Association (ISTA) [20]. For each cereal species, randomly selected 25 coated seeds, uncoated seeds, and kernels from within the coating (coated seeds of which the coating was removed) respectively were cut longitudinally using a sharp blade into halves and each half was then sliced four times. The sliced grains were immediately placed into 2 mL pre-weighed test tubes and weighed again. The samples were then transferred to an oven and dried at 103 °C for 17 h ± 15 min and later cooled in a desiccator for 1 h ± 15 min and weighed. The mass of the individual seeds was calculated by subtracting the mass of the empty test tubes from those of the test tube containing seeds. The amount of water lost was the difference between the weight in the pre-oven test tubes and post-oven test tubes. Therefore, the initial moisture content (MCi) was the moisture difference divided by the seed mass expressed in percentage.

2.3. Estimating Grain Moisture Content over Time

Twenty-five seeds were randomly selected for all cereals from all treatments and individually weighed on an electronic balance. Coated as well as uncoated seeds of each cereal were placed on moist filter paper (Ecolab-Bogen-Filterpapier, Neolab, Heidelberg, Germany) in labeled plates 19.5 cm by 19.5 cm and plates were immediately transferred to the growth chambers (Percival Intellus Ultra Controller) and maintained at 25 °C. Individual uncoated seeds and coated seeds with the coating attached and coated seeds with the coating removed were directly weighed every 3 h for the first 12 h and subsequently every 6 h until the

30th h. After each weighing time, the 25 coated seeds were quickly washed, blotted on filter paper to remove excess water, and weighed to get a weight of the seed within the coat. The difference between an individual grain weight at a given time t and its original weight at time t_0 divided by its weight at time t_0 expressed as a percentage plus MCi at time t_0 gives the moisture content (MC) at that point in time. The MC for the seeds within the coatings was estimated based on the assumption that their initial weight was the same as that of an average seed (on the thousand grain weight bases).

The moisture content in the coating was calculated as:

1. Individual coated grain (coat share >75%) – average uncoated grain = weight of coating
2. Weight of imbibed grain – weight imbibed seed within = weight of imbibed coating
3. Weight of imbibed coating – weight of coating = amount of water
4. Moisture content of coating = (amount of water/Imbibed coating) × 100

2.4. Estimating Grain Imbibition Rate and Capacity

Knowing the initial MC and MC enabled the estimation of the imbibition rate and imbibition capacity from the model postulated by Peleg [19] based on a two-parameter sorption equation, thus:

$$M_t = M_0 \pm \frac{t}{K_1 + K_2 t} \tag{1}$$

where M_t is the MC of grains at time t in percentage, M_0 is the initial MC (%), K_1 is the Peleg rate constant (min $\%^{-1}$), and K_2 is the Peleg capacity constant ($\%^{-1}$). High values of K_1 or K_2 imply low imbibition rate (IR) and low capacity, respectively, and the inverse is true for low values of K_1 and K_2. Imbibition is an absorptive process and so '±' in Equation (1) becomes '+'. The imbibition or absorption rate (R) was obtained from the first derivative of Equation (1):

$$R = \frac{dM}{dt} = \pm \frac{K_1}{(K_1 + K_2)^2} \tag{2}$$

The Peleg rate constant K_1 is the inverse of the initial imbibition rate (IR) of grains (R_0) at time t_0. The Peleg imbibition capacity constant K_2 relates to the maximum or minimum attainable grain MC. As $t \rightarrow \infty$, Equation (1) gives the relation between grain equilibrium moisture content (M_e) and K_2:

$$M|t_\infty = M_e = M_0 \pm \frac{1}{K_2} \tag{3}$$

In order to obtain K_1 and K_2, Equation (1) was linearized resulting in the following Equation:

$$\frac{t}{M_t - M_0} = K_1 + K_2 t \tag{4}$$

A plot of $t/M_t - M_0$ versus time, therefore, resulted in a straight line, with K_1 being the y-intercept and K_2 the slope, and from these constants, the equilibrium MC (M_e) and initial IR (R_0) were calculated. The water absorption capacity and IR between coated and uncoated seeds were estimated by comparing the slopes of the straight lines produced when Equation (4) was plotted. The imbibition rate was also expressed in terms of the amount of water imbibed by the grain per hour.

2.5. Measurement of Oxygen Concentrations in the Coated and Uncoated Seeds

The oxygen concentrations across a section of uncoated seeds or seeds coated with hydro-absorber were determined for barley, rye, and wheat using a less than 50 μm tip glass fiber needle-type oxygen micro-sensor (micro-sensor, Presens, Regensburg, Germany) connected to a fiber optic oxygen meter (Microx TX3, Presens) with the sensor mounted on a micro-manipulator, as described by Rolletschek et al. [21]. Before measurements began, the oxygen sensor together with the temperature sensor was calibrated in oxygen free sodium sulfite (Na_2SO_3) dissolved in 100 mL distilled water and later in pure water, as recommended by the manufacturers. Coated and uncoated seeds were randomly selected and placed intermittently in 30 min intervals at marked positions on moist filter paper and immediately transferred to the growth chambers at 25 °C. After 18 h and then every 6 h, seeds were removed in the sequence in which they were placed in the plates, the coating was quickly washed off and excess water was blotted off. Oxygen profiles were measured by fixing the kernel on a micro-manipulator stage and driving the oxygen needle tip at 50, 100, and then 200 μm distances into the embryo from the top and through the scutellum with the aid of the micro-manipulator; the readings were allowed to stabilize and were recorded, after which the needle was driven to a new position and the process was repeated (Figure 1). The path of the needle was verified and confirmed by dissecting each seed at the end of each measurement and viewing under a microscope. In order to measure the amount of oxygen in the coating, the same process was repeated but the hydro-absorber coating was not washed off and the whole grain was mounted. To protect the needled tip, the whole capsule was moved into the coating and the needle tip extended only at the sites where measurements were required. The thickness of coating, testa, and endosperm were determined by observing at least five representative seeds at 10-fold magnification under a stereo microscope (LeitzBioMeD) calibrated with a stage micrometer scale. Data are presented as means of at least six grains and oxygen was recorded as percentage saturation.

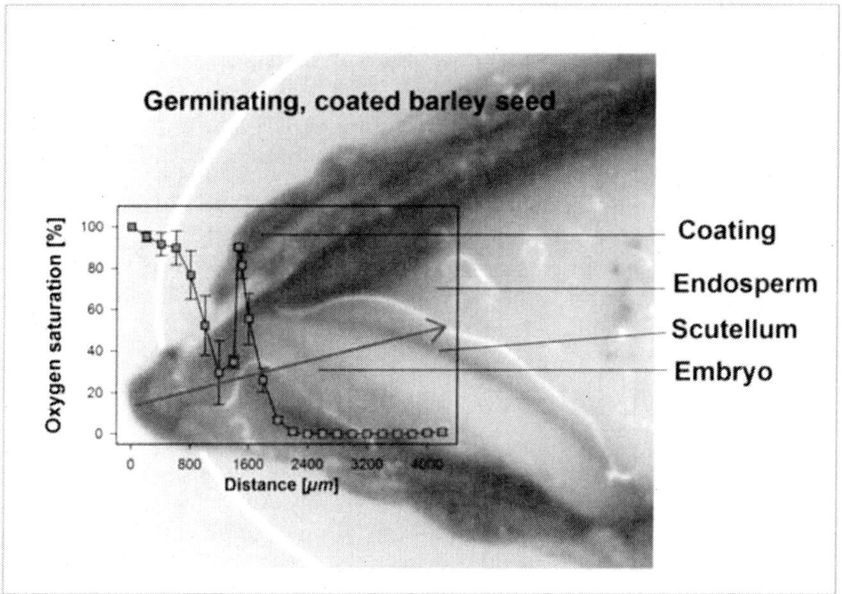

Figure 1. Exemplary illustration of an oxygen profile through the coating, the seed coat, and the embryo into the endosperm. The oxygen profile shown was taken along the path of the micro-probe, as indicated by the arrow on the image. Distances shown in the graph do not fully match dimensions in the image due to a slight skew in the longitudinal cut.

2.6. Extraction and Quantification of Sugars

Barley, rye, and wheat seeds, both coated and uncoated, were place on moist filter paper in plates, as previously reported in [7]. Every 6 hours from the 12th h, the coating was washed off the coated seed and in both coated and uncoated seeds, embryonic material was quickly excised with a stainless steel razor blade on ice, and freeze dried in 2 mL Eppendorf vials.

Ten milligrams of freeze dried embryonic material per replicate were ground in 'FastPrep Lysing matrix A' tubes at 6.0 m per second for 40 s in the FastPrep® (MP Biomedicals, Strasbourg, France)—24 homogenizer. Five hundred microliters (500 μL) 80% ethanol was added, the tubes vortex and heated at 60 °C for 30 min. Each tube was centrifuged for 10 min at 13,000× *g* and the supernatant transferred to another Eppendorf-vial. The process, from addition of ethanol to centrifugation, was repeated twice. The ethanol extract was evaporated to dryness in a vacuum centrifuge (Eppendorf Concentrator 5301, Eppendorf GmbH, Hamburg, Germany) at 45 °C for 2.5 h and re-dissolved in 900 μL water by shaking overnight. Sugars were assayed as described by Andersen et al. [22]. For glucose, 100 μL were assayed in a total volume of 200 μL water. The dye, glucose oxidase/peroxidase/ABTS-solution (400 μL), was added and samples were incubated in a water bath at 37 °C for 30 min.

For sucrose plus glucose, 50 μL were assayed in a total volume of 300 μL in NaAc-buffer containing 50 mM Na-acetate and 15 mM magnesium chloride (pH 4.6). Invertase solution, 2 μL, and the dye were added. Samples were read at 418 nm in a spectrophotometer (Beckmann, Type UV-Du640, Beckmann Coulter GmbH, Krefeld, Germany).

2.7. Calculation of the Amount of Sugars

From the linear standard curve, the amount of sugars in each sample measured per milligram dry matter or milligram reserves remobilized was calculated. For example, the amount of glucose per milligram sample (mg/mg) in an unknown sample:

$$x = \frac{[0.3\ \mathrm{mL}]\left[Y_1\left(\frac{\mathrm{mg}}{\mathrm{mL}}\right)\right][0.9\ \mathrm{mL}]}{[0.05\ \mathrm{mL}][Z\ \mathrm{mg}]} \tag{5}$$

where: x is the amount of sugar in mg/mg dry weight plant material; Y_1 (mg/mL) is the concentration of glucose obtained from the standard curve; 0.300 mL. [Y_1 (mg/mL)] is the amount of glucose in 300 μL assayed (mg); 0.300 mL. [Y_1 (mg/mL)]. [(0.900/0.05 mL)] is the amount of glucose in the total (900 μL) samples; *Z* mg can either be weight lost (∂) from grain or increase in dry matter ($\partial 1$).

2.8. Quantification of Enzymatic Activity

Crude enzyme extracts from approximately 10 mg of freeze dried and frozen material replicated three times were ground in 'FastPrep® Lysing matrix A' tubes at speeds of 6.0 $m \cdot s^{-1}$ for 40 s in the FastPrep-24 machine. Five hundred microliters (μL) of extraction buffer consisting of 50 mM HEPES-NaOH, 1 mM EDTA, and 2.5 mM dithiothreitol, (pH 7.0) was added to the samples that were vortex. The samples were then centrifuged for 10 min at 20,000× *g* to pellet insoluble material. The soluble protein extract was removed and insoluble proteins extracted with buffer containing 1 M NaCl, 50 mM HEPES-NaOH, 1 mM EDTA, and 2.5 mM dithiothreitol, pH 7.0. Soluble protein extract (400 μL) was dialyzed against extraction buffer for 20 h at 0 °C on a 14,000 MWCO dialysis membrane (Visking, Karlsruhe, Germany) to remove endogenous soluble carbohydrates. The concentration of total protein was measured in the extract as described by Bradford [23] using a bovine serum albumin (BSA) standard.

Activities of soluble acid invertase and insoluble acid invertase were measured as described by Tsai et al. [24] with minor modifications. Invertase extracts (20 μL) were assayed in a total volume of 300 μL, with an assay buffer containing 50 mM sodium acetate, 15 mM magnesium chloride, and 100 mM sucrose (pH 5.0). Assays were incubated for 1 h at 30 °C, with blank terminated immediately after addition of protein extracts. All reactions were terminated by boiling and 400 μL glucose oxidase/peroxidase/ABTS-solution added to all tubes that were incubated at 37 °C in a water bath for 30 min. Samples were read in a spectrophotometer (Beckman, Type UV-Du640) at a wavelength of 418 nm. Three samples were assayed for each seed category with duplicate quantification of each.

2.9. Statistical Analysis

The difference in moisture content and total acid invertase activity between seeds from within the coating and uncoated seeds was calculated with the aid of the Proc GLM procedure of SAS. Least significant differences between means for moisture contents were calculated for time that measurements were carried out in all cereals, meanwhile, for differences in total acid invertase activity, letters were allocated with dissimilar letters on a pair of bars denoting significant difference at alpha equals 5%. The differences in the amount of water imbibed by both seeds from the coating and uncoated seeds for a given treatment and resultant parameter and between cereals was also carried out by Proc GLM procedure in SAS and letters were allocated to denote significance at alpha equals 5%.

3. RESULTS

3.1. Effects of Coating on Imbibition

In order to germinate, seeds imbibe water. The entry of water into seeds increases their moisture linearly which later plateaus at equilibrium moisture content [18,19]. This pattern was also observed in the current study for barley, rye, and wheat seeds. Barley and wheat seeds from within the coating had higher moisture contents compared to their uncoated counterparts (Figure 2). This was reflected by the high imbibition rate constant (K_1) for seeds within the coating (SFC), while their imbibition capacity constant, K_2, was consistently lower (Table 1). The initial absorption rate (R_0) was higher in the coated seeds (entire grain) in all three cereal species as compared to the uncoated seeds, but the SFC had the lowest initial absorption rates (Table 1). The initial moisture content of the seeds within the coating was increased by coating in rye and wheat but not in barley. In comparison to the uncoated seeds, the equilibrium moisture content (M_e) was increased in the seeds from within the coats in all three cereals, with the largest increase (15.5%) in wheat and the smallest (4.5%) in barley (Table 1).

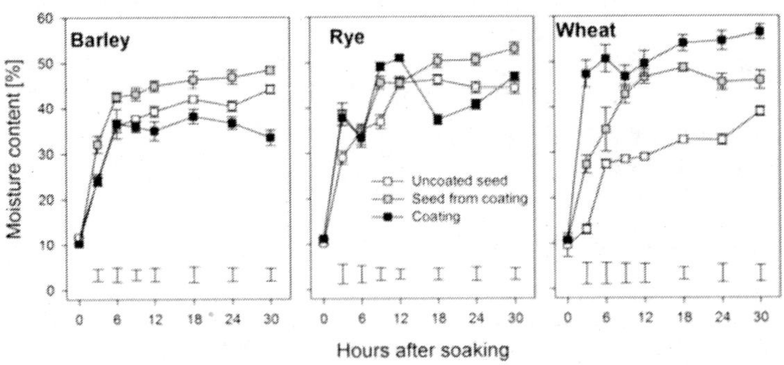

Figure 2. Moisture content of seeds within the coat, the uncoated seeds, and the coating during the first 30 h after soaking in barley, rye, and wheat. Error bars represent standard error of means and bars at the bottom of the graphs represent the least significant difference between treatments at each point in time.

Table 1. Imbibition characteristics during the first 30 h after soaking of coated and uncoated barley, rye, and wheat seeds. SFC refers to the seed within the coating; Uncoated refers to the original seed without coating. K_1 is the Peleg rate, K_2 the capacity constant, IC is the Imbibition capacity, and M_e is the equilibrium or saturation moisture content. Small letters refer to mean comparisons between SFC and the uncoated seed for each parameter analyzed at alpha equals 5%, meanwhile capital letters denote differences between cereals for the same treatment for any given parameter at alpha equals 5%.

Cereals	Treatments	$K_1 \times 10^{-2}$ (h %$^{-1}$)	$K_2 \times 10^{-2}$ (%$^{-1}$)	IC (%)	M_e (%)
Barley	SFC	5.0 (±0.8)^{aB}	2.5 (±0.1)^{aB}	40.4 (±0.8)^{aB}	51.1 (±0.8)^{aB}
	Uncoated	5.9 (±1.2)^{aB}	2.8 (±0.1)^{aB}	35.8 (±1.7)^{aA}	47.5 (±1.7)^{aA}
Rye	SFC	8.9 (±0.9)^{aA}	2.1 (±0.1)^{bC}	47.1 (±2.1)^{aA}	58.3 (±2.1)^{aA}
	Uncoated	5.4 (±0.3)^{bB}	2.6 (±0.2)^{aB}	37.9 (±0.3)^{bA}	48.1 (±0.3)^{bA}
Wheat	SFC	2.5 (±0.6)^{bB}	2.8 (±0.1)^{bA}	36.2 (±1.0)^{aB}	47.0 (±1.0)^{aB}
	Uncoated	11.8 (±2.1)^{aA}	4.1 (±0.2)^{aA}	24.3 (±0.9)^{bB}	33.9 (±0.9)^{bB}

3.2. Oxygen Profiles in Coated and Uncoated Seeds

Oxygen saturation levels were measured across coated and uncoated seeds of all three cereals through the embryonic tissue into the endosperm (Figure 1) at regular intervals during the first 48 h after soaking. The coating material was hard and could not easily be penetrated and because of this, we measured oxygen levels in two steps. The sensor was first inserted through the coating material until it reached the testa. Then, to avoid contaminations, the seed coating was washed off and the sensor was newly placed and inserted into the seed. Thus, the apparent increase of oxygen saturation between the coating and the seed in measurements from coated seeds is an artifact owed to the measuring technique (Figure 3). In general, oxygen saturation decreased rapidly in the coating of coated seeds by 80–90% of atmospheric oxygen levels and remained at levels of 1–5% in barley and rye embryos in seeds from within the coating and 8–10% in wheat embryos from seeds from within the coating (Figure 3). The three species differed substantially in the oxygen saturation levels in the embryo in the uncoated seeds during the first 30 h of germination. In barley, average oxygen saturation in the embryo varied between 50% and 80% of atmospheric levels; in rye, mean oxygen saturation was between 70% and 80%; and in wheat, oxygen saturation levels decreased with increasing distance into the embryo to levels of 20–30%, with mean values at about 40%. In barley and rye, endosperm oxygen saturation in

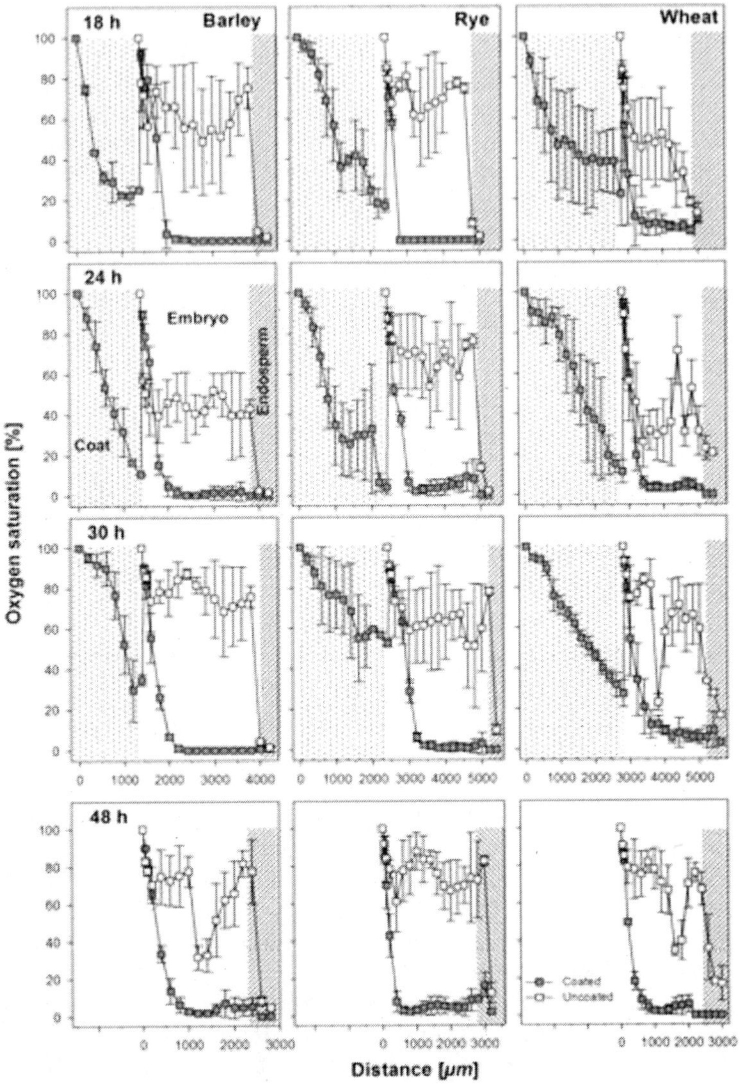

Figure 3. Oxygen profiles taken as indicated in Figure 1 at 18, 24, 30, and 48 h after soaking from coated and uncoated seeds of barley, rye, and wheat. The apparent increase of oxygen concentration after the coating material in the profiles from the coated seeds is an artifact due to the necessity to remove the coat before entering the micro-probe into the embryo. Coats where not present anymore at 48h after soaking and the radicula was visible. Error bars represent standard error of means.

coated and uncoated grains was between 1% and 10% whereas in wheat, endosperm oxygen saturation in uncoated seeds was at about 20% and in coated seeds between 1% and 5%. After 48 h, the radicle had already broken through and no more coating material was present at the site of sensor entry. Still, the embryonic oxygen saturation levels were strongly reduced to 1–5% in coated seeds as compared to uncoated seeds in all three species.

Theoretically, sucrose in either endosperm or scutellum is synthesized from starch breakdown and should thus be linearly related with endosperm mobilization, whereas glucose found in the embryo should be a product of sucrose cleavage in the embryo. If sucrose is cleaved into anything else but glucose and fructose, for example, UDP glucose, the analyses used in this study would not detect it as glucose. Figure 4 shows the relationship between mobilization of endosperm reserves and sucrose found in the respective embryos and the relationship between this sucrose and glucose found in the same tissues during the first 30 h of germination. Sucrose in the embryo was linearly related to the amount of endosperm reserves mobilized in all three species and amounts were similar with minor differences between species (Figure 4a–c). For glucose as related to sucrose, the situation was different. The method of sugar analyses used here only detects glucose but not UDP-glucose. A strong linear relationship between glucose and sucrose mobilized was found in the embryos from uncoated seed, indicating an active invertase-based cleavage pathway for sucrose, whereas less than a third of glucose was found in the embryos from coated seeds, indicating an inhibition of the invertase-based pathway (Figure 4d). For the embryos from coated seeds in rye and wheat, a similar pattern was found (Figure 4e–f), although in rye and wheat embryos from coated seeds, smaller amounts of sucrose mobilized endosperm reserves were found when compared with the uncoated seed (Figure 4b–c). However, in these species, very little glucose was found in embryos from uncoated seeds, despite the fact that sucrose was amply available (Figure 4e–f).

3.3. Effects of Coating on Soluble and Insoluble Invertase Activity

The activity of soluble and insoluble invertase in embryonic tissues of both coated and uncoated seeds did not show any specific significant trend at any specific point in time during the first 48 h. However, when combined and expressed as mean activity over 48 h, the total invertase activity was higher in barley embryos from coated seeds than in embryos from uncoated seeds, but significantly lower in embryos from coated seeds in rye and wheat (Figure 5).

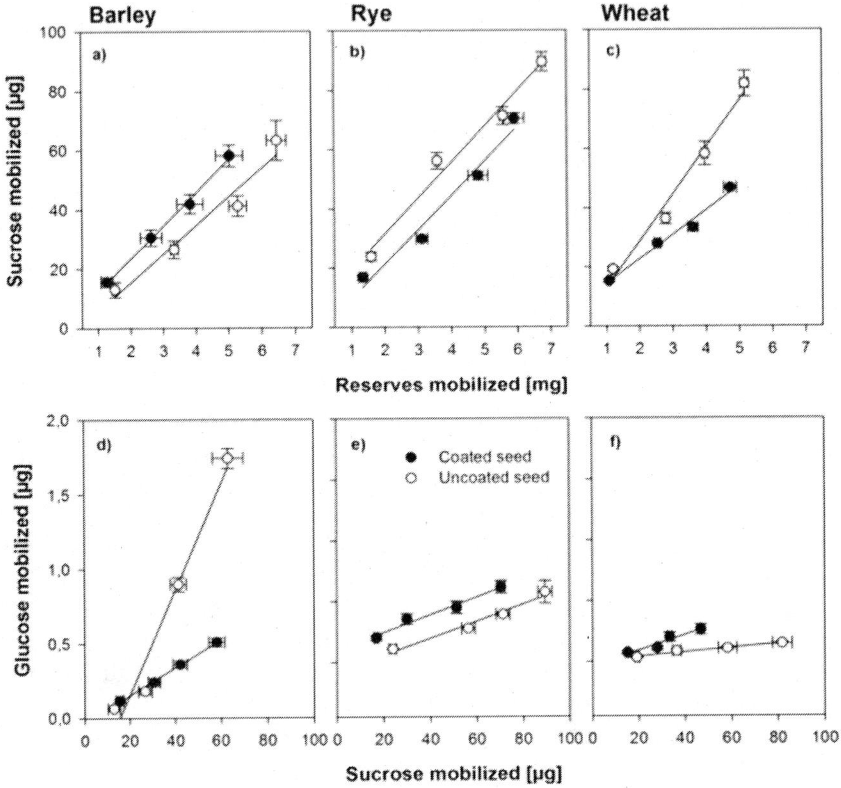

Figure 4. Relationships between the amount of reserves mobilized from the grain and the amount of sucrose found in the embryo (**a–c**) and the relationship between the amount of sucrose and glucose found in the embryonic tissues (**d–f**) in the seed from within the coat and uncoated seeds of barley, rye, and wheat. Error bars represent standard error of means.

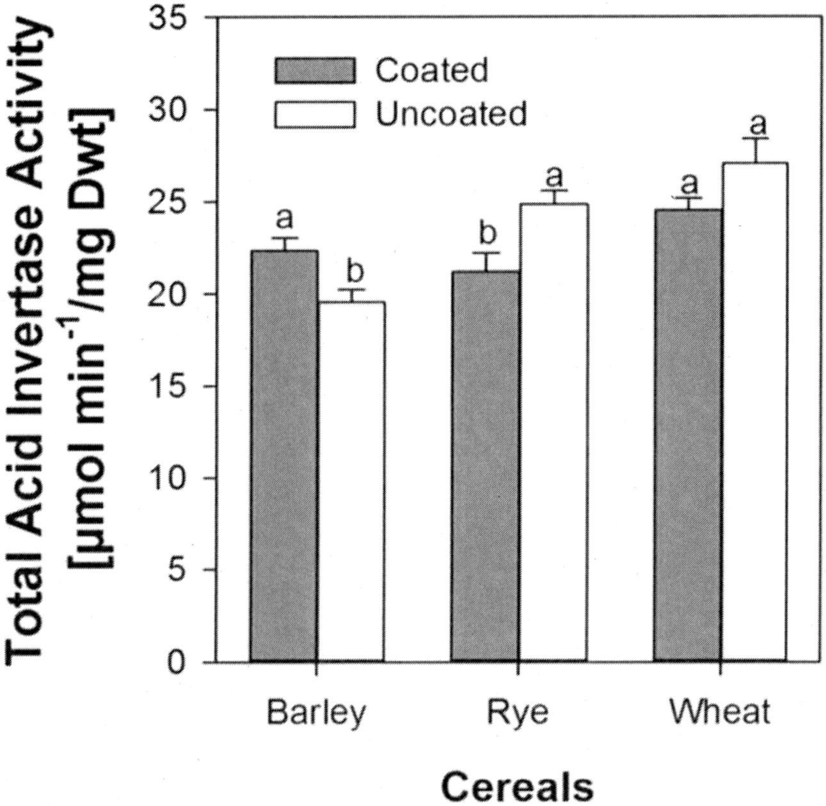

Figure 5. Total acid invertase activity averaged across the first 48 h after soaking in embryonic tissue from seeds from within the coat and uncoated seeds of barley, rye, and wheat. Error bars represent standard error of means. A bar pair with dissimilar letters is significantly different at alpha equals 5% and vice versa.

4. DISCUSSION

4.1. Effects of Coating on Water Imbibition

We studied water uptake of both coated and uncoated seeds for 30 h after imbibition. Imbibition begins during phase I of seed germination and continues through phase II [8]. Initially, there is a rapid uptake of water mainly through the micropyle of the dry seed (phase I) until all of the matrices and cell contents are fully hydrated [10,11,25]. This behavior was observed in both uncoated seeds and the seeds from within the coats of barley, rye, and wheat during imbibition; however, the seeds from within the coating showed a much steeper initial uptake, especially in wheat (Figure 2). This phase lasted for about 3 and 6 h in seeds from within the coating and uncoated barley and rye seeds, respectively, whereas duration of phase I was 6 and 12 h in wheat for uncoated seeds and seeds from within the coating, respectively. This delay in entering into phase II in wheat in the coated seeds may play a role in the delayed germination that was subsequently observed [4]. Although the moisture content of seeds from within the coating in these cereals was higher compared to that in the uncoated seeds, the imbibition rate was lower and the maximum imbibition capacity at any point was also lower in seeds from within the coating (Figure 2; Table 1). The reason for this could be that the water absorbed through the coating doted with hydro-absorber was channeled slowly to the kernel within, resulting in fewer membranes being damaged in the seeds enveloped by this coating compared to those in uncoated seeds, since it is known that rehydration imposes considerable stress upon cell components that may lead to leakage of solutes, which is indicative of temporal membrane damage [10,11].

Seed moisture content dynamics plotted to determine the imbibition rate and capacity showed similar patterns in both uncoated seeds and seeds from within coats with initial sharp increases in the amount of water absorbed which later level off at equilibrium with very high coefficients of determination (0.971–0.999), and this was in agreement with similar observations reported for other seeds [18,26]. However, the imbibition rate over time was increased in seeds for within the coat in barley and rye (Figure 2), resulting in more water absorbed into these seeds as imbibition progressed, which resulted in very high equilibrium moisture contents (Table 1). The difference in equilibrium moisture content between the seeds from within the coating and the uncoated seeds was 4.5%, 9.5%, and 15.5% for barley, rye, and wheat, respectively. The high equilibrium moisture content observed in rye and wheat could have led to oversaturation in the kernel within the hydro-absorber coating, resulting in low oxygen conditions, interruption of most metabolic processes—probably leading to tissue damage [10,25] and ultimately resulting in the poor germination rates reported in hydro-absorber coated wheat and, to a lesser extent, in coated rye [4].

4.2. Effect of Coating on Oxygen Availability

Oxygen concentrations in embryos have been investigated in maturing seeds with the aid of oxygen micro-sensors and, in some cases, reported in relation to the enzymes, invertase and sucrose synthase [15,16,21]. However, we investigated for the first time in situ the oxygen supply to the embryo in germinating seeds with the aid of oxygen micro-sensors. We measured oxygen concentrations through the center of the embryo passing through the scutellum in coated and uncoated seeds as well as in the hydro-absorber coating surrounding barley, rye, and wheat seeds at the site above the embryo. Measurements were done in the embryo because it is the site where products resulting from endosperm reserve mobilization are channeled to for seedling growth [27,28]. The presence of an imbibing hydro-absorber coating around the kernel resulted in strongly hypoxic conditions in the embryo. Relative to atmospheric conditions, oxygen concentration decreased in the hydro-absorber coating by 70% to 90%. Since the coating material was completely saturated with water, atmospheric oxygen was strongly inhibited to diffuse through this barrier to reach the seed within the coat. Thus, the coating and the water contained within effectively blocked the oxygen supply to the seed. Oxygen in the embryos of coated seeds was greatly reduced by 90–99% relative to atmospheric concentrations across species and time and about 45–75%, 60–70%, and 35–65% in barley, rye, and wheat, respectively, compared to their uncoated counterparts (Figure 3). However, oxygen measurements were carried out mostly in seeds, coated or uncoated, that would have germinated. Lower oxygen concentrations in the embryo do not imply that oxygen was completely lacking but that its amount was much reduced and with mobilization actively progressing, the amount available was quickly consumed. This was evident from oxygen profiles for tissues surrounding the embryo center which suggest that oxygen supply to and around the embryo vary depending on the position with respect to the embryo center and this also varied as imbibition progressed in both coated and uncoated seeds. The micropyle has been shown to be the main gateway through which water enters seeds [29] and may also serve as supplier of oxygen at the same time, since the endosperm of maturing cereal grains has been shown to be almost anoxic [15], as well as the endosperm of germinating grains, as shown here (Figure 3).

4.3. Effects of Coating on Sugar Metabolism and Enzyme Activity

Sucrose stored in the aleurone layer, and also produced from oil catabolism, is believed to be actively secreted into the endosperm and taken up by the embryo as an early energy source before starch breakdown becomes the dominant source of carbohydrate supply [30]. The oxygen concentration in the embryo affects the pathway of sucrose cleavage with hypoxic conditions resulting in inhibition of acid invertases and activation of sucrose synthase [13,14,31], whose level has been reported to be elevated in both barley and wheat embryos only but not at other sites [14]. This could easily show the importance of the pathway switch, since the reserves mobilized from the endosperm were proportional to the sucrose accumulation found in the embryo, thereby indicating no inhibition of the starch breakdown during germination resulting from low oxygen levels. However, the argument would hold only for both cereals (i.e., barley and wheat) if the glucose accumulated in the embryos of the uncoated seeds was proportional to the sucrose accumulated in the embryo, whereas no proportionality would be observed between glucose and sucrose accumulation in the embryonic tissues of seed from within the coat, which was the case for barley, but not for wheat (Figure 4). This may have resulted from different durations of the phasing during early germination, particularly imbibition and phase I [10,11], as described above, so the early germination phases may have not been fully aligned in relation to time, and consequently, sucrose transport or glycolysis in the embryonic tissue may have been delayed.

4.4. Effect of Hydro-Absorber Coating on Enzyme Activity

We determined the activities of soluble and insoluble invertase during the first 48 h after imbibition of all three cereal species for coated and uncoated seeds. Both activity as well as amounts present in the embryonic tissue did not vary systematically between species or coatings. Mean total invertase activity did show significant differences between species and coatings (Figure 5), however, it was not conclusive with regard to glucose and sucrose availability in the embryo. It is possible that we were not able to capture the depression of invertase activity that may have happened in vivo under strongly hypoxic conditions in the embryo, as the activity measurements were performed in vitro under laboratory conditions with oxygen present during the analyses. The mean values we found are in contrast to the report by Guglielminetti et al. [32], who reported very low invertase activity in barley and rye under anoxic conditions. We suggest that based on our knowledge of oxygen supply in the embryo, invertase was present but its activity was depressed, or, alternatively, both invertase and sucrose synthase might have been active at different times during imbibition.

5. CONCLUSIONS

As postulated earlier [7] we investigated moisture uptake to the uncoated seed and the seed within the coating as well as the oxygen profiles in the embryos of the respective seeds. We found that seeds within coatings absorbed significantly more moisture than uncoated seeds. Using a micro probe technology, we provided evidence that coating results in near anoxic oxygen concentrations in the developing embryonic tissues in all three species. In barley, sucrose was not cleaved via the invertase pathway anymore, despite the fact that invertase activity levels in coated seeds were increased. In rye and wheat, clear evidence from the sugar availability in the embryo could not be found, but invertase activities were significantly lower in embryos from coated seeds in these two species. In addition to genetic differences between the species, differences in the timing of imbibition and progressing germination may also have interfered with the measurements during the first 30 h of germination. The data on the enzymatic activities remained elusive, since within the experimental set-up it was not possible to measure enzymatic activities or abundance in situ. We could show that the functionality of the invertases was not compromised by the low oxygen environment. However, whether or not the activity is actually depressed within the living tissue remains the objective of further studies.

ACKNOWLEDGMENTS

This research was in part funded by the German Ministry for Science and Education (BMBF) under the reference Fkz: 01LS05103 and Freudenberger Feldsaaten GmbH. Magdeburger Strasse 2. 47800, Krefeld. Deutschland.

AUTHOR CONTRIBUTIONS

Folkard Asch and Linda Gorim conceived and designed the experiments; Linda Gorim performed the experiments; Folkard Asch and Linda Gorim analyzed the data.

CONFLICTS OF INTEREST

The authors declare no conflict of interest.

REFERENCES

1. Berdahl, J.D.; Barker, R.E. Germination and emergence of Russian wildrye seeds coated with hydrophilic materials. Agron. J. 1980, 72, 1006–1008.
2. Schneider, A.; Renault, P. Effects of coating on seed imbibitions: I. Model estimates of water transport coefficient. Crop Sci. 1997, 37, 1841–1849.
3. Willenborg, C.J.; Gulden, R.H.; Johnson, E.N.; Shirtliffe, S.J. Germination characteristics of polymer-coated canola (Brassica napus L.) seeds subjected to moisture stress at different temperatures. Agron. J. 2004, 96, 786–791.
4. Gorim, L.; Asch, F. Effects of composition and share of seed coatings on the mobilization efficiency of cereal seeds during germination. J. Agron. Crop Sci. 2012, 198, 81–91.
5. Baxter, L.; Waters, L. Effect of a hydrophilic polymer seed coating on the imbibitions, respiration and germination of sweet corn at four matrix potentials. J. Am. Soc. Hortic. Sci. 1986, 111, 517–520.
6. Klein, J.D.; Sachs, M. Measurement of water uptake and volatile production by coated wheat seeds and subsequent seedling growth. Seed Sci. Technol. 1992, 20, 299–305.
7. Gorim., L.; Asch, F. Seed coating reduces respiration losses and affects sugar metabolism during germination in cereals. Funct. Plant Biol. 2015, 42, 209–218.
8. Zeng, Y.; Wu, Y.; Avigne, T.W.; Koch, E.K. Differential regulation of sugar-sensitive sucrose synthases by hypoxia and anoxia indicate complementary transcriptional and posttranscriptional responses. Plant Phys. 1998, 116, 1573–1583.
9. Howell, K.A.; Narsai, R.; Carroll, A.; Ivanova, A.; Lohse, M.; Usadel, B.; Millar, A.H.; Whelan, J. Mapping metabolic and transcript temporal switches during germination in rice highlights specific transcription factors and the role of RNA instability in the germination process. Plant Phys. 2009, 149, 961–980.
10. Nonogaki, H.; Bassel, G.W.; Bewley, J.D. Germination-still a mystery. Plant Sci. 2010, 179, 574–581.
11. Weitbrecht, K.; Müller, K.; Leubner-Metzger, G. First off the mark: Early seed germination. J. Exp. Bot. 2011, 62, 3289–3309.
12. Sreenivasulu, N.; Usadel, B.; Winter, A.; Radchuk, V.; Scholz, U.; Stein, N.; Weschke, W.; Strickert, M.; Close, T.J.; Stitt, M.; et al. Barley grain maturation and germination: Metabolic pathway and regulatory network commonalities and differences highlighted by new MapMan/PageMan profiling tools. Plant Phys. 2008, 146, 1738–1758.
13. Koch, K. Sucrose metabolism: Regulatory mechanisms and pivotal roles in sugar sensing and plant development. Curr. Opin. Plant Biol. 2004, 7, 235–246.

14. Perata, P.; Guglielminetti, L.; Alpi, A. Mobilization of endosperm reserves in cereal seeds under anoxia. Ann. Bot.1997, 79, 49–56.
15. Rolletschek, H.; Koch, K.; Wobus, U.; Borisjuk, L. Positional cue for the starch/lipid balance in maize kernels and resource partitioning to the embryo. Plant J. 2005, 42, 69–83.
16. Rolletschek, H.; Stangelmayer, A.; Borisjuk, L. Methodology and significance of micro-sensor-based oxygen mapping in plant seeds—An overview. Sensors 2009, 9, 3218–3227.
17. Maharaj, V.; Sankat, C.K. Rehydration characteristics and quality of dehydrated dasheen leaves. Can. Agric. Eng. 2000, 42, 81–85.
18. Turhan, M.; Sayar, S.; Gunasekaran, S. Application of Peleg model to study water absorption in chickpea during soaking. J. Food Eng. 2002, 53, 153–159.
19. Peleg, M. An empirical model for the description of moisture sorption curves. J. Food Sci. 1988, 53, 1216–1217.
20. International Seed Testing Association (ISTA). International Rules for Seed Testing; The International Seed Testing Association (ISTA): Bassersdorf, Switzerland, 2010.
21. Rolletschek, H.; Borisjuk, L.; Koschorreck, M.; Wobus, U.; Weber, H. Legume embryo develop in a hypoxic environment. J. Exp. Bot. 2002, 53, 1099–1107.
22. Andersen, N.M.; Asch, F.; Wu, Y.; Jensen, R.C.; Naested, H.; Mogensen, O.V.; Koch, E.K. Soluble invertase expression is an early target of drought stress during the critical, abortion-sensitive phase of young ovary development in maize. Plant Phys. 2002, 130, 591–604.
23. Bradford, M.M. A rapid and sensitive method for the quantification of microgram quantities of protein utilizing the principle of protein dye binding. Ann. Biochem. 1976, 72, 248–254.
24. Tsai, C.Y.; Salamini, F.; Nelson, O.E. Enzymes in carbohydrate metabolism in the developing endosperm of maize. Plant Phys. 1970, 46, 299–306.
25. Leopold, A.C. Volumetric components of seed imbibition. Plant Phys. 1983, 73, 677–680.
26. Meyer, C.; Steudle, E.; Peterson, C.A. Patterns and kinetics of water uptake by soybean seeds. J. Exp. Bot. 2007, 58, 717–732.
27. Fincher, G.B. Molecular and cellular biology associated with endosperm mobilization in germinating cereal grains. Annu. Rev. Plant Phys. Plant Mol. Biol. 1989, 40, 305–346.
28. Aoki, N.; Scofield, G.N.; Wang, X.-D.; Offler, C.E.; Patrick, J.W.; Furbank, R.T. Pathway of sugar transport in germinating wheat seeds. Plant Phys. 2006, 141, 1255–1263.
29. Rathjen, J.R.; Strounina, E.V.; Mares, D.J. Water movements into dormant and non dormant wheat (Triticum aestivum L.) grains. J. Exp. Bot. 2009, 60, 1619–1631.
30. Chrispeels, M.J.; Tenner, A.J.; Johnson, K.D. Synthesis and release of sucrose by the aleurone layer of barley: Regulation by gibberelic acid. Planta 1973, 113, 35–46.
31. Zeng, Y.; Wu, Y; Avigne, W.T.; Koch, K.E. Rapid repression of maize invertases by low oxygen. Invertase/sucrose synthase balance, sugar signaling potential, and seedling survival. Plant Phys. 1999, 121, 599–608.
32. Guglielminetti, L.; Perata, P.; Alpi, A. Effect of anoxia on carbohydrate metabolism in rice seedlings. Plant Phys. 1995, 108, 735–741.

CHAPTER 15

Proteomic Analysis of Tung Tree (*Vernicia fordii*) Oilseeds during the Developmental Stages

Zhiyong Zhan [1,2,3,†], Yicun Chen [1,2,†], Jay Shockey [4], Xiaojiao Han [1,2] and Yangdong Wang [1,2,*]

[1]*State Key Laboratory of Tree Genetics and Breeding, Chinese Academy of Forestry, Beijing 100091, China*

[2]*Research Institute of Subtropical Foresty, Chinese Academy of Forestry, Fuyang 311400, Zhejiang, China*

[3]*Jiangxi Academy of Forestry, Nanchang 330013, Jiangxi, China*

[4]*United States Department of Agriculture-Agricultural Research Service, Southern Regional Research Center, New Orleans, LA 70124, USA*

ABSTRACT

The tung tree (*Vernicia fordii*), a non-model woody plant belonging to the Euphorbiaceae family, is a promising economic plant due to the high content of a novel high-value oil in its seeds. Many metabolic pathways are active during seed development. Oil (triacylglycerols (TAGs)) accumulates in oil bodies distributed in the endosperm cells of tung tree seeds. The relationship between oil bodies and oil content during tung tree seed development was analyzed using ultrastructural observations, which confirmed that oil accumulation was correlated with the volumes and numbers of oil bodies in the endosperm cells during three different developmental stages. For a deeper understanding of seed development, we carried out proteomic analyses. At least 144 proteins were differentially expressed during three different developmental stages. A total of 76 proteins were successfully identified using matrix-assisted laser desorption/ionization time-of-flight mass spectrometry/mass spectrometry (MALDI-TOF/MS/MS). These proteins were grouped into 11 classes according to their functions. The major groups of differentially expressed proteins were associated with energy metabolism (25%), fatty acid metabolism (15.79%) and defense (14.47%). These results strongly suggested that a very high percentage of gene expression in seed development is dedicated to the synthesis and accumulation of TAGs.

Keywords: tung tree; seed; oil content; oil body; development; proteomics

1. INTRODUCTION

Seed development and germination have been studied using transcriptome and proteomic analyses, either independently or in combination [1], and these methods have had an increased impact on the study of the seed proteome [2]. Most of the papers related to seed proteomics have focused on the construction of reference proteome maps at specific stages [3,4] including dormancy [5], germination [2,6,7], post-germination [8,9], development and maturation [10,11]. During this period, comparative proteomics was used as a basic method to provide insight into the identification of functional gene products and their expression levels in seeds [12]. Based on changes in protein expression in oilseeds, which contain an abundance of oils compared to starch or protein, seeds proteomics has emphasized the importance of energy and metabolism during seed development and germination. Several proteins related to oil mobilization were successfully identified in *Jatropha curcas* [6,11] and *Brassica napus* [13,14]. The tung (*Vernicia fordii* H. and *V. montana* L., previously classified as *Aleurites fordii*) tree is a multipurpose perennial plant belonging to the genus *Vernicia* of the spurge family (Euphorbiaceae). Tung is adaptable to several soil types, provided that proper drainage and aeration conditions are met [15]. Individual tung fruits typically contain multiple (usually 4–5) seeds surrounded by a thick verrucose seed coat. Tung oil, extracted from the seeds, contains 80% (mole %) α-eleostearic acid, a conjugated trienoic 18-carbon fatty acid ($18{:}3\Delta^{9cis,\ 11trans,\ 13trans}$) that imparts useful drying and blending properties to the oil [16]. Currently, tung oil is widely used in industries as a drying oil and is also used in paints, high-quality printing, plasticizers, and certain types of medicines and chemical reagents

[17,18,19]. China collectively produces approximately 80,000 tons of oil per year, which represents 70%–80% of the world market [19]. However, the output of tung oil in China still will not meet the projected requirements of the international market. The ultimate objective of the tung tree breeding program in China is to create a new hybrid species or find improved varieties with enhanced oil yield, quality and resistance.

Most of the previous research on tung oil focused on a single or a few genes that are related to the regulation of tung oil synthesis [20,21,22,23], therefore, the results of the studies have not provided a comprehensive picture of metabolism during seed development. In this study, a powerful proteomic approach was applied to study the differential protein expression in tung trees during seed development of the tung trees to provide a critical analysis that will assist in the development of new and improved breeding strategies in China.

2. RESULTS

2.1. Tung Endosperm Cell Ultrastructure

The most obvious subcellular structures (Figure 1) were the protein storage vacuoles (PSV), vacuoles (V) and oil bodies (O). The PSV and V from the first time point (Figure 1A1) had average diameters of approximately 8 μm (±0.05, n = 30) and 10 μm (±0.07, n = 30), respectively. The diameters of the PSV and V decreased during seed growth. The structure of the PSV from the second and third time points was approximately 4 μm (±0.03, n = 30) (Figure 1B1) and 3 μm (±0.05, n = 30) (Figure 1C1), respectively. The structure of V largely disappeared during the last stage (Figure 1C1). The oil bodies were distributed throughout the entire tissue area in all three developmental stages. An obvious increase in the number and volume of the oil bodies was observed in the second and third developmental stages, as the PSV and V declined (Figure 1A2,B2,C2). In addition, the single oil bodies appeared smoother in the third developmental stage compared to the earlier time points (Figure 1A3,B3,C3).

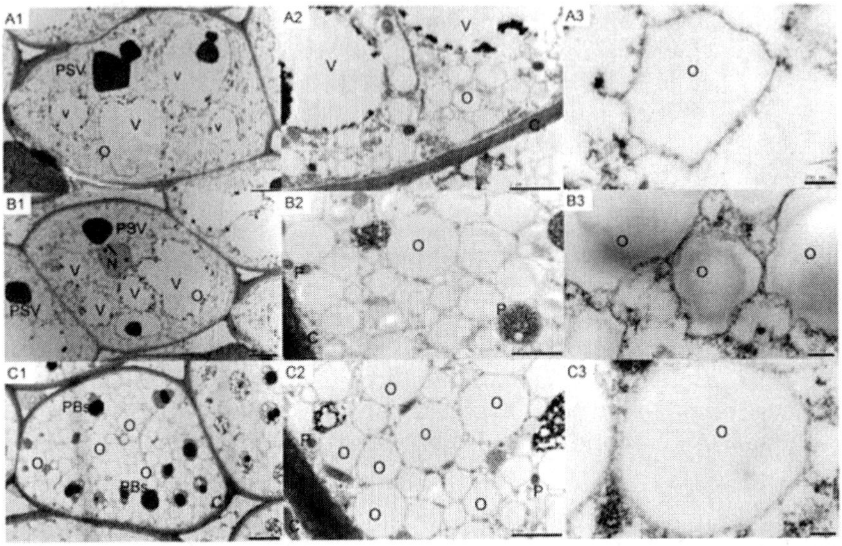

Figure 1. Ultrastructural observations of tung tree seeds across three development stages. Tung tree seeds prepared on: 25 August (**A**); 9 September (**B**); and 26 September (**C**). Images of: the whole tissue sample (**1**); cell sections (**2**); and single oil bodies (**3**). O, oil body; PSV, protein storage vacuole; PBs, protein bodies; P, protein particle; V, vacuole; C, cell wall; and N, nucleus.

2.2. Oil Content and Fatty Acid (FA) Composition

The oil content from three independent endosperm samples from each of the three different developmental stages was measured. The oil content significantly increased during seed development (Figure 2). Eleostearic acid was the most abundant fatty acid, which was followed by linoleic acid, and oleic acid. The eleostearic acid percentage of the total fatty acids increased from 72.28% to 77.5% (Table 1). Linolenic acid was undetectable at the two later time points. No significant decreases were observed for any of the other fatty acids. These results indicated that polyunsaturated fatty acids, particularly eleostearic acid, accumulate at high levels in tung seeds during all three developmental stages.

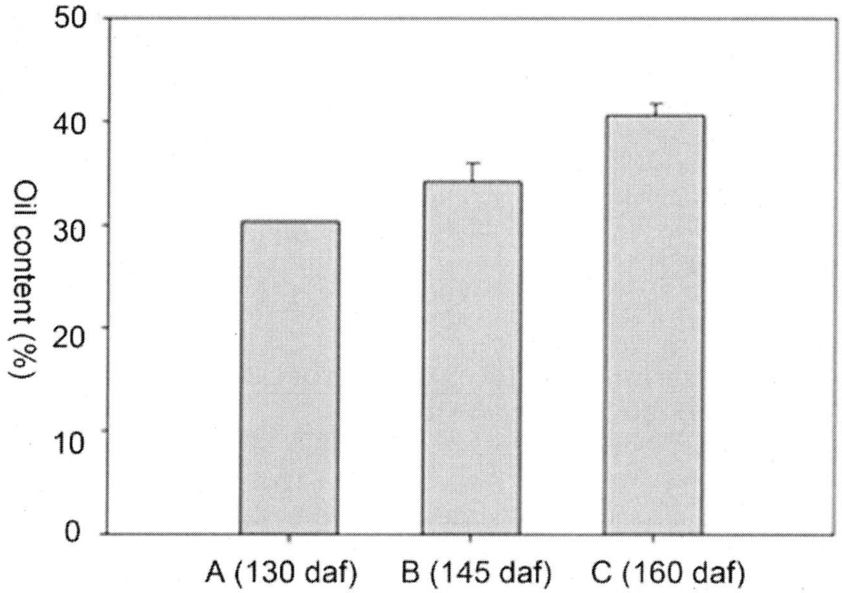

Figure 2. The oil content (%) in tung seeds of three developmental stages. A, B and C represent the seeds harvested on 25 August, 9 September and 26 September, respectively.

Table 1. Changes in the fatty acid composition (%) of tung tree seed oil in the three developmental stages.

Time	Fatty Acid					
	Palmitic Acid	Stearic Acid	Oleic Acid	Linoleic Acid	Linolenic Acid	Eleostearic Acid
8.25 (A)	3.74 ± 0.02	2.75 ± 0.03	7.42 ± 0.08	13.82 ± 0.12	0.21 ± 0.15	72.28 ± 0.06
9.9 (B)	2.50 ± 0.05	1.95 ± 0.01	7.99 ± 0.06	9.11 ± 0.01	-	77.47 ± 0.05
9.26 (C)	2.35 ± 0.06	2.17 ± 0.02	8.03 ± 0.08	7.85 ± 0.05	-	77.49 ± 0.02

The content of each fatty acid was calculated as the percentage of the total measured fatty acids. The "-" denotes that the fatty acid data were undetectable. The values are the means of three biological replicates (±SD).

2.3. Comparison of the Different Protein Extraction Methods

Recalcitrant plant tissues, such as tung seeds and other plant seeds, contain many compounds which impede protein extraction. Therefore, three different methods, including trichloroacetic acid (TCA)-acetone extraction (M1), phenol extraction (M2), and TCA-acetone combined with phenol extraction (M3), were compared to identify the most suitable method for proteomic analysis (Supplementary Materials Figure S1). The protein extraction yield using M1 was the highest, followed by M2 and M3; inverse results were observed for protein quality (data not shown). The highest number of protein spots was visualized in the M1 2-DE gel followed by M3 and M2. M3 gave the best resolution of the 2-DE gels and the clearest gel background (Supplementary Materials Figure S1). These results indicated that although M3 is a complicated extraction protocol, it was the most suitable for the comprehensive proteomic analysis of tung tree seeds.

2.4. Identification of Differentially Expressed Proteins in Different Functional Categories

At least 144 protein spots changed in abundance between the three different seed developmental stages. Using MALDI-TOF/MS/MS (Figure 3), 76 protein spots were successfully identified and these protein spots were analyzed in greater detail (Supplementary Materials Figure S2, Table S1). The proteins could be sorted into eleven functional groups as follows: energy metabolism (25.01%), fatty acid metabolism (15.79%), defense-related (14.47%), protease (11.84%), unknown (9.21%), peroxidase (6.58%), signal transduction (5.26%), cell construction (5.26%), transcription-related (2.63%), protein modification (2.63%) and storage (1.32%) (Figure 4). The percentage in parentheses represents the proportion of each protein group.

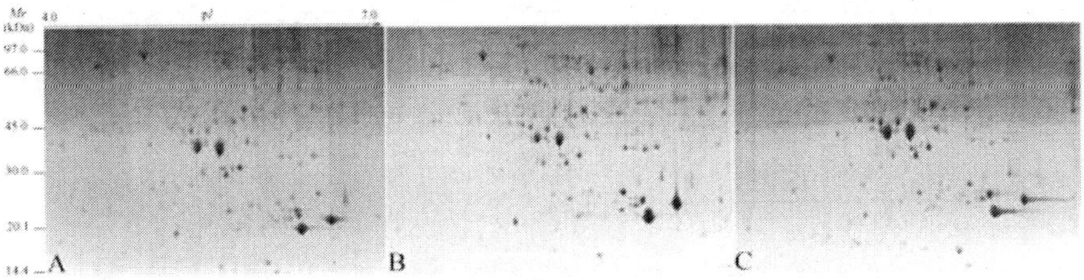

Figure 3. Images from the 2-DE gel maps of tung oilseeds from the three developmental stages tung tree seeds collected at three time-points: 25 August (**A**); 9 September (**B**); and 26 September (**C**).

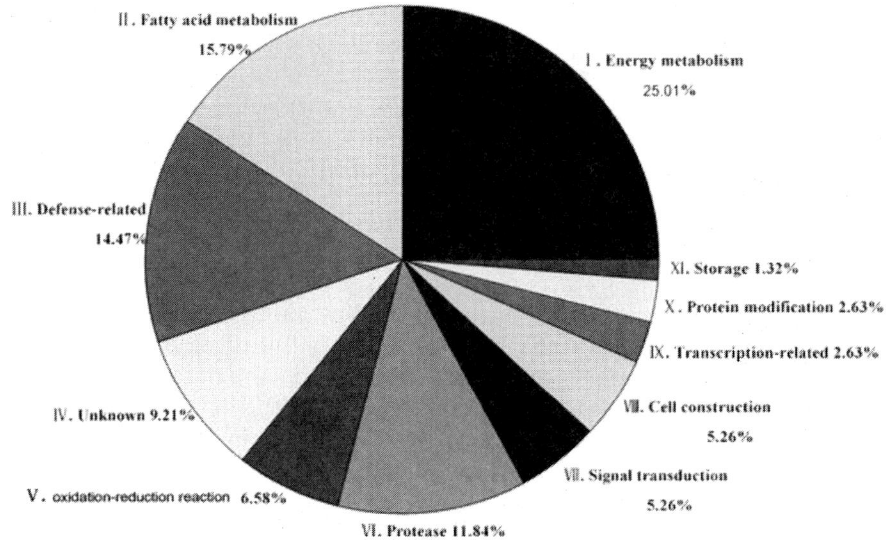

Figure 4. The functional proportion of the identified proteins expressed in the tung tree seeds during the three developmental stages. A total of 76 identified differential proteins were assigned to functional categories. The Roman numeral in each category corresponds to the functional category described in Supplementary Materials Table S1. The percentage represents the proportion of each category.

2.4.1. Energy Metabolism

The proteins in this group are involved in energy metabolism during seed development and included proteins with potential roles in glycolysis (S68), the tricarboxylic acid cycle (S29, S54 and S63), citric acid cycle (S45 and S56), energy-related (S11, S18 and S70) and other metabolic processes (S19, S28, S30, S33, S43, S55, S62, S72 and S74). Most of the proteins in this group, (10 out of 19 spots), were continuously up-regulated during seed development (i.e., the protein level increased in each successive time point). Only the protein of 6-phosphogluconate dehydrogenase (S18), which is a participant in the pentose phosphate pathway, was continuously down-regulated, whereas proteins S11 and S70, which are involved in other energy-related processes, were up-regulated during seed development (Supplementary Materials Table S1). In plants, glutamine synthetase (S33) (Supplementary Materials Figure S2) participates in primary metabolic processes, such as the biosynthesis of amino acids and nitrogen metabolism, and is up-regulated during tung seed development. Caffeoyl-

CoA *O*-methyltransferase (S19), *S*-adenosylmethionine synthetase (S55) and UTP-glucose-1-phosphate uridylyltransferase (S72) were identified as continuously up-regulated proteins in this data set (Supplementary Materials Figure S2, Table S1), which is not surprising given their likely roles in secondary metabolic processes, such as the biosynthesis of lignin (S19) and the phenylpropanoid pathway (S55).

2.4.2. Fatty Acid Metabolism

The proteins related to fatty acid metabolism in this group were mainly enzymes involved in the downstream mobilization and metabolism of oil-derived products. These enzymes function in the glyoxylate cycle, the citric acid cycle, glycolysis and the pentose phosphate pathway. Pyruvate dehydrogenase E1 component subunit alpha (S50), which participates in the citric acid cycle, showed a three-fold increase in abundance during seed development. Other up-regulated proteins included stearoyl-acyl carrier protein desaturase (S31, S57), acyl-transferase (S46), plastid 3-ketoacyl-ACP synthase (KAS) (S58) and 3-oxoacyl-[acyl-carrier-protein] reductase (S76), which likely participate in the biosynthesis of tung oil and other storage or membrane lipids in tung seeds. This result indicated that the observed increase in the accumulation of tung oil between 25 August and 26 September required increased levels of gene expression from this group. The abundance of NAD(P)-binding Rossmann-fold-containing protein (S4), esterase precursor (S17) and acetyl-CoA carboxylase (ACCase, S64) were decreased in the seeds harvested on 26 September compared with the seeds of 25 August (Figure 5). These data, particularly for ACCase, suggested that fatty acid synthesis begins to decline as the seeds approach complete maturation. Noticeably, the fatty acid metabolism group shares some proteins with the energy metabolism group, including malate dehydrogenase (S63), which plays an important role in the glyoxylate cycle.

2.4.3. Defense-Related

The proteins in this group are mainly involved in the stress-response (S3, S69), protection (S13, S14, S32, S34, S47) and detoxification (S23, S37, S59, S61). Most of the proteins in this group were up-regulated and their likely functions involved responses to abiotic stresses given their roles in cold, heat, salt and drought tolerance in many other plant species [24]. Nucleoside diphosphate kinase 2 (NDPK2) (S3), which is a member of NDP kinases family, enhances salt stress tolerance [25]. Glycine-rich proteins (GRP) (S69) improve tolerance against freezing stresses [26]. Defense-related proteins, similar to GRP (S10), are associated with signal transduction pathway. Heat shock proteins (HSP) (S13, S14, S32, S34, and S47) were also up-regulated, likely to assist in proper protein folding and to guard against the deleterious effects of the high temperatures encountered during the summer months as tung seeds mature [27]. Superoxide dismutase (SOD) (S23 and S37) may act as a first line of defense against reactive oxygen species (ROS), a process in which MnSOD (S37) is the principal antioxidant enzyme [28]. Cysteine protease inhibitor (S59) and alcohol dehydrogenase (S61) were up-regulated during seed development and likely participate in detoxification pathways. Other functional categories included protein modification, storage, peroxidase, signal transduction, cell construction, transcription-related, and proteins with unknown functions (Supplementary Materials Table S1).

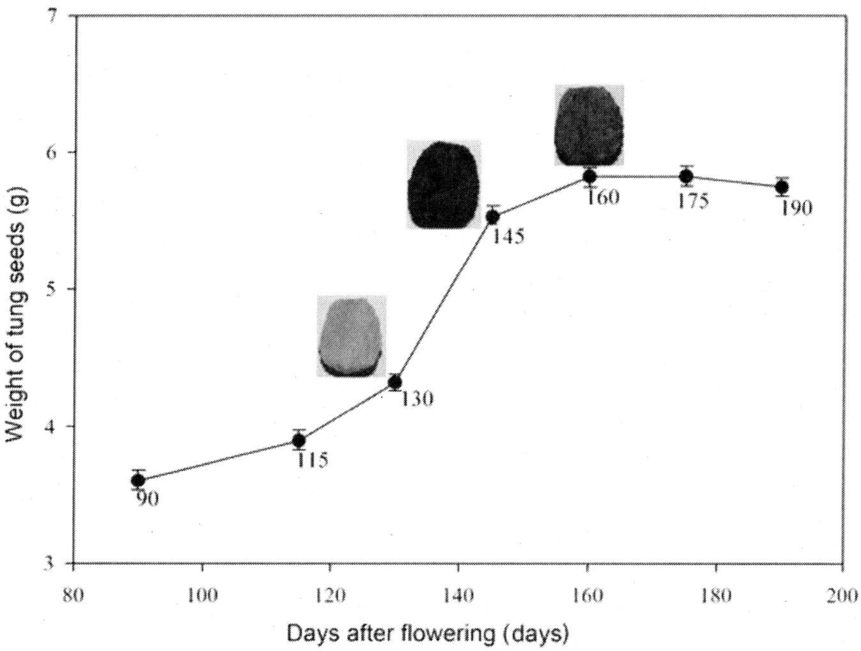

Figure 5. The weight of the tung seeds during development. The *x*-axis units indicate the number of days after flowering. The *y*-axis units indicate the average fresh weight (in grams) of six randomly selected seeds.

3. DISCUSSION

3.1. Efficient Extraction Methods for Tung Tree Seed Proteins

Isaacson et al. [29] recently compared the following two basic protocols for plant proteomics protein preparation: one is based on TCA–acetone extraction and the second relies on phenol precipitation. As the most commonly used protein extraction technique, TCA–acetone extraction results in protein pellets that are hard to re-solubilize and always retains some insoluble material, which results in protein losses. However, a high number of plant samples can be processed using TCA–acetone extraction. This method effectively inhibits proteases as well as phenoloxidases and peroxidases [30]. Phenol extraction protocols were first reported by Hurkman and Tanaka [31]. Phenol interacts with proteins through hydrogen bonding and causes proteins to become denatured and soluble in the organic phase. Phenol also acts as a dissociating agent that decreases molecular interactions between proteins and other materials [32]. Previous studies have indicated that use of phenol results in less streaking on both the horizontal and vertical dimensions of two-dimensional protein gels compared to samples prepared using TCA–acetone [30,33].

However, both strategies are usually applied to relatively "easy" tissues, such as etiolated shoots and young root tips [29]. An efficient strategy for protein extraction from recalcitrant plant tissues (e.g., seeds) using a combination of TCA–acetone and phenol extraction was presented here, which resulted in high-quality protein samples according to definitions presented by Wang et al. [34]. In this report, three protein extraction methods for tung tree seeds were compared with respect to the protein yield, quality and resolution of the 2-DE gel. The protein yield obtained by the combination of TCA–acetone and phenol preparation (M3) was relatively low, but resulted in the best protein quality and gel resolution (Supplementary Materials Figure S1). Thus, method M3 was deemed the most suitable for the proteomic analysis of the tung tree seeds.

3.2. The Relationship between Oil Bodies and Oil Content in Seeds During Developmental Stages

Previous studies have indicated that there is a close relationship between oil content and oil bodies in various plant species. The differences in oil body abundance in high oil versus low oil cultivars of *Brassica napus* indicated that larger (and presumably fewer) oil bodies exist in the lower oil content cultivars [35]. Large oil bodies exist in high oil content cultivars of maize [36], which is influenced, at least in part, by the level of seed oleosin proteins. Oil bodies develop during embryo development in higher-oil sunflower cultivars, and then stop developing before the complete growth of the embryo is achieved [37].

In this study, the changes in the cellular ultrastructure, oil body number, and oil content during development of tung seed endosperm were analyzed. According to the results, the oil content in the seeds increased from 25 August to 26 September. In the 25 August seeds, the average oil content was 30.3%, and increased to 34.2% and 40.6%, during the second and third time points, respectively (Figure 2). The ultrastructural observations of the endosperm cells showed that the volumes of the oil bodies decreased, whereas the number of oil bodies increased through progressive developmental stages (Figure 1, Supplementary Materials Figure S3).

3.3. Seed Reserve Synthesis during Development

Seed reserves are mainly accumulated during seed development for the next generation of germination and seedling growth. The most important reserves are oil, proteins and starches, and the seeds of different species can be categorized according to the relative amounts of these three metabolites. In the seeds of the tung tree, oil is the most abundant reserve. From 25 August to 26 September, the oil content in the seeds increased substantially (Figure 2).

At the protein level, malate dehydrogenase (S29 and S63) was down-regulated from 25 August to 9 September, whereas the enzymes related to oil synthesis (S31, S40, S57, S58 and S64) were up-regulated across these the three developmental stages (Supplementary Materials Table S1, Figure 5), which correlated with the increase in oil content and oil body abundance in the 26 September seeds (Figure 3 and Figure S3). These results provide strong molecular-level support for the notion that oil, as the primary reserve in tung seeds, actively accumulates throughout a large portion of seed development and that much of the gene expression in these organs is dedicated to the pathways necessary for oil biosynthesis.

3.4. Some Enzymes Were Undetectable during Seed Development

Lipid metabolism in plant seeds is complex and requires over 100 enzymatic reactions and 600 proteins [38]. An oil body consists of a neutral lipid core enclosed by a membrane lipid monolayer that is primarily coated with structural proteins, such as oleosins and caleosin [39,40]. It was surprising that oleosins, the most abundant oil body-associated proteins, were not found in this study. This may be because the predicted isoelectric point for oleosins is 9.7 and the pH gradient strips (IPGs) used in present study ranged from pH 4 to 7. Consequently, all the oleosins may have been lost during the first dimension IEF [41]. Caleosin, which is another important protein related to oil bodies, is a Ca^{2+}-binding oil body surface protein that participates in oil-vacuole interactions that affect the breakdown of oil bodies during germination [8,42]. Storage proteins accumulate during seed development, particularly at the late stages of development [43]. However, few storage proteins were identified other than a nutrient reservoir protein (S67), which was found to be continuously up-regulated in all three developmental stages (Supplementary Materials Figure S2) and indicated its role in nutrient deposition during seed development.

This data set represents a significant step forward in oilseed proteomics; however, we identified fewer proteins than expected. A significant limitation to MS identification could be the lack of currently available genomic resources for the tung tree. In the study, the tung tree proteins were identified using homology to the sequences of other oil or woody plants including *Arabidopsis* and *Ricinus communis*. Nonetheless, the results shown here provide a clear basis to better understand the interplay between different metabolic pathways in developing oilseeds and will help to guide the rational design of future experiments.

4. MATERIALS AND METHODS

4.1. Materials

The seeds were harvested approximately 130 days (25 August), 145 days (9 September) and 160 days (26 September) after flowering, and were marked A, B and C in this paper, respectively (Figure 5). Tung tree fruits were harvested from trees in the State-owned Dongfanghong Forest Farms located in Jinhua City (E119°14'–E120°46', N28°32'–N29°41'), Zhejiang Province, China. The seeds were immediately removed from the fruit, and the seed fractions were stored at −80 °C after quick-freezing in liquid nitrogen and used for protein extraction and two-dimensional gel electrophoresis (2-DE). Other seed fractions were stored at −4 °C, and used for oil content and fatty acid composition analysis.

4.2. Tissue Preparation for Transmission Electron Microscopy

For the ultrastructural observation studies, seed endosperm tissues were cut from the seeds as blocks (2 mm length × 1 mm width × 1 mm height). The tissue blocks were fixed in 2.5% glutaraldehyde in 100 mM phosphate buffer (pH 7.0) for 4 h at room temperature, and then washed with 100 mM phosphate buffer (pH 7.0). Subsequently, the blocks were treated with 1% osmic acid (OsO_4) and rinsed with phosphate buffer (3 × 15 min) and then dehydrated using a graded series of ethanol treatments (50%, 70%, 85%, 95%, and 100% *v*/*v*). The infiltration process was completed using a graded ethanol/Epon/Spurr's epoxy resin series, followed by embedding the blocks in 100% (*w*/*v*) Spurr's epoxy resin and polymerization at 60 °C for 24 h. The samples were cut into thin sections (70 nm) using an ultramicrotome (Leica MZ6, LeiMicrosystems, Wetzlar, Germany), and then collected onto copper grids, and poststained with saturated uranyl acetate and 0.4% lead citrate. After the sections were rinsed with dH_2O (6 × 15 s), they were viewed with a H-7650 transmission electron microscope (Hitachi company, Koka, Japan).

4.3. Oil Content and Fatty Acid Composition Analysis

Endosperm tissues collected at different developmental stages were milled into a fine powder and the oil content was analyzed using Soxhlet extraction. Fatty acid composition analysis was performed according to the methods described by Yang [8].

4.4. Protein Extraction, Two-Dimensional Gel Electrophoresis and Data Analysis

Effective separation and identification of proteins depends upon efficient protein isolation methodology. TCA–acetone extraction [44], phenol extraction [45] and TCA–acetone/phenol extraction [46] with a minimal modification were used to prepare the protein samples of tung seeds and the effects of these extraction methods were compared by 2-DE and sodium dodecyl sulfate polyacrylamide gel electrophoresis (SDS-PAGE) in pre-experiment. The results of pre-experiment indicated that TCA–acetone combined with phenol extraction method was suitable for the protein extraction from tung seeds [47]. The detailed protocols of TCA–acetone combined with phenol extraction method for protein extraction and 2-DE in this study are as follows.

Seed endosperms were ground into a fine power under liquid nitrogen. During grinding, 10% (*w/w*) polyvinyl-polypyrrolidone (PVPP) was added. Dry powdered tissue (0.5 g/sample) was transferred into centrifugation tubes, and suspended in 2 mL of extraction buffer I [10% (*w/v*) TCA, 0.07 (*v/v*) β-mercaptoethanol (β-ME) in cold acetone and 18 μL phenylmethylsulfonyl fluoride (PMSF, 200 mM)], for 2 h at −20 °C. After centrifugation at 12,000 rpm for 15 min at 4 °C, the supernatants were removed and the precipitated pellets were washed in 1 mL cold acetone containing 0.07% β-ME for 5 min. After centrifugation, the pellets were rinsed again with 0.1 mM ammonium acetate in methanol for 5 min. Then, the pellets were homogenized using Tris-saturated phenol and an equal volume of extraction buffer II [500 mM Tris-HCl (pH 8.65), 50 mM ethylenediaminetetraacetic (EDTA), 2% (*v/v*) 100 mM KCl] at −20 °C for 2 h. The phenol phase was collected, and protein was precipitated overnight at −20 °C after adding 0.1 mM ammonium acetate in methanol. After centrifugation, the proteins were rinsed once each in 0.1 mM ammonium acetate in methanol and 100% cold acetone, respectively. The pellets were vacuum-dried for 1 h and stored at −80 °C for later use.

The dried protein pellets were dissolved in solution buffer containing 7 M urea, 2 M thiourea, 4% 3-[(3-cholamidopropyl)-dimethylammonio]-1-propane sulfonate (CHAPS), 65 mM dithiothreitol (DTT), and 2% (*v/v*) ampholine pH 3–10 (GE Healthcare Bio-Science, UK Ltd., Buckinghamshire, UK). After centrifugation at 12,000 rpm for 15 min, the supernatant was collected for protein quantification.

The immobilized pH gradient strips (IPGs, pH 4–7, 24 cm, GE Healthcare Bio-Science, UK Ltd.) were used in two-dimensional gel electrophoresis on the Multiphor™ II Electrophoresis System (GE Healthcare Bio-Science, UK Ltd.). A total of 400 μL sample containing 850 μg proteins were loaded in IPGs for rehydration at 50 V and 20 °C constant temperature. Isoelectric focusing (IEF) was performed at 500 V for 30 min, 1000 V for 1.5 h and 8000 V for 4 h, and finished at 70,000 Vh, all at 50 μA/strip. After the IEF run, the IPGs were equilibrated in equilibration buffer [6 M urea, 2% (*w/v*) SDS, 20% (*w/v*) glycerol, 375 mM tris-HCl (pH 8.8) and 2% (*w/v*) DTT] at room temperature for 15 min. The second equilibration was carried out in the same buffer, except that DTT was replaced with 2.5% (*w/v*) iodoacetamide. Second-dimension protein separation and visualization was achieved by 12.5% SDS-PAGE electrophoresis and post-staining with 0.1% Coomassie Brilliant Blue (CBB) R-250.

The images of 2-DE gels were analyzed using Image Master 2D Platinum Version 5.0 Analysis Software (GE Healthcare Bio-Science, UK Ltd.). Only the protein spot abundances which changed more than 1.2-fold across the three different developmental stages.

4.5. In-Gel Protein Digestion and MALDI-TOF/MS Analysis

Protein digestion and MALDI-TOF/MS analyses were performed as follows. Each protein gel piece was destained with 100 mM NH_4HCO_3 in 30% (*v*/*v*) acetonitrile (ACN) for 2 h at 40 °C. The gel pieces were minced, lyophilized and digested in 25 mM NH_4HCO_3 with 10 ng sequencing-grade modified trypsin (Promega, Madison, WI, USA) at 37 °C overnight. After digestion, peptides were extracted by three washes with 0.1% trifluoroacetic acid (TFA) in 60% ACN. The peptides were desalted by ZipTipC-18 pipet tips (Millipore, Bedford, MA, USA) according to the product manual. Tryptic peptide samples were analyzed using a 4800 Plus MALDI TOF/TOFTM Analyzer (Applied Biosystems, Foster, CA, USA).

ACKNOWLEDGMENTS

This work was supported financially by Zhejiang Province Natural Science Foundation of China (LZ13C160001).

AUTHOR CONTRIBUTIONS

Y.C., Z.Z., and Y.W. conceived the original screening and research plans; Y.W. supervised the experiments; Z.Z. and Y.C performed most of the experiments; J.S. and X.H. provided technical assistance; and Z.Z., J.S., Y.C. and Y.W. supervised and complemented the writing.

CONFLICTS OF INTEREST

The authors declare no conflict of interest.

REFERENCES

1. Sghaier-Hammami, B.; Valledor, L.; Drira, N.; Jorrin-Novo, J.V. Proteomic analysis of the development and germination of date palm (Phoenix dactylifera L.) zygotic embryos. Proteomics 2009, 9, 2543–2554.

2. Miernyk, J.A.; Hajduch, M. Seed proteomics. Proteomics 2011, 74, 389–400.
3. Huang, H.; Møller, I.M.; Song, S.Q. Proteomics of desiccation tolerance during development and germination of maize embryos. Proteomics 2012, 75, 1247–1262.
4. Schiltz, S.; Gallardo, K.; Huart, M.; Negroni, L.; Sommerer, N.; Burstin, J. Proteome reference maps of vegetative tissues in Pea. An investigation of nitrogen mobilization from leaves during seed filling. Plant Physiol. 2004, 135, 2241–2260.
5. Chibani, K.; Ali-Rachedi, S.; Job, C.; Job, D.; Jullien, M.; Grappin, P. Proteomic analysis of seed dormancy in Arabidopsis. Plant Physiol. 2006, 142, 1493–1510.
6. Sheoran, I.S.; Olson, D.J.H.; Ross, A.R.S.; Sawhney, V.K. Proteome analysis of mbryo and endosperm from germination tomato seesds. Proteomics 2005, 5, 3752–3764.
7. Yang, P.; Li, X.; Wang, X.; Chen, H.; Chen, F.; Shen, S. Proteomic analysis of rice (Oryza sativa) seeds during germination. Proteomics 2007, 7, 3358–3368.
8. Yang, M.F.; Liu, Y.J.; Liu, Y.; Chen, H.; Chen, F.; Shen, S.H. Proteomic analysis of oil mobilization in seed germination and postgermination development of Jatropha curcas. Proteome Res. 2009, 8, 1441–1451.
9. Wong, P.F.; Abubakar, S. Post-germination changes in Hevea brasiliensis seeds proteome. Plant Sci. 2005, 169, 303–311.
10. Finnie, C.; Melchior, S.; Roepstorff, P.; Svensson, B. Proteome analysis of grain filling and seed maturation in barly. Plant Physiol. 2002, 129, 1308–1319.
11. Liu, H.; Yang, Z.; Yang, M.; Shen, S. The differential proteome of endosperm and embryo from mature seed of Jatropha curcas. Plant Sci. 2011, 181, 660–666.
12. Nouri, M.Z.; Komatsu, S. Comparative analysis of soybean plasma membrane proteins under osmotic stress using gel-based and LC MS/MS-based proteomics approaches. Proteomics 2010, 10, 1930–1945.
13. Sheoran, I.S.; Pedersen, E.J.; Ross, A.R.S.; Sawhney, V.K. Dynamics of protein expression during pollen germination in canola (Brassica napus). Planta 2009, 230, 779–793.
14. Hajduch, M.; Casteel, J.E.; Hurrelmeyer, K.E.; Song, Z.; Agrawal, G.K.; Thelen, J.J. Proteomic analysis of seed filling in Brassica napus. Developmental characterization of metabolic isozymes using high-resolution two-dimensional gel electrophoresis. Plant Physiol. 2006, 141, 32–46.
15. Potter, G. The domestic tung industry. I. Production and improvement of the tung tree. Econ. Bot. 1959, 13, 328–342.
16. Sonntag, N. Composition and characteristics of individual fats and oils. Bailey's Ind. Oil Fat Prod. 1979, 1, 289–477.
17. Park, J.; Kim, D.; Wang, Z. Production and characterization of biodiesel from tung oil. Appl. Biochem. Biotechnol. 2008, 148, 109–117.
18. Shang, Q.; Jiang, W.; Lu, H.; Liang, B. Properties of tung oil biodiesel and its blends with 0# diesel. Bioresour. Technol.2010, 101, 826–828. [PubMed]
19. Chen, Y.; Chen, J.; Chang, C.; Chang, C. Biodiesel production from tung (Vernicia montana) oil and its blending properties in different fatty acid compositions. Bioresour. Technol. 2010, 101, 9521–9526.
20. Cao, H.; Chapital, D.C.; Shockey, J.M.; Klasson, K.T. Expression of tung tree diacylglycerol acyltransferase 1 in E. coli. BMC Biotechnol. 2011, 11, 73–85.
21. Dyer, J.M.; Chapital, D.C.; Kuan, J.C.W.; Mullen, R.T.; Turner, C.; McKeon, T.A.; Pepperman, A.B. Molecular analysis of a bifunctional fatty acid conjugase/desaturase from tung. Implications for the evolution of plant fatty acid diversity. Plant Physiol. 2002, 130, 2027–2038.
22. Shockey, J.M.; Gidda, S.K.; Chapital, D.C.; Kuan, J.C.; Dhanoa, P.K.; Bland, J.M.; Rothstein, S.J.; Mullen, R.T.; Dyer, J.M. Tung tree DGAT1 and DGAT2 have nonredundant functions in triacylglycerol biosynthesis and are localized to different subdomains of the endoplasmic reticulum. Plant Cell 2006, 18, 2294–2313.
23. Turchetto-Zolet, A.C.; Maraschin, F.S.; de Morais, G.L.; Cagliari, A.; Andrade, C.M.B.; Margis-Pinheiro, M.; Margis, R. Evolutionary view of acyl-CoA diacylglycerol acyltransferase (DGAT), a key enzyme in neutral lipid biosynthesis. BMC Evol. Biol. 2011, 11, 263–276.

24. Li, X.J.; Yang, M.F.; Chen, H.; Qu, L.Q.; Chen, F.; Shen, S.H. Abscisic acid pretreatment enhances salt tolerance of rice seedlings: Proteomic evidence. Biochim. Biophys. Acta 2010, 1804, 929–940.
25. Tang, L.; Kim, M.D.; Yang, K.S.; Kwon, S.Y.; Kim, S.H.; Kim, J.S.; Yun, D.J.; Kwak, S.S.; Lee, H.S. Enhanced tolerance of transgenic potato plants overexpressing nucleoside diphosphate kinase 2 against multiple environment stresses. Transgenic Res. 2008, 17, 705–715.
26. Kim, J.Y.; Park, S.J.; Jang, B.; Jung, C.H.; Ahn, S.J.; Goh, C.H.; Cho, K.; Han, O.; Kang, H. Functional characterization of a glycine-rich RNA-binding protein 2 in Arabidopsis thaliana under abiotic stress conditions. Plant J. 2007, 50, 439–451.
27. Han, F.; Chen, H.; Li, X.J.; Yang, M.F.; Liu, G.S.; Shen, S.H. A comparative proteomic analysis of rice seedlings under various high-temperature stresses. Biochim. Biophys. Acta 2009, 1794, 1625–1634.
28. Kwon, S.I.; An, C.S. Cloning and expression of mitochondrial MnSOD from the small radish (Raphanus sativus L.). Mol. Cells 2003, 16, 194–200. [PubMed]
29. Isaacson, T.; Damasceno, C.M.B.; Saravanan, R.S.; He, Y.H.; Catalá, C.; Saladié, M.; Rose, J.K.C. Sample extraction techniques for enhanced proteomic analysis of plant tissues. Nat. Protocol. 2006, 1, 769–774.
30. Saravanan, R.S.; Rose, J.K.C. A critical evaluation of sample extraction techniques for enhanced proteomic analysis of recalcitrant plant tissues. Proteomics 2004, 4, 2522–2532.
31. Hurkman, W.J.; Tanaka, C.K. Solubilisation of plant membrane proteins for analysis by two-dimensional gel electrophoresis. Plant Physiol. 1986, 81, 802–806.
32. Sheoran, I.; Ross, A.; Olson, D.; Sawhney, V. Compatibility of plant protein extraction methods with mass spectrometry for proteome analysis. Plant Sci. 2009, 176, 99–104.
33. De la Fuente, M.; Borrajo, A.; Bermúdez, J.; Lores, M.; Alonso, J.; López, M.; Santalla, M.; de Ron, A.M.; Zapata, C.; Alvarez, G. 2-DE-based proteomic analysis of common bean (Phaseolus vulgaris L.) seeds. Proteomics 2011, 74, 262–267.
34. Wang, W.; Vignani, R.; Scali, M.; Cresti, M. A universal and rapid protocol for protein extraction from recalcitrant plant tissues for proteomic analysis. Electrophoresis 2006, 27, 2782–2786.
35. Hu, Z.Y.; Wang, X.F.; Zhang, G.M.; Liu, G.H.; Hua, W.; Wang, H.Z. Unusually large oilbodies are highly correlated with lower oil content in Brassica napus. Plant Cell Rep. 2009, 28, 541–549.
36. Ting, J.T.L.; Lee, K.Y.; Ratnayake, C.; Platt, K.A.; Balsamo, R.A.; Huang, A.H.C. Oleosin genes in maize kernels having diverse oil contents are constitutively expressed independent of oil contents. Size and shape of intracellular oil-bodies are determined by the oleosins/oils ratio. Planta 1996, 199, 158–165.
37. Mantese, A.L.; Medan, D.; Hall, A.J. Achene structure development and lipid accumulation in sunflower cultivars differing in oil content at maturity. Ann. Bot. 2006, 97, 999–1010.
38. Nguyen, H.M.; Baudet, M.; Cuine, S.; Adriano, J.; Barthe, D.; Billon, E.; Bruley, C.; Beisson, F.; Peltier, G.; Ferro, M.; et al. Proteomic profiling of oil bodies isolated from the unicellular green microalga Ch-lamydomonas reinhardtii: With focus on proteins involved in lipid metabolism. Proteomics 2011, 11, 4266–4273.
39. Murphy, D.J. Structure, function and biogenesis of storage lipid bodies and oleosins in plants. Prog. Lipid Res. 1993, 2, 247–280.
40. Huang, A.H.C. Oil bodies and oleosins in seeds. Annu. Rev. Plant Physiol. 1992, 43, 177–200.
41. Tnani, H.; López, I.; Jouenne, T.; Vicient, C.M. Protein composition analysis of oil bodies from maize embryos during germination. Plant Physiol. 2011, 168, 510–513.
42. Poxleitner, M.; Rogers, S.W.; Samuels, A.L.; Browse, J.; Rogers, J.C. A role for caleosin in degradation of oil-body storage lipid during seed germination. Plant J. 2006, 47, 917–933.
43. Bewley, J.D. Seed germination and dormancy. Plant Cell 1997, 9, 1055–1066.

44. Granier, F. Extraction of plant proteins for two-dimensional electrophoresis. Electrophoresis 1988, 11, 712–718.
45. Wang, W.; Scali, M.; Vignani, R.; Spadafora, A.; Sensi, E.; Mazzuca, S.; Cresti, M. Protein extraction for two-dimensional electrophoresis from olive leaf, a plant tissue containing high levels of interfering compounds. Electrophoresis 2003, 14, 2369–2375.
46. Wang, W.; Tai, F.; Chen, S. Optimizing protein extraction from plant tissues for enhanced proteomics analysis. J. Sep. Sci. 2008, 31, 2032–2039.
47. Zhan, Z.Y.; Wang, Y.D.; Chen, Y.C.; Han, X.J.; Cui, Q.Q. Comparison of three protein extraction methods for tung trees (Vernicia fordii) seed. Bull. Bot. Res. 2013, 33, 91–97.

CHAPTER 16

Designing Novel Breeding Strategies for Producing High-Oil Crops Based on a Molecular Understanding of Triacylglycerol Metabolism

Masatake Kanai[1, 2], Shoji Mano[1, 2, 3], Makoto Hayashi[4] and Mikio Nishimura[1*]

[1] *Department of Cell Biology, National Institute for Basic Biology, Okazaki, Japan*

[2] *Laboratory of Biological Diversity, Department of Evolutionary Biology and Biodiversity, National Institute for Basic Biology, Okazaki, Japan*

[3] *Department of Basic Biology, SOKENDAI (The Graduate University for Advanced Studies), Okazaki, Japan*

[4] *Department of Bioscience, Nagahama Institute of Bio-Science and Technology, Nagahama, Japan*

ABSTRACT

Seeds are storage organ in plants and main resource of plant oils to human civilization and the demand of plant oils are increasing yearly and expansion of the production capacity is an urgent issue worldwide. Thus, it is necessary to improve oil yields per unit area and generation of crops with high-oil content is needed. *Arabidopsis thaliana* plays a vital role in advancement of genetics and molecular biology in plant sciences. The forward and reverse genetic approaches with Arabidopsis have provided an overview of triacylglycerol metabolism. The elucidation of the overview contributes to understanding of spatiotemporal regulation of a metabolic flow of triacylglycerol metabolism in plant cell. This understanding sheds light on bottlenecks in triacylglycerol biosynthesis and provides novel clues for increasing seed triacylglycerol content. Recent advance in metabolic engineering approaches demonstrate several evidences that triacylglycerol metabolism is coordinated with other metabolisms. Most notably, triacylglycerol biosynthesis competes with biosynthesis of starch or seed storage proteins. These studies indicate that alterations of the metabolic pathways to avoid the competitions could be a novel concept for increasing seed oil content.

Keywords: Seed, oil, Triacylglycerol, Metabolic engineering, *Arabidopsis thaliana*

1. INTRODUCTION

Seeds are storage organs in plants that accumulate massive quantities of carbohydrates, proteins, and oils, which are collectively referred to as seed storage reserves. Seed storage reserves are utilized to supply energy and nutrients to the embryo during postgerminative growth until the plant acquires the ability to perform photosynthesis. Hence, seed storage reserves are an easily degradable source of organic matter for organisms.

Seeds are widely used as industrial materials. For example, plant oils are mainly derived from seeds. In 2014, the overall consumption of plant oils was 170,980 kt, which is nearly double that consumed in 2004. This increase is attributed to a greater demand for dietary oil, as well as industrial materials for use as carbon-neutral oils, representing an alternative to petroleum.

To improve oil productivity in plants, it is essential to increase seed yields in crops. However, the agricultural area worldwide has been flat for 40 years. Therefore, improving seed yield per plant has become increasingly important. Since increasing seed yield is one of the major issues in plant science, effective strategies for increasing yield have been investigated. Considering the many successes in the field of metabolic engineering of microorganisms, metabolic engineering represents a promising approach for increasing oil yields in seeds. Lipid metabolism has been extensively studied, and its metabolic pathways and regulatory systems have been elucidated. Additionally, in-depth analysis of crop genomes has been greatly expedited by recent advances in life science technologies (next-generation sequencing technology, genome editing, and so on). Therefore, numerous translational studies of model plant species have been performed, which have shed light on crop species. This chapter summarizes recent advances in our understanding of oil metabolism in seeds and introduces promising strategies for increasing oil production in crops.

2. MAIN COMPONENTS OF SEEDS

Seeds accumulate carbohydrates, proteins, and oils as seed storage reserves. The ratios of these components dramatically differ among plant species, representing a major factor that determines the usage and application of seeds [1].

2.1. Cereal Seeds

Starch consists of carbohydrates, which comprise polysaccharides composed of glucose [2–4]. Major cereal seeds, such as rice, wheat, and maize, contain large amounts of starch. Barley and rye seeds also accumulate starch as their main storage reserves. The starch contents in these cereal seeds are over 70%, whereas the contents of proteins and oils are less than 20% (**Table 1**). Therefore, cereal seeds represent an important energy source for animals.

2.2. Oil Seeds

Oil seeds mainly accumulate lipids. Oil seeds are one of the major sources of plant oils. These seeds are compressed and heated to extract oils from their cells. Oil crops, which produce oil seeds, are cultivated worldwide. Oil palm, soybean, and rapeseed, the "big three", represent three-quarters of total plant oil production (**Table 2**). Palm oil and olive oil are extracted from fruits, whereas many other oils are extracted from seeds (**Table 2**).

Table 1. Carbohydrate, protein, and fat contents in cereal seeds. Values (%/w) in this table were extracted from the Food Composition Database of the Ministry of Education, Culture, Sports, Science, and Technology in Japan (http://fooddb.mext.go.jp/index.pl).

Grain	Carbohydrates	Proteins	Oils
wheat	72.2	10.6	3.1
rice	73.8	6.8	2.7
maize	70.6	8.6	5.0
barley	72.1	10.9	2.1
rye	70.7	12.7	2.7

Table 2. Major oil crops and worldwide oil production in 2013–2014. The rates of total production and % total plant oil production are summarized for major oil crops; the values were obtained from FAOSTAT (http://faostat.fao.org/).

	Plant species	Production (kilotons, 2013–2014)	Rate (% total plant oil production)	Storage organ
1	oil palm	59,360	34.09	fruit
2	soybean	44,439	25.52	seed
3	rapeseed	27,029	15.52	seed
4	sunflower	16,169	9.29	seed
5	cotton	4,859	2.79	seed
6	peanut	3,548	2.04	seed
7	olive	3,416	1.96	fruit
8	corn	3,148	1.80	seed

Table 3. Oil and protein concentrations in major oil seed crops. The oil and protein concentrations were measured following the methods of Kanai et al. (2016). The variety of sesame used in this experiment was not identified.

Species	Concentration (%)		Variety
	Oil	Protein	
soybean	20.1	41.4	Kariyutaka
rapeseed	41.0	25.6	Kizakinonatane
sunflower	33.8	30.4	Mammoth Russian
peanut	42.6	25.8	Omasari
sesame	49.9	21.6	–

2.2.1. Oil contents of oil seeds

Of the major oil crops, the oil content of oil palm fruit is over 50%, whereas that of soybean and rapeseed is 20.1% and 41%, respectively (**Table 3**) [5, 6]. The seed oil content in sunflower, peanut, and sesame is 33.8%, 42.6%, and 49.9%, respectively (**Table 3**). Thus, the seed oil contents of typical oil crops range from approximately 20–50%. These values are relatively low compared to the starch contents of rice, wheat, maize, and so on, suggesting that there is strong potential for increasing oil contents in seeds.

2.2.2. Protein contents of oil seeds

The protein content of major cereal seeds is less than 15% (**Table 1**), whereas the protein content of soybean and rapeseed, which are major suppliers of plant oils, is 41.4% and 25.6%, respectively (**Table 3**). Most oil seeds accumulate more proteins, i.e., seed storage proteins (hereafter referred to as SSPs), than cereal seeds.

3. TRIACYLGLYCEROL METABOLISM IN SEEDS

Lipids are essential components not only for animals, but also for plants. Various species of lipids are biosynthesized throughout plant organs. Some lipids, such as membrane lipids [7], cuticular waxes [8], and volatile oils, comprising fatty acids [9–11], alcohols, terpenes, and so on, are produced in plants in response to changes in the external environment. On the other hand, seed oils, the primary subject of this chapter, are composed of triacylglycerol (referred to hereafter as TAG) [12–14]. Except for palm oil, major seed oils are liquids at room temperature (**Figure 1A**). One molecule of TAG is composed of a glycerol backbone and three fatty acids (**Figure 1B**).

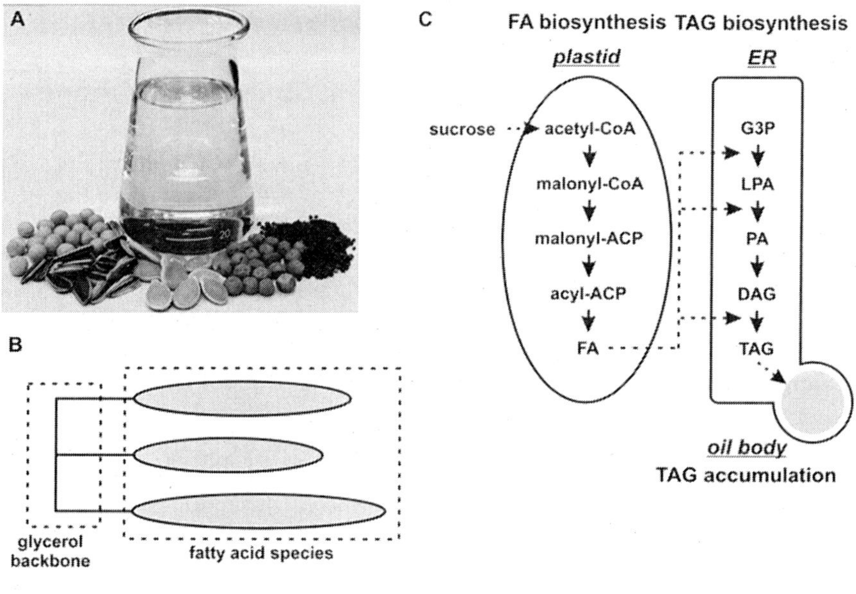

Figure 1. Triacylglycerol in plant seeds. (A) Plant oil extracted from seeds. (B) Structure of TAG. TAG is produced by the assembly of one glycerol with three fatty acids. (C) Schematic diagram of TAG biosynthesis in plant cells.

3.1. Triacylglycerol Biosynthesis

TAG biosynthesis primarily involves two steps, i.e., fatty acid biosynthesis and the assembly of fatty acids with a glycerol backbone (**Figure 1C**).

3.2. Fatty Acid Biosynthesis

In plant cells, fatty acids are synthesized in plastids. Fatty acid biosynthesis begins with the carboxylation of acetyl-CoA to malonyl-CoA by acetyl-CoA carboxylase (reaction (1) in **Figure 2**). The malonyl group in malonyl-CoA is transferred to acyl carrier protein (hereafter referred to as ACP), producing malonyl-ACP (reaction (2) in **Figure 2**). Next, 3-keto-butyryl-ACP is synthesized via the condensation reaction of malonyl-ACP with acetyl-CoA (reaction (3) in **Figure 2**). The 3-keto-butyryl-ACP molecule is converted to butyryl-ACP via reduction and dehydration (reaction (4) in **Figure 2**). The provision of C2 units, reduction, and dehydration are repeated, leading to the elongation of the carbon chain of acyl-ACP (carbon chain elongation in **Figure 2**). Synthesized acyl-ACP is catalyzed by thioesterase to form free fatty acids, which are again converted to acyl-CoA and imported to the endoplasmic reticulum (hereafter referred to as ER).

3.3. Assembly of Fatty Acids with The Glycerol Backbone

Acyl-CoA is a donor of the acyl-group for TAG biosynthesis. *De novo* synthesis of TAG begins with the transfer of the acyl-group to the sn-1 position of glycerol 3-phosphate, leading to the production of lysophosphatidic acid (reaction (1) in **Figure 3**). Subsequently, phosphatidic acid is produced by the transfer of the acyl group to the sn-2 position (reaction (2) in **Figure 3**). Next, the sn-3 position of phosphatidic acid is dephosphorylated and converted to diacylglycerol (reaction (3) in **Figure 3**), and, finally, TAG is synthesized by the transfer of the acyl-group to the sn-3 position of diacylglycerol (hereafter referred to as DAG) (reaction (4) in **Figure 3**). In addition to the *de novo* synthesis pathway, TAG is produced via an alternative pathway through phosphatidylcholine (hereafter referred to as PC) [15, 16]. PC, one of the main components of membrane lipids, is present in pools in the ER [16]. PC pools affect *de novo* TAG synthesis and the acyl-CoA pool in the cytosol to supply DAG as a precursor for TAG [16].

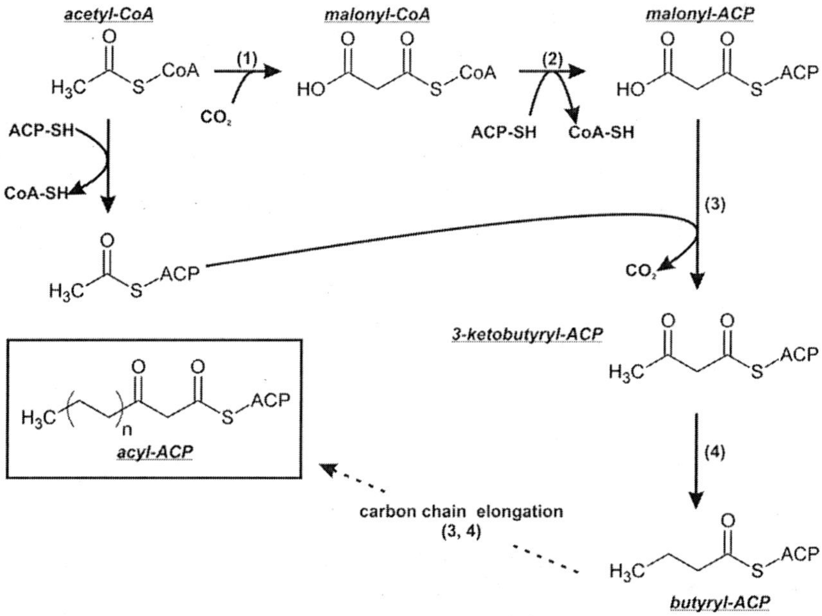

Figure 2. Fatty acid biosynthesis in plastids: (1) acetyl-CoA carboxylase, (2) malonyl CoA:ACP transacylase, (3) 3-ketoacyl-ACP synthase, and (4) 3-ketoacyl-ACP reductase, 3-hydroxyacyl ACP dehydrase, and enoyl-ACP reductase.

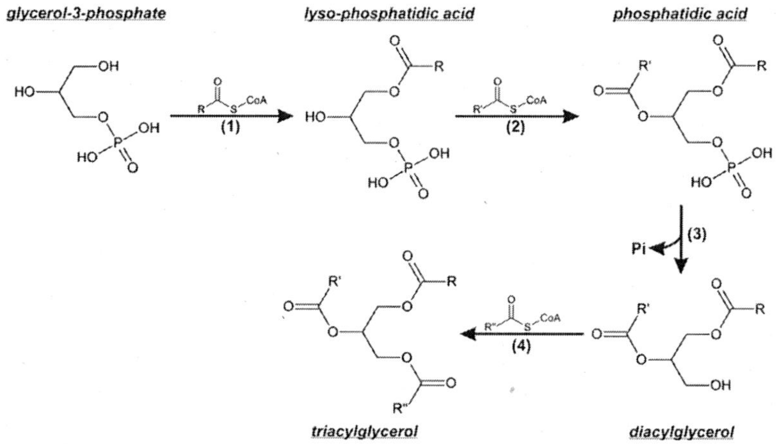

Figure 3. Triacylglycerol formation in the ER: (1) acyl-CoA:G3P acyltransferase, (2) acyl-CoA:LPA acyltransferase, (3) PA phosphatase, and (4) acyl-CoA:DAG acyltransferase.

3.4. Triacylglycerol Accumulation

Synthesized TAGs accumulate in compartments in the ER (**Figure 1C**), which are converted to vesicles through budding. The vesicles then develop into oil-accumulating organelles, i.e., oil bodies [16–19]. Oil bodies are TAG storage organelles with a single layer membrane, whose major membrane protein is oleosin [18, 20, 21]. Oleosin blocks adhesion between neighboring oil bodies, which allows small oil bodies to be packed tightly together without adhesion in the cells of oil seeds [22].

4. STRATEGIES FOR IMPROVING OIL CONTENT BY MODIFYING TRIACYLGLYCEROL METABOLISM IN OIL SEED CROPS

Since plant oils are commercially important, improving oil seed crops has long been a focus of breeders. Such breeding efforts, which began in ancient times, have led to the improvement of oil contents in several crops. In addition, the oil contents of the seeds of modern cultivars are significantly higher than those of wild species [23, 24]. Currently, many plant breeders have undertaken the challenge of further increasing seed oil contents. However, recent studies using quantitative trait loci analyses revealed that seed oil contents are controlled by many genes with additive effects [25–28], suggesting that traditional breeding methods based on cross-fertilization may be inadequate for further increasing seed oil contents.

By contrast, genetic transformation may be an effective tool for increasing seed oil contents. The model plant *Arabidopsis thaliana*, a close relative of the major oil crop *Brassica napus*, produces typical oil seeds (**Figure 4A** and **B**). The use of Arabidopsis drastically simplifies both the processes of screening mutants with abnormal phenotypes and generating transformants, in which genes of interest are introduced. The use of Arabidopsis has accelerated the process of uncovering TAG metabolic pathways in plants. In fact, most genes encoding enzymes involved in TAG metabolism have been identified based on characterization of Arabidopsis mutants defective in TAG metabolism [29, 30]. Elucidating TAG metabolic pathways has revealed the limiting factors and key enzymes in this process and has led to the development of novel strategies for improving seed TAG contents. In this section, we review important strategies for improving TAG contents in seeds based on metabolic engineering approaches.

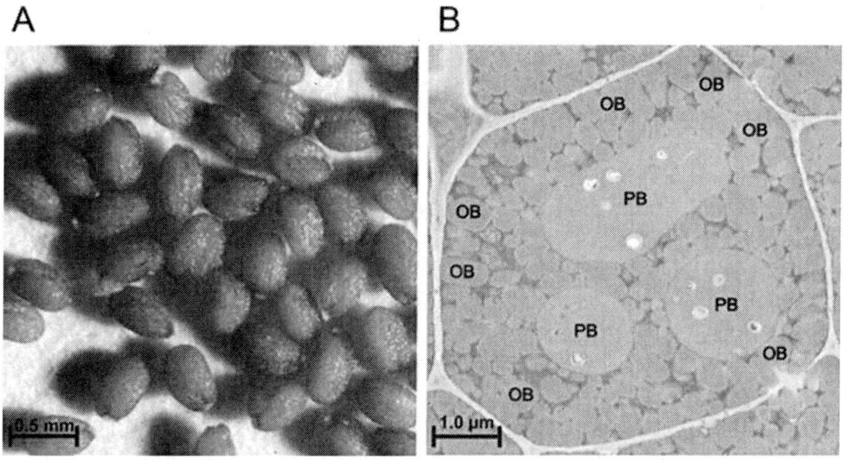

Figure 4. Seeds of *Arabidopsis thaliana*. (A) Microscopic image of Arabidopsis seeds. (B) Electron microscopic image of Arabidopsis seed cells. OB: oil body, an oil storage organelle in seeds. PB: protein body, an organelle that accumulates seed storage protein.

4.1. Enhancement of Tag Biosynthesis

The TAG biosynthetic pathway, which extends across plastids, the cytosol, and the ER, involves many enzymes [30]. Additionally, DAG biosynthetic pathways include the *de novo* synthetic pathway and the phosphatidylcholine-derived pathway (hereafter referred to as PC-derived pathway), whose activities greatly differ among plant species [15, 16]. Thus, increasing the activity of enzymes in the *de novo* or PC-derived pathway does not always increase seed oil contents in every plant species. On the other hand, TAG production from DAG by acyl-CoA:DAG acyltransferase is a common pathway among plants. Overexpressing acyl-CoA:DAG acyltransferase significantly increases seed TAG contents in Arabidopsis [31] and in other plants species [32–37]. These findings indicate that the conversion of DAG to TAG is one of the rate-limiting steps in the production of TAG in seeds, and they suggest that the acyl-CoA:DAG acyltransferase gene would be a promising target gene for increasing TAG contents.

4.2. Suppression of Tag Degradation

Synthesized TAG is stored in oil bodies and degraded during germinative growth [17, 38, 39]. The TAG degradation pathway has also been uncovered, and most genes encoding enzymes in this pathway have been identified [17, 39]. The expression of these genes is upregulated after seed imbibition, and TAG degradation activity rapidly increases in imbibed seeds [40–42]. These genes are also expressed during seed development in several plants [43, 44]. In fact, TAG degradation occurs in developing seeds [45–47]. Therefore, the TAG degradation pathway is activated during seed development, and seeds lose some of the TAG synthesized during seed development. This finding suggests that suppressing TAG degradation would be a promising strategy for improving seed oil contents. Oil degradation begins with TAG hydrolysis via TAG lipase. TAG lipase was genetically identified as *SUGAR DEPENDENT 1* in Arabidopsis [48] and was subsequently identified in rapeseed and Jatropha [47, 49]. Suppressing *SUGAR DEPENDENT 1* expression significantly increases seed oil contents [47, 49]. These reports indicate that suppressing TAG degradation via suppressing *SUGAR DEPENDENT 1* expression may represent an effective strategy for increasing seed oil contents in oil seed crops.

5. NOVEL STRATEGIES FOCUSED ON CARBON FLUX IN SEEDS

In production systems that utilize microorganisms and cultured cells for metabolic engineering, individual cells synthesize organic compounds, providing a limitless source of carbon in the medium. By contrast, the seed, representing only one of many plant organs, synthesizes storage reserves using a limited carbon source provided by photosynthesis. This limited carbon source is thought to represent another limiting factor to the use of seeds as production systems, in addition to the limited activities of metabolic enzymes. To utilize seeds for oil production, it is important to obtain a comprehensive understanding of the factors that regulate TAG production, including carbon flow, metabolite transport, TAG metabolism, compartmentation, competition between other reserves, and so on. Recent studies identifying other factors that limit oil production in seeds have opened up the possibility of developing novel strategies for improving seed oil contents.

5.1. Effective Utilization of "The Window" During Seed Development

Fertilized embryos grow into mature seeds through sequential development. Seed storage oils are produced only during a short period of seed development; high TAG biosynthesis activity only occurs for a short period of time. The master transcription factor of TAG biosynthesis, WRINKLED1 (hereafter referred to as WRI1), is expressed during this short period and induces the expression of genes related to fatty acid biosynthesis [50–54]. Although WRI1 positively regulates TAG biosynthesis, the overexpression of *WRI1* fails to increase TAG contents in seeds [51, 53]. These findings suggest that, in order to increase seed TAG contents, it is necessary to reconsider the timing and duration of *WRI1* expression. Arabidopsis seeds initially accumulate TAG, followed by proteins [1]. A detailed analysis of seed development revealed that there is a time lag between the termination of TAG biosynthesis and the initiation of protein biosynthesis, i.e., there is a "window" period in the middle phase of seed development during which the activities of the TAG and protein synthesis pathways are low (**Figure 5**) [54]. Additionally, strong expression of *WRI1* during this window extends the duration of TAG biosynthesis and increases TAG contents in seeds [54], indicating that *WRI1* expression during this window effectively induces the TAG biosynthesis pathway. This finding has important implications for developing novel strategies for increasing seed TAG contents. The timing and duration of TAG and seed storage protein synthesis differ greatly among plant species. Therefore, identifying the window phase and fine tuning of *WRI1* expression are essential for generating oil crops with increased TAG contents in seeds.

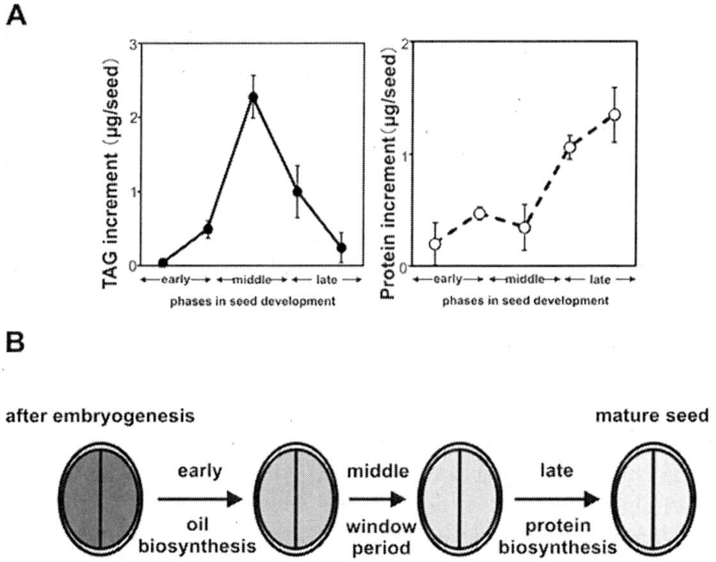

Figure 5. Oil and protein biosynthesis during seed development. (A) Phases of oil and protein biosynthesis during seed development in *Arabidopsis thaliana*. (B) Schematic diagram of oil and protein biosynthesis during seed development in *Arabidopsis thaliana*.

5.2. Concentrating The Carbon Source Into Triacylglycerol Biosynthesis by Reducing The Levels of Other Organic Materials

Seeds accumulate various organic materials, including carbohydrates and proteins, in addition to TAG. Therefore, reducing carbohydrate and protein levels would help direct the carbon source into TAG biosynthesis.

5.2.1. Reducing polysaccharide biosynthesis

Seeds are generally covered with seed coats. Members of the Brassica family, including Arabidopsis, rapeseed, and so on, accumulate large amounts of mucilage consisting of pectin [55, 56]. Therefore, a considerable volume of the carbon source is consumed by mucilage production. Shen et al. reported that knockout of *GLABRA2*, which encodes a WRKY transcription factor controlling epidermal development, increases seed oil content in Arabidopsis [57]. Subsequent research demonstrated that increasing oil contents by suppressing *GLABRA2* expression reduces mucilage biosynthesis [58]. These findings indicate that reducing mucilage biosynthesis causes the carbon source to be directed into TAG biosynthesis. This transfer of the carbon source via reduction of polysaccharide levels has been verified in leaves. Starch comprises one of the main storage reserves in leaves. On the other hand, TAG biosynthesis activity is quite low in leaves. Therefore, to accumulate TAG in leaves, it is essential to both suppress starch biosynthesis and activate TAG biosynthesis. Sanjaya et al. found that overexpressing *WRI1* while suppressing the production of a small subunit of adenosine diphospho (ADP)-glucose pyrophosphorylase, a key enzyme for starch biosynthesis, significantly increases TAG contents in leaves [59]. These results indicate that reducing polysaccharide biosynthesis leads to the funneling of the carbon source into TAG biosynthesis, representing a possible novel strategy for increasing seed TAG contents in oil crops.

5.2.2. Reducing seed storage protein biosynthesis

As described above, oil seeds accumulate large amounts of seed storage proteins in addition to TAG. Crop breeders have reported a negative correlation between TAG and protein contents, especially in soybean [60, 61], suggesting that reducing the levels of seed storage proteins increases TAG contents in seeds. However, in Arabidopsis, knockout of genes encoding major seed storage proteins has little effect on TAG contents in seeds, although the protein content is reduced [54, 62]. These results may be due to the time lag between TAG and protein biosynthesis (**Figure 5**): little of the surplus carbon source derived from the suppression of protein biosynthesis is utilized for TAG biosynthesis because the TAG biosynthesis activity is quite low when protein synthesis is active in the late phase of seed development [54]. Therefore, simultaneously overexpressing *WRI1* and reducing protein synthesis greatly increases seed TAG contents due to effective utilization of the surplus carbon source [54]. This finding indicates that reducing protein synthesis indeed provides the surplus carbon source required for TAG production which, when combined with the simultaneous activation of TAG biosynthesis, leads to increased TAG production. This finding also demonstrates that the combination of these functional strategies has an additive effect on seed TAG content (**Figure 6**) [54, 63]. Breeding lines with reduced SSP content have been established in several crops [64, 65]. Introduction of *WRI1* into these breeding lines represents a potentially important strategy for high-oil seed production.

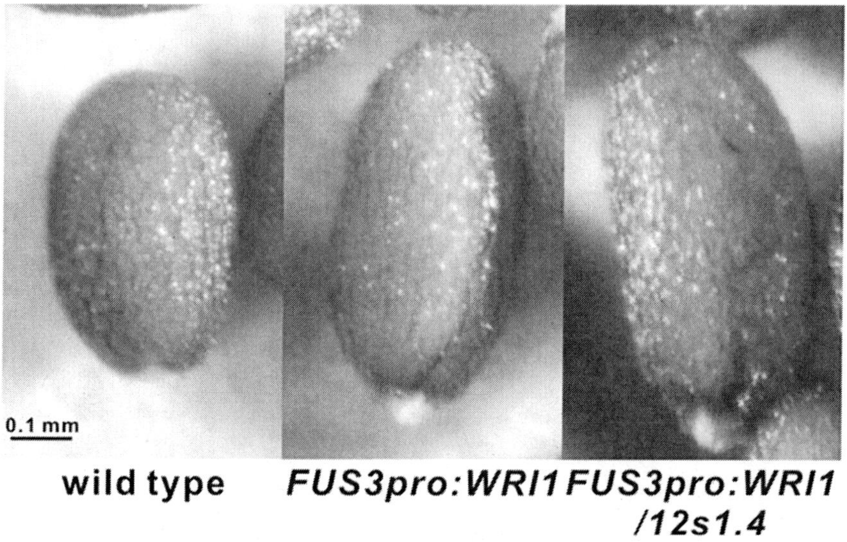

Figure 6. During the enlargement of seeds, the activation of TAG biosynthesis and the reduction of protein synthesis occur simultaneously. *FUS3pro:WRI1*; seeds from transgenic plants harboring *WRI1* under the control of the *FUS3* promoter in wild-type plants (Col-0): *WRI1* is expressed during the window phase. *FUS3pro: WRI1/12s1.4*; seeds from transgenic plants harboring *WRI1* under the control of the *FUS3* promoter in a double knockout mutant of *12S1* and *12S4*, encoding major seed storage proteins in *Arabidopsis thaliana*.

6. CONCLUSION

Molecular genetic analysis of Arabidopsis has led to the incredibly rapid elucidation of the mechanisms underlying the metabolism and regulation of seed storage reserves. The results from these basic studies provide powerful clues to help solve important issues in crop breeding. Optimizing the strategies developed based on the results of basic studies for use in crops will lead to crucial innovations for improving crop yields.

REFERENCES

1. Hills MJ: Control of storage-product synthesis in seeds. Current Opinion in Plant Biology. 2004;7:302–308.
2. Santos-Mendoza M, Dubreucq B, Baud S, Parcy F, Caboche M, Lepiniec L: Deciphering gene regulatory networks that control seed development and maturation in Arabidopsis. The Plant Journal. 2008;54:608–620.
3. Smith AM, Zeeman SC, Smith SM: Starch degradation. Annual Review of Plant Biology. 2005;56:73–98.
4. Zeeman SC, Kossmann J, Smith AM: Starch: Its metabolism, evolution, and biotechnological modification in plants. Annual Review of Plant Biology, Vol 61. 2010;61:209–234.
5. Bourgis F, Kilaru A, Cao X, Ngando-Ebongue GF, Drira N, Ohlrogge JB, Arondel V: Comparative transcriptome and metabolite analysis of oil palm and date palm mesocarp that differ dramatically in carbon partitioning. Proceedings of the National Academy of Sciences of the United States of America. 2011;108:18186–18186.
6. Dussert S, Guerin C, Andersson M, Joet T, Tranbarger TJ, Pizot M, Sarah G, Omore A, Durand-Gasselin T, Morcillo F: Comparative transcriptome analysis of three oil palm fruit and seed tissues that differ in oil content and fatty acid composition. Plant Physiology. 2013;162:1337–1358.
7. Wang XM: Plant phospholipases. Annual Review of Plant Physiology and Plant Molecular Biology. 2001;52:211–231. DOI: 10.1146/annurev.arplant.52.1.211.
8. Samuels L, Kunst L, Jetter R: Sealing plant surfaces: Cuticular wax formation by epidermal cells. Annual Review of Plant Biology. 2008;59:683–707.
9. Dudareva N, Negre F, Nagegowda DA, Orlova I: Plant volatiles: Recent advances and future perspectives. Critical Reviews in Plant Sciences. 2006;25:417–440.
10. Schwab W, Davidovich-Rikanati R, Lewinsohn E: Biosynthesis of plant-derived flavor compounds. The Plant Journal. 2008;54:712–732.
11. Weber H: Fatty acid-derived signals in plants. Trends in Plant Science. 2002;7:217–224.
12. Baud S, Lepiniec L: Physiological and developmental regulation of seed oil production. Progress in Lipid Research. 2010;49:235–249.
13. Durrett TP, Benning C, Ohlrogge J: Plant triacylglycerols as feedstocks for the production of biofuels. The Plant Journal. 2008;54:593–607.
14. Ohlrogge JB, Jaworski JG: Regulation of fatty acid synthesis. Annual Review of Plant Physiology and Plant Molecular Biology. 1997;48:109–136.
15. Bates PD, Browse J: The significance of different diacylgycerol synthesis pathways on plant oil composition and bioengineering. Frontiers in Plant Science. 2012;3
16. Bates PD, Stymne S, Ohlrogge J: Biochemical pathways in seed oil synthesis. Current Opinion in Plant Biology. 2013;16:358–364.

17. Graham IA: Seed storage oil mobilization. Annual Review of Plant Biology. 2008;59:115–142.
18. Napier JA, Beaudoin F, Tatham AS, Alexander LG, Shewry PR: The seed oleosins: Structure, properties and biological role. Advances in Botanical Research, Vol 35. 2001;35:111–138.
19. Shimada TL, Hara-Nishimura I: Leaf oil bodies are subcellular factories producing antifungal oxylipins. Current Opinion in Plant Biology. 2015;25:145–150.
20. Napier JA, Stobart AK, Shewry PR: The structure and biogenesis of plant oil bodies: The role of the er membrane and the oleosin class of proteins. Plant Molecular Biology. 1996;31:945–956.
21. Tzen JTC, Cao YZ, Laurent P, Ratnayake C, Huang AHC: Lipids, proteins, and structure of seed oil bodies from diverse species. Plant Physiology. 1993;101:267–276.
22. Siloto RMP, Findlay K, Lopez-Villalobos A, Yeung EC, Nykiforuk CL, Moloney MM: The accumulation of oleosins determines the size of seed oilbodies in Arabidopsis. The Plant Cell. 2006;18:1961–1974.
23. Skoric D: Achievements and future-directions of sunflower breeding. Field Crops Research. 1992;30:231–270.
24. Zheng P, Allen WB, Roesler K, Williams ME, Zhang S, Li J, Glassman K, Ranch J, Nubel D, Solawetz W, Bhattramakki D, Llaca V, Deschamps S, Zhong GY, Tarczynski MC, Shen B: A phenylalanine in dgat is a key determinant of oil content and composition in maize. Nature Genetics. 2008;40:367–372.
25. Li HW, Zhao TJ, Wang YF, Yu DY, Chen SY, Zhou RB, Gai JY: Genetic structure composed of additive QTL, epistatic QTL pairs and collective unmapped minor QTL conferring oil content and fatty acid components of soybeans. Euphytica. 2011;182:117–132.
26. Hwang EY, Song QJ, Jia GF, Specht JE, Hyten DL, Costa J, Cregan PB: A genome-wide association study of seed protein and oil content in soybean. Bmc Genomics. 2014;15
27. Burns MJ, Barnes SR, Bowman JG, Clarke MHE, Werner CP, Kearsey MJ: QTL analysis of an intervarietal set of substitution lines in *Brassica napus*: seed oil content and fatty acid composition. Heredity. 2003;90:39–48.
28. Jiang CC, Shi JQ, Li RY, Long Y, Wang H, Li DR, Zhao JY, Meng JL: Quantitative trait loci that control the oil content variation of rapeseed (*Brassica napus L.*). Theoretical and Applied Genetics. 2014;127:957–968.
29. McGlew K, Shaw V, Zhang M, Kim RJ, Yang W, Shorrosh B, Suh MC, Ohlrogge J: An annotated database of Arabidopsis mutants of acyl lipid metabolism. Plant Cell Reports. 2015;34:519–32.
30. Li-Beisson Y, Shorrosh B, Beisson F, Andersson MX, Arondel V, Bates PD, Baud S, Bird D, Debono A, Durrett TP, Franke RB, Graham IA, Katayama K, Kelly AA, Larson T, Markham JE, Miquel M, Molina I, Nishida I, Rowland O, Samuels L, Schmid KM, Wada H, Welti R, Xu C, Zallot R, Ohlrogge J: Acyl-lipid metabolism. The Arabidopsis book / American Society of Plant Biologists. 2013;11:e0161.
31. Jako C, Kumar A, Wei YD, Zou JT, Barton DL, Giblin EM, Covello PS, Taylor DC: Seed-specific over-expression of an Arabidopsis cDNA encoding a diacylglycerol acyltransferase enhances seed oil content and seed weight. Plant Physiology. 2001;126:861–874.
32. Vigeolas H, Waldeck P, Zank T, Geigenberger P: Increasing seed oil content in oil-seed rape (*Brassica napus L.*) by over-expression of a yeast glycerol-3-phosphate dehydrogenase under the control of a seed-specific promoter. Plant Biotechnology Journal. 2007;5:431–441.
33. Lardizabal K, Effertz R, Levering C, Mai J, Pedroso MC, Jury T, Aasen E, Gruys K, Bennett K: Expression of Umbelopsis ramanniana DGAT2A in seed increases oil in soybean. Plant Physiology. 2008;148:89–96.
34. Andrianov V, Borisjuk N, Pogrebnyak N, Brinker A, Dixon J, Spitsin S, Flynn J, Matyszczuk P, Andryszak K, Laurelli M, Golovkin M, Koprowski H: Tobacco as a

production platform for biofuel: Overexpression of *Arabidopsis* DGAT and LEC2 genes increases accumulation and shifts the composition of lipids in green biomass. Plant Biotechnology Journal. 2010;8:277–287.

35. Wang ZK, Huang WJ, Chang JM, Sebastian A, Li YG, Li HY, Wu XX, Zhang BB, Meng FL, Li WB: Overexpression of SiDGAT1, a gene encoding acyl-CoA:diacylglycerol acyltransferase from Sesamum indicum L. Increases oil content in transgenic Arabidopsis and soybean. Plant Cell Tissue and Organ Culture. 2014;119:399–410.
36. Taylor DC, Zhang Y, Kumar A, Francis T, Giblin EM, Barton DL, Ferrie JR, Laroche A, Shah S, Zhu W, Snyder CL, Hall L, Rakow G, Harwood JL, Weselake RJ: Molecular modification of triacylglycerol accumulation by over-expression of DGAT1 to produce canola with increased seed oil content under field conditions. Botany-Botanique. 2009;87:533–543.
37. Weselake RJ, Shah S, Tang MG, Quant PA, Snyder CL, Furukawa-Stoffer TL, Zhu WM, Taylor DC, Zou JT, Kumar A, Hall L, Laroche A, Rakow G, Raney P, Moloney MM, Harwood JL: Metabolic control analysis is helpful for informed genetic manipulation of oilseed rape (Brassica napus) to increase seed oil content. Journal of Experimental Botany. 2008;59:3543–3549.
38. Mano S, Nishimura M: Plant peroxisomes. Plant Hormones. 2005;72:111–154.
39. Theodoulou FL, Eastmond PJ: Seed storage oil catabolism: A story of give and take. Current Opinion in Plant Biology. 2012;15:322–328.
40. Carrera E, Holman T, Medhurst A, Peer W, Schmuths H, Footitt S, Theodoulou FL, Holdsworth MJ: Gene expression profiling reveals defined functions of the ATP-binding cassette transporter comatose late in phase ii of germination. Plant Physiology. 2007;143:1669–1679.
41. Kamada T, Nito K, Hayashi H, Mano S, Hayashi M, Nishimura M: Functional differentiation of peroxisomes revealed by expression profiles of peroxisomal genes in Sesamum indicum L. Plant and Cell Physiology. 2003;44:1275–1289.
42. Kanai M, Nishimura M, Hayashi M: A peroxisomal ABC transporter promotes seed germination by inducing pectin degradation under the control of ABI5. The Plant Journal. 2010;62:936–947.
43. Turley RB, Trelease RN: Development and regulation of three glyoxysomal enzymes during cotton seed maturation and growth. Plant Molecular Biology. 1990;14:137-46.
44. Comai L, Dietrich RA, Maslyar DJ, Baden CS, Harada JJ: Coordinate expression of transcriptionally regulated isocitrate lyase and malate synthase genes in *Brassica-napus L.* The Plant Cell. 1989;1:293–300.
45. Chia TYP, Pike MJ, Rawsthorne S: Storage oil breakdown during embryo development of Brassica napus (L.). Journal of Experimental Botany. 2005;56:1285–1296.
46. Chen H, Wang FW, Dong YY, Wang N, Sun YP, Li XY, Liu L, Fan XD, Yin HL, Jing YY, Zhang XY, Li YL, Chen G, Li HY: Sequence mining and transcript profiling to explore differentially expressed genes associated with lipid biosynthesis during soybean seed development. Bmc Plant Biology. 2012;12
47. Kelly AA, Shaw E, Powers SJ, Kurup S, Eastmond PJ: Suppression of the *SUGAR-DEPENDENT1* triacylglycerol lipase family during seed development enhances oil yield in oilseed rape (*Brassica napus L.*). Plant Biotechnology Journal. 2013;11:355–361.
48. Eastmond PJ: SUGAR-DEPENDENT1 encodes a patatin domain triacylglycerol lipase that initiates storage oil breakdown in germinating Arabidopsis seeds. The Plant Cell. 2006;18:665–675.
49. Kim MJ, Yang SW, Mao HZ, Veena SP, Yin JL, Chua NH: Gene silencing of *Sugar-dependent 1 (JcSDP1)*, encoding a patatin-domain triacylglycerol lipase, enhances seed oil accumulation in Jatropha curcas. Biotechnology for Biofuels. 2014;7
50. Maeo K, Tokuda T, Ayame A, Mitsui N, Kawai T, Tsukagoshi H, Ishiguro S, Nakamura K: An AP2-type transcription factor, WRINKLED1, of Arabidopsis thaliana binds to

the AW-box sequence conserved among proximal upstream regions of genes involved in fatty acid synthesis. The Plant Journal. 2009;60:476–487.

51. Baud S, Wuilleme S, To A, Rochat C, Lepiniec L: Role of WRINKLED1 in the transcriptional regulation of glycolytic and fatty acid biosynthetic genes in Arabidopsis. The Plant Journal. 2009;60:933–947.

52. Baud S, Mendoza MS, To A, Harscoet E, Lepiniec L, Dubreucq B: WRINKLED1 specifies the regulatory action of LEAFY COTYLEDON2 towards fatty acid metabolism during seed maturation in Arabidopsis. The Plant Journal. 2007;50:825–838.

53. Cernac A, Benning C: *WRINKLED1* encodes an AP2/EREB domain protein involved in the control of storage compound biosynthesis in Arabidopsis. The Plant Journal. 2004;40:575–585.

54. Kanai M, Mano S, Kondo M, Hayashi M, Nishimura M: Extension of oil biosynthesis during the mid-phase of seed development enhances oil content in *Arabidopsis seeds*. Plant Biotechnology Journal. 2016;14:1241–1250.

55. Western TL, Skinner DJ, Haughn GW: Differentiation of mucilage secretory cells of the Arabidopsis seed coat. Plant Physiology. 2000;122:345-355.

56. Haughn GW, Western TL: *Arabidopsis*seed coat mucilage is a specialized cell wall that can be used as a model for genetic analysis of plant cell wall structure and function. Frontiers in Plant Science. 2012;3

57. Shen B, Sinkevicius KW, Selinger DA, Tarczynski MC: The Homeobox gene *GLABRA2 affects seed oil content in Arabidopsis*. Plant Molecular Biology. 2006;60:377–387.

58. Shi L, Katavic V, Yu YY, Kunst L, Haughn G: Arabidopsis *glabra2* mutant seeds deficient in mucilage biosynthesis produce more oil. The Plant Journal. 2012;69:37–46.

59. Sanjaya, Durrett TP, Weise SE, Benning C: Increasing the energy density of vegetative tissues by diverting carbon from starch to oil biosynthesis in transgenic Arabidopsis. Plant Biotechnology Journal. 2011;9:874–883.

60. Eskandari M, Cober ER, Rajcan I: Genetic control of soybean seed oil: Ii. QTL and genes that increase oil concentration without decreasing protein or with increased seed yield. Theoretical and Applied Genetics. 2013;126:1677–1687.

61. Mao TT, Jiang ZF, Han YP, Teng WL, Zhao X, Li WB: Identification of quantitative trait loci underlying seed protein and oil contents of soybean across multi-genetic backgrounds and environments. Plant Breeding. 2013;132:630–641.

62. Lin YM, Pajak A, Marsolais F, McCourt P, Riggs CD: Characterization of a cruciferin deficient mutant of Arabidopsis and its utility for overexpression of foreign proteins in plants. Plos One. 2013;8

63. van Erp H, Kelly AA, Menard G, Eastmond PJ: Multigene engineering of boosts seed oil content in Arabidopsis. Plant Physiology. 2014;165:30-6.

64. Schmidt MA, Barbazuk WB, Sandford M, May G, Song ZH, Zhou WX, Nikolau BJ, Herman EM: Silencing of soybean seed storage proteins results in a rebalanced protein composition preserving seed protein content without major collateral changes in the metabolome and transcriptome. Plant Physiology. 2011;156:330–345.

65. Kawakatsu T, Hirose S, Yasuda H, Takaiwa F: Reducing rice seed storage protein accumulation leads to changes in nutrient quality and storage organelle formation. Plant Physiology. 2010;154:1842–1854.

CHAPTER 17

Effect of Sowing Method and N Application on Seed Yield and N Use Efficiency of Winter Oilseed Rape

Klaus Sieling, Ulf Böttcher and Henning Kage

Agronomy and Crop Science, Institute of Crop Science and Plant Breeding, Christian-Albrechts-University of Kiel, Hermann-Rodewald-Str. 9, D-24118 Kiel, Germany

ABSTRACT

In northern Europe, replacing winter barley with winter wheat as the preceding crop for winter oilseed rape (*Brassica napus* L.; WOSR) often results in a delayed WOSR sowing and poor autumn growth. Based on data from a field experiment running in 2009/2010, 2010/2011, and 2012/2013, this study aims (i) to investigate how a delayed sowing method affects seed yield, N offtake with the seeds, and apparent N use efficiency (NUE) of WOSR; (ii) to test the ability of autumn and spring N fertilization to compensate for the negative effects of a delayed sowing method; and (iii) to estimate the minimum autumnal growth for optimal seed yield. In order to create sufficiently differentiated canopies, a combination of four sowing methods (first week of August until the third week of September) and four autumn N treatments (0, 30, 60, and 90 kg·N·ha^{-1}) was established. Each of these 16 different canopies was fertilized with 5 N amounts (0/0, 40/40, 80/80, 120/120, 140/140 kg·N·ha^{-1}) in spring in order to estimate separate N response curves.

Above-ground N accumulation in autumn and seed yield and N offtake by the seeds were determined. Plant establishment after mid-September significantly decreased seed yield. Autumn N fertilization of at least 30 kg·N·ha^{-1}increased seed yield and N offtake by the seeds without any significant interaction with sowing method and spring N supply. However, the pathway(s) remain(s) unclear. Spring N fertilization up to 130 kg·N·ha^{-1} (estimated by a Linear-Plateau N response curve) increased seed yield. NUE decreased with increasing N supply, where WOSR used autumn N to a lesser extent than spring N. An above-ground N uptake of at least 10–15 kg·N·ha^{-1} at the end of autumn growth was required to achieve high seed yields. From an environmental point of view, optimal autumn growth should be attained by choosing an adequate sowing method, not by applying additional N in autumn.

Keywords: oilseed rape; seed yield; sowing method; N fertilization; autumn N uptake; N use efficiency

1. INTRODUCTION

Farmers in northern Europe often replace winter barley with winter wheat as the preceding crop for winter oilseed rape (*Brassica napus* L.; WOSR) causing a delayed WOSR sowing due to the later wheat harvest. Luteman and Dixon [1] and Sieling et al. [2] observed a reduced autumn and winter growth by delayed plant establishment, especially if it was later than 10 September, increasing the risk of winter kill [3,4,5]. While Luteman and Dixon [1] found only small yield penalties, Uzan et al. [6] reported a decline in seed yield of canola cultivars with a delay in sowing date under Mediterranean environment conditions due to shortening the length of the reproductive period and consequently the potential grain-filling period. Also, Scott et al. [7] showed large yield penalties if sowing occurred after mid-September.

In order to compensate for the poor autumn growth, an additional nitrogen (N) supply of about 30–50 kg·N·ha^{-1} in autumn is intended to ensure crop N supply, adequate crop growth before winter and, consequently, a good overwinter survival. Autumn N, especially in early-sown oilseed rape, leads to a better above-ground growth [2] and enhances N accumulation before winter [8]. It is, however, debatable whether this increases seed yield and overwintering. Results of Ogilvy and Bastiman [9] revealed that, although plots receiving nitrogen in the seedbed or at the two-leaf stage were more vigorous before winter compared with unfertilized plants, neither the number of plants established, the survival over winter, nor the seed yield were affected by this treatment. According to Engström et al. [10] no effects on seed yield occurred after applying 30 or 60 kg·N·ha^{-1} at sowing to WOSR in Sweden. In addition, Sieling and Kage [11] observed that WOSR yield increased by 0.2 t·ha^{-1} after an autumn N supply of 40 kg·N·ha^{-1}. In experiments in France, severe N deficiencies in autumn, described in terms of the nitrogen nutrition index, reduced shoot biomass, tap root biomass, leaf area index, and radiation-use efficiency compared with N sufficient treatments [12]. However, no difference in seed yield occurred. The authors assumed that sufficient N release due to natural mineralization provided enough growth in autumn to ensure sufficient regrowth in spring. In other experiments, autumn N gave a small yield response where the preceding cereal straw was baled or incorporated instead of burning [13,14]. In addition, N application to boost crop growth rate may lead to plants being more susceptible to freezing in winter due to vigorous vegetative growth prior to the first killing frost or to increased leaching risk of N not taken up by the crop.

Spring N fertilization clearly increases WOSR seed yield. WOSR demands high amounts of N fertilizer often exceeding 200 kg N·ha^{-1} to achieve maximum yields in high yielding environments with yield levels >4 t·ha^{-1}. Scott et al. [15] and Zhao et al. [16] suggested a maximum yield response to a N rate around 200 kg·N·ha^{-1}, which was in good agreement with the results of Bilsborrow et al. [17] who obtained >85% of the maximum recorded yield with an application of 150 kg·N·ha^{-1}. Several approaches from, for example, France and Germany take the amount of N in the WOSR canopy at the end of autumn growth and/or at the beginning of spring growth into account [18,19,20]. In one study of insignificant leaf losses over winter, a well-developed rapeseed canopy showed a higher leaf area index at the beginning of spring growth than a poor canopy, which, in consequence, allows for reduced N fertilization rates [21]. Henke et al. [19] showed that even if canopy N is lost over winter due to frost, optimum N requirements in spring decreased with increasing canopy N in late autumn. These results assumed a negative interaction between autumn N supply, promoting above-ground dry matter accumulation and N uptake, and spring N application [2].

WOSR has the ability to take up more N than winter cereals between sowing and spring [22]; however, relatively little of this N ends up in the seed [23]. Compared to the N requirement, N offtake by the WOSR seeds is comparatively low, leading to low nitrogen harvest indices and high N balance surpluses (fertilizer N minus N offtake by the seeds), which can lead to high N leaching rates during periods of heavy rainfall [23,24,25,26,27]. Thus, along with an increasing rate of N-fertilization, the rate of fertilizer-N recovery (apparent N use efficiency (NUE)) declines substantially [26]. Due to only small yield effects, N fertilization in autumn is less N efficient in spring, thus most of the autumn N remains in the system. In field experiments reported by Sieling and Kage [11], 40 kg·N·ha^{-1} in autumn increased N offtake by the seeds by about 4 kg·N·ha^{-1}, thus 36 kg·N·ha^{-1} were left in the soil changing the N balance.

In general, it is assumed that WOSR requires at least six to eight leaves, a root collar diameter of 1 cm, a tap root length of 20 cm, and about 1 g of dry matter (DM) per plant to secure optimal winter survival [28]. Under the climate conditions of northern Europe, sowing date and N availability in autumn are the main factors influencing WOSR growth before winter, which can be managed by the farmers [2]. However, no data are available on which to base thresholds for optimal autumn growth.

Based on data from a three-year field experiment, this study aims (i) to investigate how delayed sowing affects seed yield, N offtake with the seeds, and N use efficiency of WOSR; (ii) to test the ability of autumn and spring N fertilization to compensate for the negative effects of a delayed sowing method; and (iii) to estimate the minimum autumnal growth for optimal seed yield. We hypothesize that delaying sowing reduces seed yield, N offtake, and N use efficiency, while autumn N supply may increase these factors, especially in late sown canopies. In addition, spring N fertilization may be reduced if additional N was applied in autumn.

2. RESULTS

The experimental design provided an increased seed density at later sowing in order to compensate for a decreased emergence rate, thus confounding both effects. However, in considering the average of the three years, plant densities were quite similar for all sowing methods ranging between 37 and 47 plants·m^{-2} at the end of autumn growth and between 36 and 40 plants·m^{-2} at the beginning of spring growth. Consequently, winter survival also remained unaffected. Therefore, no significant interaction between sowing date and seed density occurred.

2.1. Seed Yield

In considering the average of all three years and all other treatments, sowing in September significantly decreased WOSR seed yield ($p < 0.05$) (Table 1). While both sowing methods in August resulted in similar seed yields of about 5 t·ha^{-1}, drilling in the first and third September week caused a yield loss of 0.3 t·ha^{-1} and 1.2 t·ha^{-1}, respectively. Autumn N application significantly increased seed yield by 0.6 t·ha^{-1}, in comparison to the unfertilized and the 90 kg·N·ha^{-1}treatments (Table 1). The sowing method by autumn N interaction was not significant at $p = 0.05$.

Table 1. Sowing method by autumn N application interaction effects on winter oilseed rape (*Brassica napus* L.; WOSR) seed yield (t·ha^{-1}) (based on the average of 2009/2010, 2010/2011, 2012/2013 and all spring treatments).

	Autumn N Application (kg·N·ha^{-1})				
Sowing Method ‡	0	30	60	90	Mean
SD 1	4.72	4.87	5.02	5.12	4.93 ^{a,†}
SD 2	4.74	4.94	5.16	5.18	5.01 ^{a}
SD 3	4.05	4.74	4.86	4.98	4.66 ^{b}
SD 4	3.42	3.71	3.84	4.06	3.76 ^{c}
Mean	4.23 ^{c}	4.56 ^{b}	4.72 ^{a,b}	4.83 ^{a}	

‡—SD 1—first week of August; SD 2—third week of August; SD 3—first week of September; SD 4—third week of September. †—Different letters indicate significant differences within a factor at $p = 0.05$. The sowing method by autumn N interaction was not significant ($p > 0.05$). SD—sowing method.

Spring N supply boosted seed yield (Figure 1a; for estimated model parameters, see Table 2). However, N amount that exceeded 116–136 kg·N·ha^{-1} depending on the sowing method failed to show any further increase. The comparison of the function parameter revealed that the sowing method only influenced the level of the yield plateau ranging between 4.23 t·ha^{-1} if sown in the third week of September (SD 4) and 5.56 t·ha^{-1} at the third week of August (SD 2). In contrast, the N application rate at the intersection of the linear model and the plateau yield (Nopt), as well as the linear slope, remained unaffected (Table 2) indicating that the sowing method (SD) by spring N interaction was not significant.

A similar pattern occurred when comparing spring N effects at different autumn N levels (Figure 1b; for estimated model parameters, see Table 2). Applying N in autumn increased the plateau yield from 4.48 t·ha^{-1} in the unfertilized control up to 5.12, 5.20, and 5.32 t·ha^{-1} in the 30, 60, and 90 kg·N·ha^{-1} treatment, respectively, but did not affect Nopt and slope (Table 2). No significant interactions between both N application dates occurred.

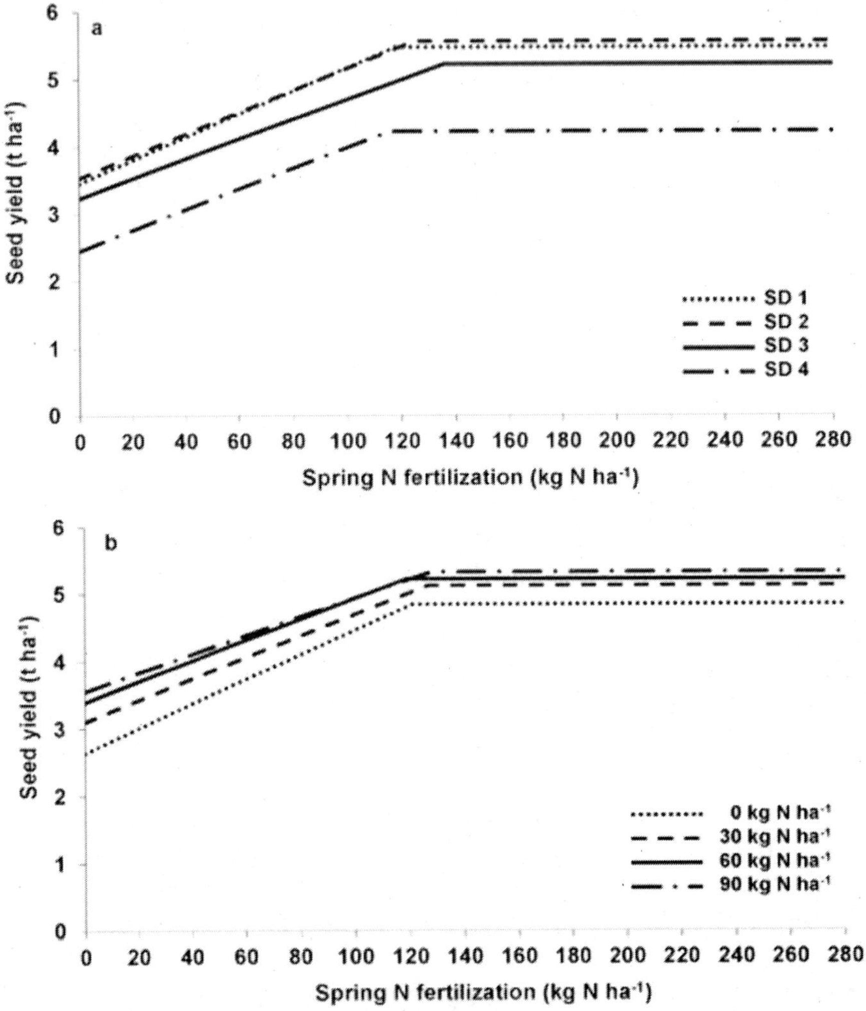

Figure 1. Effect of sowing method (SD) by spring N fertilization interaction (**a**) and autumn N by spring N fertilization (**b**) on seed yield of oilseed rape (2010, 2011, 2013; for function parameters, see Table 2). (SD 1—first week of August; S—third week of August; S—first week of September; S—third week of September).

Table 2. Estimated model parameters of the linear-plateau functions quantifying the relationship between spring N fertilization (kg·N·ha^{-1}) and seed yield (t·ha^{-1}) or N offtake with the seeds (kg·N·ha^{-1}) and between N uptake at the end of autumn growth and seed yield (t·ha^{-1}) (2009/2010, 2010/2011, 2012/2013).

Sowing Method §	Autumn N (kg·N·ha^{-1})	Spring N (kg·N·ha^{-1})	n	R^2	p	Plateau †	Nopt ‡	Slope
Seed yield (Figure 1a)								
SD 1			240	0.65	<0.0001	5.47 $^{a,b\$}$	117.7 ^{ns}	0.0172 ^{ns}
SD 2			239	0.66	<0.0001	5.56 ^{a}	125.2 ^{ns}	0.0162 ^{ns}
SD 3			240	0.49	<0.0001	5.22 ^{b}	136.2 ^{ns}	0.0147 ^{ns}
SD 4			240	0.27	<0.0001	4.23 ^{c}	116.4 ^{ns}	0.0153 ^{ns}
Seed yield (Figure 1b)								
	0		240	0.41	<0.0001	4.48 ^{b}	121.4 ^{ns}	0.0185 ^{ns}
	30		240	0.44	<0.0001	5.12 ^{a}	127.3 ^{ns}	0.0159 ^{ns}
	60		239	0.39	<0.0001	5.20 ^{a}	119.1 ^{ns}	0.0152 ^{ns}
	90		240	0.41	<0.0001	5.32 ^{a}	127.7 ^{ns}	0.0138 ^{ns}
N offtake by the seeds (Figure 2a)								
SD 1			240	0.73	<0.0001	162.8 ^{b}	156.5 ^{ns}	0.5344 ^{a}
SD 2			239	0.75	<0.0001	171.1 ^{a}	195.3 ^{ns}	0.4360 ^{b}
SD 3			240	0.65	<0.0001	159.6 ^{b}	182.2 ^{ns}	0.4257 ^{b}
SD 4			240	0.39	<0.0001	131.3 ^{c}	176.1 ^{ns}	0.3428 ^{c}
N offtake by the seeds (Figure 2b)								
	0		240	0.56	<0.0001	148.5 ^{b}	181.4 ^{ns}	0.4568 ^{ns}
	30		240	0.62	<0.0001	157.7 ^{a}	190.4 ^{ns}	0.4189 ^{ns}
	60		239	0.56	<0.0001	158.3 ^{a}	177.2 ^{ns}	0.4056 ^{ns}
	90		240	0.63	<0.0001	164.6 ^{a}	190.8 ^{ns}	0.3909 ^{ns}
Seed yield (Figure 4)								
		0	190	0.42	<0.0001	3.99 ^{c}	64.9 ^{a}	0.0245 ^{c}
		80	190	0.41	<0.0001	4.80 ^{b}	13.1 ^{b}	0.1329 ^{b}
		160	192	0.44	<0.0001	5.40 ^{a}	6.8 ^{c}	0.3487 ^{a}
		240	191	0.43	<0.0001	5.47 ^{a}	6.7 ^{c}	0.3697 ^{a}
		280	191	0.42	<0.0001	5.52 ^{a}	6.4 ^{c}	0.4319 ^{a}

†—Plateau—plateau yield and N offtake, respectively. ‡ Nopt—N application rate (kg·N·ha^{-1}) and N uptake at the end of autumn growth, respectively, at the intersection of the linear model and the plateau (i.e., minimum N amount required to achieve maximum yield). §—SD 1—first week of August; SD 2—third week of August; SD 3—first week of September; SD 4—third week of September. $—Different letters indicate significant differences within one parameter and one factor at $p = 0.05$; ns-not significant at $p = 0.05$.

2.2. N Offtake by the Seeds

Considering the N offtake by the seeds revealed similar results as observed for seed yield. Delaying sowing reduced N offtake ranging between 112 kg·N·ha^{-1} sown at the latest date and 141 kg·N·ha^{-1}, if crop establishment was made in the third week of August. The additional 30 kg·N·ha^{-1} applied in autumn increased N offtake by 9 kg·N·ha^{-1}, thus leaving 21 kg·N·ha^{-1} (= 70% of the applied N amount) in the system (Table 3). A further increase in the autumn N supply up to 60 and 90 kg·N·ha^{-1} increased N offtake by 14 and 18 kg·N·ha^{-1}, thus boosting the N surplus by 46 (= 77%) and 72 kg·N·ha^{-1} (= 80%), respectively. Again, no significant interactions could be identified.

Table 3. Sowing method by autumn N application interaction effects on N offtake by WOSR seed yield (kg·N·ha^{-1}) (based on average of 2009/2010, 2010/2011, 2012/2013 and all spring treatments).

	Autumn N Application (kg·N·ha^{-1})				
Sowing Method §	0	30	60	90	Mean
SD 1	133	156	140	143	138 a,b,†
SD 2	134	138	145	146	141 a
SD 3	116	136	140	143	133 b
SD 4	100	110	114	123	112 c
Mean	121 c	130 b	135 a,b	139 a	

§—SD 1—first week of August; SD 2—third week of August; SD 3—first week of September; SD 4—third week of September. †—Different letters indicate significant differences within a factor at $p = 0.05$. The sowing method by autumn N interaction was significant at $p = 0.05$.

Since spring N fertilization increased not only seed yield, but also seed N concentration (data not shown), the N rate required to achieve maximum yield (Nopt) was higher for N offtake (160–195 kg·N·ha^{-1}) than for seed yield (about 116–136 kg·N·ha^{-1}) (Figure 2a; for estimated model parameters, see Table 2) at all sowing methods. In addition, the linear slope decreased with delayed sowing.

In combination with spring N, autumn N application was positively correlated with only the level of the plateau ranging between 149 and 166 kg·N·ha^{-1} (Figure 2b; for estimated model parameters, see Table 2), while Nopt and the slope were similar. Autumn N did not affect seed N concentration; in consequence, increased N offtake was mainly due to higher seed yields.

2.3. Seed Oil Concentration and Oil Yield

Seed oil concentration was highest at early sowing and fell if sowing was delayed (Figure 3). Likewise, spring N significantly decreased oil concentration while autumn N had no effects. The slope was steepest if plant establishment occurred in August. In consequence, oil yield showed a similar pattern as seed yield with significant effects of sowing method (2.17, 2.19, 2.03, and 1.61 t·ha^{-1} in SD 1, SD 2, SD 3, and SD 4, respectively), autumn N (1.84, 1.99, 2.06, and 2.10 t·ha^{-1} in the 0, 30, 60, and 90 kg·N·ha^{-1} treatment, respectively) and spring N application (1.44, 1.99, 2.20, 2.18, and 2.18 t·ha^{-1} in the 0, 80, 160, 240, and 280 kg·N·ha^{-1} treatment, respectively). No significant interactions were observed.

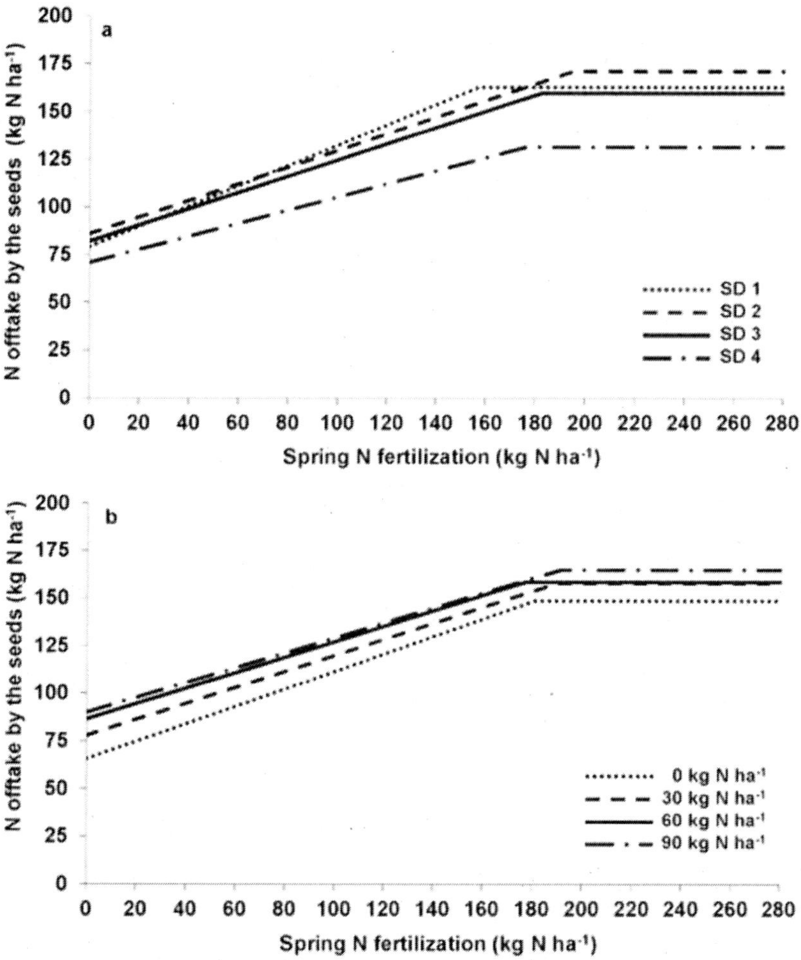

Figure 2. Effect of sowing method (SD) by spring N fertilization interaction (**a**) and autumn N by spring N fertilization (**b**) on N offtake by the seeds of oilseed rape (2010, 2011, 2013; for function parameters, see Table 2). (SD 1—first week of August; SD 2—third week of August; SD 3—first week of September; SD 4—third week of September).

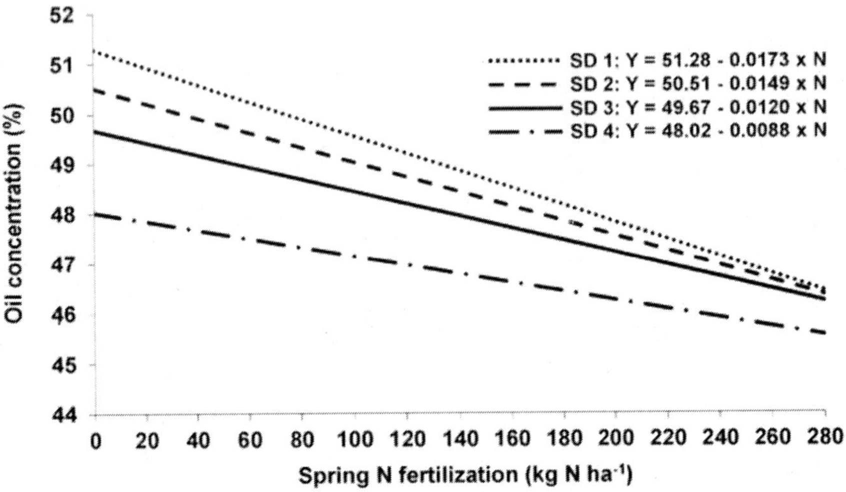

Figure 3. Effect of sowing method (SD) by spring N fertilization interaction on seed oil concentration of the seeds of oilseed rape (2010, 2011, 2013). (SD 1—first week of August; SD 2—third week of August; SD 3—first week of September; SD 4—third week of September).

2.4. Apparent N Use Efficiency

The apparent N use efficiency (NUE) describes the additional N offtake by the seeds due to the fertilization in relation to the respective N supply. Since no significant interaction between autumn and spring N application occurred, NUE of autumn N was presented based on the average of the spring N treatments and vice versa (Table 4). NUE of autumn N supply decreased with increasing N amount from 25% in the 30 $kg{\cdot}N{\cdot}ha^{-1}$ treatment down to 20% in the plots which had received 90 $kg{\cdot}N{\cdot}ha^{-1}$; however, this reduction was not significant at $p = 0.05$. The date of plant establishment significantly affected NUE, being lowest at SD 1 (11%) and highest at SD 3 (38%). In contrast, the early sown plots utilized spring applied N to a larger extent than the later sown ones (Table 4). Increasing spring N application significantly reduced the respective NUE from 46% (80 $kg{\cdot}N{\cdot}ha^{-1}$) to 28% (280 $kg{\cdot}N{\cdot}ha^{-1}$). The sowing method by autumn N interaction and the sowing method by spring N interaction were not significant ($p > 0.05$).

Table 4. Sowing method by N application interaction effects on apparent N use efficiency (%) (2009/2010, 2010/2011, 2012/2013).

	Sowing Method §				
	SD 1	SD 2	SD 3	SD 4	Mean
Autumn N application (kg·N·ha^{-1})					
0	-	-	-	-	-
30	8.4	13.0	44.4	32.6	24.6 $^{ns,+}$
60	14.0	18.5	40.0	22.5	23.7 ^{ns}
90	11.1	13.2	30.3	25.0	19.9 ^{ns}
Mean	11.1 ^{c}	14.9 ^{c}	38.2 ^{a}	26.7 ^{b}	
Spring N application (kg·N·ha^{-1})					
0	-	-	-	-	-
80	53.1	47.8	41.3	41.9	46.0 ^{a}
160	47.0	43.6	42.6	34.3	41.9 ^{a}
240	35.9	35.1	31.6	26.2	32.2 ^{b,c}
280	31.9	31.5	28.1	22.1	28.4 ^{c}
Mean	42.0 ^{a}	39.5 ^{a,b}	35.9 ^{b}	31.1 ^{c}	

§—SD 1—first week of August; SD 2—third week of August; SD 3—first week of September; SD 4—third week of September. +—ns—not significant ($p > 0.05$). Different letters indicate significant differences within one parameter and one factor at $p = 0.05$. The sowing method by autumn N interaction and the sowing method by spring N interaction were not significant ($p > 0.05$).

2.5. Relation between Above-Ground N Uptake in Autumn and Seed Yield

The combination of four sowing methods and four autumn N treatments resulted in 16 different WOSR canopies before winter. The estimation of the N uptake at the end of autumn growth allowed to determine the least required N application needed for optimal seed yield. Presuming a Linear-Plateau approach for each spring N treatment separately, the minimum autumn N uptake depended on the amount of spring N fertilization (Figure 4). A comparison of the estimated model parameters (Table 4) revealed similar curves if spring N supply was at least 160 kg·N·ha^{-1}. In this case, WOSR canopies should have taken up at least about 7 kg·N·ha^{-1} to achieve maximum seed yield. In contrast, canopies without spring N supply required 65 kg·N·ha^{-1} accumulated in the above-ground biomass. Larger canopies did not result in higher seed yield.

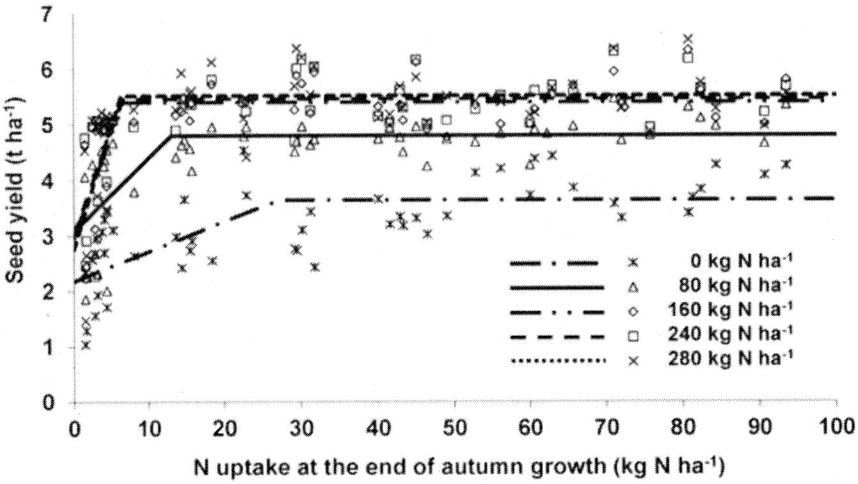

Figure 4. Relationship between N uptake of oilseed rape at the end of autumn growth and seed yield at different spring N fertilization levels (2010, 2011, 2013; for function parameters see Table 2).

3. DISCUSSION

The experimental design allowed for an investigation of the effects of sowing method, autumn N, and spring N supply, as well as their interactions, on yield and N offtake by the seeds of WSOR. Due to varying sowing methods and autumn N supplies, 16 different WOSR canopies were established in autumn, each combined with increasing spring N fertilization in order to estimate N response curves separately for each canopy. The results confirmed the current recommendation that under the climatic conditions of northern Europe WOSR should ideally be sown between the middle and end of August, whereas plant establishment after mid-September clearly decreased seed yield, which was in agreement with [7]. It should be noted that, if the weather conditions in the following spring are favorable, even an WOSR crop sown in the third week of September with two to four leaves at the end of autumn growth was able to achieve more than 5 t·ha^{-1} [29]; however, since long-term weather forecasts are not yet available, the risk of yield penalties increases with shifting WOSR sowing to the end of September.

In combination with late sowing, farmers often apply additional N in autumn to ensure adequate WOSR growth before winter. No significant interaction between the sowing method and autumn as well as spring N fertilization on seed yield could be observed in our experiments (Table

1). Applying at least 30 kg·N·ha^{-1} in autumn increased WOSR yield at all sowing methods averaged over three years, although unfertilized WOSR canopies established in August already accumulated about 100 g·DM·m^{-2} and 34 kg·N·ha^{-1} at the end of autumn growth compared to 3 g·DM·m^{-2} and <2 kg·N·ha^{-1} in the late sown ones [29]. In addition, the N amount in early sown canopies clearly exceeded the threshold for optimal autumn N uptake, as shown in Figure 4. However, the pathway(s) which allowed autumn N to significantly increase seed yield for the early sowing method treatment require(s) further study. On the one hand, it can be argued that despite the often relatively small N uptake before winter, the additional N supply in autumn enables the plants to better utilize (N) resources in spring, thus indicating a physiological pathway. On the other hand, autumn N might act as advanced spring N fertilization if N is not leached out of the potential rooting zone during winter. In this case, an interaction between autumn and spring N treatments should be expected; however, it was not observed in our experiments (Figure 1b, Figure 2b, and Table 2). In contrast, Henke et al. [19] provided an approach which took autumn N uptake into account when estimating spring N fertilization, thus assuming an autumn N by spring N interaction. The authors suggested reducing the spring N amount if N in the canopy exceeds 50 kg·N·ha^{-1} at the end of autumn growth. With only small increases in N offtake by the seeds (Table 3) and the resulting small NUE (Table 4) in mind, autumn N should not exceed 30 kg·N·ha^{-1}.

In order to describe the N effects, a "Linear Response and Plateau" approach (LRP) was chosen, since using quadratic N response curves resulted in quite high N amounts required to achieve yield maximum due to a less steep slope compared to wheat [30,31]. Although the quadratic approach seems to be safer in terms of preventing financial losses under conditions of uncertainty, the risk of high positive N balances rises and the introduction of penalty functions may therefore favor the LPR model, especially against the background of high N surpluses in WOSR growing systems [31].

Based on the LRP approach, the minimum amount of spring N required for maximum seed yield ranged, based on average of the three experimental years, between 120 and 140 kg·N·ha^{-1}, regardless of the sowing method and the autumn N supply, which was lower than local recommendation according to the "Düngeverordnung" [32], the national implementation of the EU Nitrate Directive, as well as results from other authors [26,33]. Since N fertilization also increased seed N concentration, optimal N supply for the maximum N offtake by the seeds was higher (about 180–190 kg·N·ha^{-1}, Figure 2a,b and Table 2). In addition, delayed sowing methods also led to a higher seed N concentration, whereas autumn N had no effects [29].

N-use efficiency (NUE) for autumn (spring) N was calculated by comparing the seed yield according to the average of the spring (autumn) N treatments in relation to the amount of applied N in autumn (spring). In general, NUE of WOSR is lower compared to wheat or barley [23,30,34], which often has been associated with a low foliar N remobilization, thus with a high N concentration in WOSR leaves dropping off during the growth period, especially after flowering [35,36,37]. NUE decreased with increasing N amounts (Table 4). WOSR utilized autumn N to a lower extent than spring applied N [9,11]. Autumn NUE was highest when WOSR was sown in the first week of September, which was mainly due to the lower N offtake of the unfertilized plots compared to earlier sowing, while the autumn N fertilized ones showed similar values (Table 3). Therefore, NUE alone is not suitable for evaluating the sustainability of a cropping system as the N offtake and, in consequence, the N balance (N surplus) have to be taken into account. From an environmental point of view, N application in autumn, if any, should not exceed 30 kg·N·ha^{-1}, since most of the N remained in the system, increasing the risk of losses in the subsequent crops [27]. According to the presented results using the LPR approach, applying 120–130 kg·N·ha^{-1} in spring were sufficient to achieve the highest seed yields. However, using quadratic N response curves, 180–200 kg·N·ha^{-1}were required to get the economic optimal seed yield [29], representing an application rate that was similar to that used by the farmers.

The estimation of the minimum amount of N taken up by the canopy at the end of autumn in order to achieve maximum seed yield was based on the LPR approach, where only the spring N fertilization was additionally considered. Including autumn N supply did not improve the relationship (except if the canopy remained unfertilized in spring), indicating that the above-ground N accumulation at the end of autumn growth adequately covered the effects of autumn N application. Optimal autumn N uptake depended on the amount of spring N. If there are plans for 160 kg·N·ha^{-1} or more to be applied in spring, an N uptake of 10 kg·N·ha^{-1} before winter should enable the canopies to produce optimal seed yield. Even under the less favorable conditions occurring in autumn 2010, 13 kg·N·ha^{-1} seemed to be sufficient [29]. Assuming a quadratic relationship, about 40 kg·N·ha^{-1} were required to achieve 95% of the yield maximum [29] being similar to the 50 kg·N·ha^{-1}suggested by Henke et al. [19] to represent an average rapeseed canopy which will be fertilized in spring according to the official recommendation. However, a threshold of 1 g single plant DM for minimum autumn growth which is often used by the farmers results in about 20 kg·N·ha^{-1} assuming 40 plants·m^{-2} and 5% N concentration in the above-ground biomass.

4. MATERIALS AND METHODS

4.1. Site and Soil

The field experiment was carried out on a pseudogleyic sandy loam (Luvisol: 100 g·kg^{-1} clay, pH 6.6, 86 mg·kg^{-1} P, 79 mg·kg^{-1} K, 150 g·kg^{-1} Mg, 13.8 g·kg^{-1} C_{org}, 1.1 g·kg^{-1} N_{org}) at the Hohenschulen Experimental Farm (10.0° E, 54.3° N, 30 m above see level) of the Kiel University, located in northern Germany 15 km west of Kiel (Schleswig-Holstein).

The climate of NW Germany can be described as humid. Total annual rainfall averages 750 mm at the experimental site, with ca. 400 mm received during April–September, the main growing season, and ca. 350 mm during October–March. In all three years, the mean air temperature in October was below the long-term average. Especially in autumn 2010, the low temperature in August until November, in combination with very high rainfall in August and September, resulted in poor autumn growth in all treatments (Table 5).

Table 5. Monthly rainfall (mm) and mean air temperature (°C) at Hohenschulen, Germany.

	Total Rainfall (mm)				Mean Air Temperature (°C)			
	2009/2010	2010/2011	2012/2013	Long-Term Mean	2009/2010	2010/2011	2012/2013	Long-Term Mean
August	69	105	56	59	17.8	16.1	16.7	17.9
September	28	123	59	61	14.3	12.6	12.7	14.2
October	63	55	71	75	7.7	8.7	8.6	9.8
November	109	113	39	58	7.3	3.8	5.1	4.9
December	49	29	72	62	0.2	−4.6	0.0	1.6
January	21	37	68	48	−4.1	0.4	0.5	1.2
February	43	39	24	50	−1.7	−0.1	−0.7	2.0
March	56	19	22	44	3.6	3.2	−1.1	3.4
April	17	3	18	44	7.5	10.4	6.1	8.0
May	50	36	141	62	8.8	12.2	11.4	11.9
June	50	93	100	69	14.3	15.2	13.9	14.8
July	24	150	40	100	19.5	15.7	17.6	16.9

4.2. Treatments

In 2009/2010, 2010/2011, and 2012/2013, a field experiment with winter oilseed rape (WOSR) cv. Visby was established to test the effects of sowing method (SD) and N application in autumn and spring on crop growth in autumn, seed yield, and plant N recovery. Heavy rainfall in August 2011 prevented a proper crop establishment, therefore, no oilseed rape was grown in that year. In order to create sufficiently differentiated canopies, the four sowing dates ranged between the first week of August and the third week of September (Table 6). The seed density was increased with delaying the sowing dates resulting in 35, 40, 50, and 70 seeds·m^{-2}, respectively, according to common farmers' practice. Since sowing date was confounded by seed rate, the term 'sowing method' is used throughout the paper. In addition, each sowing method received four autumn N treatments (0, 30, 60, and 90 kg·N·ha^{-1} as calcium ammonium nitrate with 27% N; for application dates, see Table 6), resulting in 16 sowing method by N supply combinations. Each of these 16 different canopies was fertilized with five N amounts (0/0, 40/40, 80/80, 120/120, and 140/140 kg·N·ha^{-1} as calcium ammonium nitrate with 27% N) in spring in order to estimate separate N response curves. N was applied in split-dressings at the beginning of spring growth and at stem elongation (for application dates, see Table 6).

Table 6. Dates of sowing, autumn, and spring N application of winter oilseed rape.

		Year		
		2009/2010	2010/2011	2012/2013
Sowing	SD 1	5 August	12 August	11 August
	SD 2	18 August	28 August	23 August
	SD 3	1 September	7 September	5 September
	SD 4	21 September	21 September	17 September
N application in autumn	SD 1	27 August	24 August	3 September
	SD 2	27 August	8 September	3 September
	SD 3	2 September	8 September	10 September
	SD 4	22 September	30 September	19 September
First N application in spring		23 March	3 March	4 March
Second N application in spring		12 April	8 April	15 April

The sowing methods as main plots were randomized with four replicates, while the N treatments were randomized within the main plots. This experimental design allowed for analyzing the sowing method effects separately for each year. The single plot size was 3 m × 12 m. Winter barley was the preceding crop; its straw remained on the plots. In general, the plots were ploughed within one day before sowing. Crop management not involving the treatments (e.g., P and K supply, soil tillage, pesticide application) were applied according to local recommendations to achieve optimal yield.

4.3. Measurements

In all sowing methods plant samples were taken at the end of autumn growth. Plants were dug up from an area of 1 m^2. After washing, above-ground biomass was analyzed for dry matter (DM) and N concentration using the near infrared spectroscopy (NIRS) analysis. Autumn N uptake was calculated by multiplying above-ground DM and the respective N concentration.

In order to minimize border effects, only a core of 1.75 m × 6 m (= 10.5 m^2) was combine harvested at maturity to determine seed yield (standardized to $t \cdot ha^{-1}$ with 91% DM based on the moisture content of a seed subsample). Oil and N concentration of the seeds (at 100% DM) were determined by NIRS. N offtake by the seeds results from the product of seed yield DM and N concentration.

Apparent N use efficiency (NUE) was defined as the difference between the N offtake by the seeds in the fertilized and unfertilized treatment in relation to the N amount. Since no interaction between autumn and spring N fertilization occurred, NUE of the autumn (spring) N application was calculated based on the average of the spring (autumn) N treatments.

4.4. Statistical Analysis

First, the effects of sowing methods (SD), autumn and spring N fertilization were analyzed using a mixed model approach (SAS 9.4, Proc MIXED, SAS Institute Inc., Cary, NC, US) with "year" and "replication" as random terms. Since the SD by autumn N by spring N interaction was not significant, only the interactions between two factors as average on the remaining factor are presented.

Second, the effects of spring N application on seed yield and N offtake by the seeds were estimated by using a "Linear-Plateau" approach (Proc NLIN) separately for the sowing methods and the autumn N treatments, respectively.

The 'Linear-Plateau' model is specified by three parameters a, b, and P as follows:

$$Y = a + bX + \varepsilon \rightarrow \text{if } X < \text{Nopt} \tag{1}$$

$$Y = P + \varepsilon \rightarrow \text{if } X \geq \text{Nopt} \tag{2}$$

where Y denotes the DM yield ($t \cdot ha^{-1}$) or the N offtake by the seeds ($kg \cdot N \cdot ha^{-1}$), X the application rate of N ($kg \cdot N \cdot ha^{-1}$) or the N uptake in autumn, a the intercept, b the linear coefficient, and ε the error term. Nopt is the nitrogen application rate at the intersection of the linear model and the plateau yield P (Table 2). Function parameters were compared by a modified t-test based on Zar [38].

Seed oil concentration was analyzed by performing an analysis of covariance (Proc GLM).

A *p*-value of 0.05 was considered to be significant in the statistical analyses.

5. CONCLUSIONS

The presented results confirmed that the sowing of WOSR in the third week of August until the first week of September as being the best sowing method in northern Germany. Even in canopies which were well developed before winter, autumn N application increased seed yield. However, although economically suitable, autumn N should be critically discussed from an environmental point of view, since NUE is low and the pathway(s) of autumn N on yield is/are still unidentified.

Autumn N uptake of at least 10 $kg \cdot N \cdot ha^{-1}$ was required to achieve high seed yields, if no severe stress (severe frost during winter, drought in spring) occurred.

ACKNOWLEDGMENTS

We thank Kirsten Schulz and Cordula Weise for performing the plant sampling and the laboratory analyses.

AUTHOR CONTRIBUTIONS

Klaus Sieling analyzed the data and wrote the paper. All authors discussed the results and commented on the manuscript.

CONFLICTS OF INTEREST

The authors declare no conflict of interest.

REFERENCES

1. Luteman, P.J.W.; Dixon, F.L. The effect of drilling date on the growth and yield of oil-seed rape (Brassica napus L.). J. Agric. Sci. 1987, 108, 195–200.
2. Sieling, K.; Böttcher, U.; Kage, H. Sowing date and N application effects on tape root and above-ground dry matter of winter oilseed rape in autumn. Eur. J. Agric. 2017, 83, 40–46.
3. Lääniste, P.; Jõudu, J.; Eremeev, V.; Mäeorg, E. Sowing date influence on winter oilseed rape overwintering in Estonia. Acta Agric. Scand. 2007, B 57, 342–348.
4. Balodis, O.; Gaile, Z. Winter oilseed rape (Brassica napus L.) autumn growth. In Proceedings of the Annual 17th International Scientific Conference Research for Rural Development, Jelgava, Latvia, 18–20 May 2011; Latvia University of Agriculture: Jelgava, Latvia; Volume 1, pp. 6–12.
5. Waalen, W.; Øvergaard, S.I.; Åssveen, M.; Gusta, L.V. Winter survival of winter rapeseed and winter turnip rapeseed in field trials as explained by PPLS regression. Eur. J. Agron. 2013, 51, 81–90.
6. Uzun, B.; Zengin, Ü.; Furat, S.; Akdesir, Ö. Sowing date effects on growth, flowering, oil content and seed yield of canola cultivars. Asian J. Chem. 2009, 21, 1957–1965.
7. Scott, R.K.; Ogunremi, E.A.; Ivins, J.U.D.; Mendham, N.J. The effect of sowing date and season on growth abd yield of oilseed rape (Brassica napus). J. Agric. Sci. 1973, 81, 277–285.
8. Dejoux, J.-F.; Meynard, J.-M.; Reau, R.; Roche, R.; Saulas, P. Evaluation of environmentally-friendly crop management systems based on very early sowing dates for winter oilseed rape in France. Agronomie 2003, 23, 725–736.
9. Ogilvy, S.E.; Bastiman, B. The effect of rate and timing of autumn nitrogen on the pre-flowering dry matter production and seed yield of winter oilseed rape. Asp. Appl. Biol. 1992, 30, 413–416.
10. Engström, L.; Stenberg, M.; Aronsson, H.; Lindén, B. Reducing nitrate leaching after winter oilseed rape and peas in mild and cold winters. Agron. Sustain. Dev. 2011, 31, 337–347.
11. Sieling, K.; Kage, H. Autumnal N fertilization of late sown oilseed rape after minimum tillage. In Proceedings of the 12th International Rapeseed Congress, Wuhan, China, 26–30 March 2007; pp. 375–378.
12. Colnenne, C.; Meynard, J.-M.; Roche, R.; Reau, R. Effects of nitrogen deficiencies on autumnal growth of oilseed rape. Eur. J. Agron. 2002, 17, 11–28.
13. Chalmers, A.G. Autumn and spring fertiliser nitrogen requirements for winter oilseed rape. Asp. Appl. Biol. 1989, 23, 125–133.

14. Chalmers, A.G.; Darby, R.J. Nitrogen application to oilseed rape and implications for potential leaching loss. Asp. Appl. Biol. 1992, 30, 425–430.
15. Scott, R.K.; Ogunremi, E.A.; Ivins, J.U.D.; Mendham, N.J. The effect of fertilizers and harvest date on growth and yield of oilseed rape sown in autumn and spring. J. Agric. Sci. 1973, 81, 287–293.
16. Zhao, F.J.; Evans, E.J.; Bilsborrow, P.E.; Syers, J.K. Influence of sulphur and nitrogen on seed yield and quality of low glucosinolate oilseed rape (Brassica napus L.). J. Sci. Food Agric. 1993, 63, 29–37.
17. Bilsborrow, P.E.; Evans, E.J.; Zhao, F.J. The influence of spring nitrogen on yield, yield components and glucosinolate content of autumn-sown oilseed rape (Brassica napus). J. Agric. Sci. 1993, 120, 219–224.
18. Reau, R.; Wagner, D.; Palleau, J.P. End of winter diagnosis: Winter rapeseed and nitrogen fertilization. In Proceedings of the Third Congress of the European Society for Agronomy, Podova, Italy, 18–22 September 1994; European Society for Agrnonomy: Podova, Italy; pp. 220–221.
19. Henke, J.; Sieling, K.; Sauermann, W.; Kage, H. Analysing soil and canopy factors affecting optimum nitrogen fertilization rates of oilseed rape (Brassica napus). J. Agric. Sci. 2009, 147, 1–8.
20. Makowski, D.; Maltas, A.; Morison, M.; Reau, R. Calculating N fertilizer doses for oil-seed rape using plant and soil data. Agron. Sustain. Dev. 2005, 25, 159–161.
21. Mendham, N.J.; Shipway, P.A.; Scott, R.K. The effects of delayed sowing and weather on growth, development and yield of winter oil-seed rape (Brassica napus L.). J. Agric. Sci. 1981, 96, 389–416.
22. Barraclough, P.B. Root growth, macro-nutrient uptake dynamics and soil fertility requirements of a high-yielding winter oilseed rape crop. Plant Soil 1989, 119, 59–70.
23. Sieling, K.; Schröder, H.; Hanus, H. Mineral and slurry nitrogen effects on yield, N uptake, and apparent N use efficiency of oilseed rape (Brassica napus). J. Agric. Sci. 1998, 130, 165–172.
24. Shepherd, M.A.; Sylvester-Bradley, R. Effect of nitrogen fertilizer applied to winter oilseed rape (Brassica napus) on soil mineral nitrogen after harvest and on the response of a succeeding crop of winter wheat to nitrogen fertilizer. J. Agric. Sci. 1996, 126, 63–74.
25. Beaudoin, N.; Saad, J.K.; van Laethem, C.; Machet, J.M.; Maucorps, J.; Mary, B. Nitrate leaching in intensive agriculture in Northern France: Effect of farming practices, soils and crop rotations. Agric. Ecosyst. Environ. 2005, 111, 292–310.
26. Rathke, G.W.; Behrens, T.; Diepenbrock, W. Integrated nitrogen management strategies to improve seed yield, oil content and nitrogen efficiency of winter oilseed rape (Brassica napus L.): A review. Agric. Ecosyst. Environ. 2006, 117, 80–108.
27. Sieling, K.; Kage, H. N balance as an indicator of N leaching in an oilseed rape—Winter wheat—Winter barley rotation. Agric. Ecosyst. Environ. 2006, 115, 261–269.
28. Cramer, N. Raps—Züchtung, Anbau und Vermarktung von Körnerraps; Verlag Eugen Ulmer: Stuttgart, Germany, 1990.
29. Sieling, K. Unpublished work. 2017.
30. Sieling, K. Growth stage-specific application of slurry and mineral N to oilseed rape, wheat and barley. J. Agric. Sci.2004, 142, 495–502.
31. Henke, J.; Breustedt, G.; Sieling, K.; Kage, H. Impact of uncertainty on the optimum nitrogen fertilization rate and agronomic, ecological and economic factors in an oilseed rape based crop rotation. J. Agric. Sci. 2007, 145, 455–468.
32. Bundesgesetzblatt. Düngeverordnung. Bekanntmachung der Neufassung der Düngeverordnung, Bundesgesetzblatt. Available online: https://www.bgbl.de/xaver/bgbl/start.xav?start=//*%5B@attr_id=%27bgbl107s0221.pdf%27%5D#__bgbl__%2F%2F*%5B%40attr_id%3D%27bgbl107s0221.pdf%27%5D__1487932661132 (assessed on 11 November 2007).
33. Sieling, K.; Kage, H. Efficient N management using winter oilseed rape: A review. Agron. Sustain. Dev. 2010, 30, 271–279.

34. Sieling, K.; Schröder, H.; Finck, M.; Hanus, H. Yield, N uptake, and apparent N-use efficiency of winter wheat and winter barley grown in different cropping systems. J. Agric. Sci. 1998, 131, 375–387.
35. Rossato, L.; Lainé, P.; Ourry, A. Nitrogen storage and remobilisation in Brassica napus L. during the growth cycle: Nitrogen fluxes within the plant and changes in soluble protein patterns. J. Exp. Bot. 2001, 52, 1655–1663.
36. Ulas, A.; Behrens, T.; Wiesler, F.; Horst, W.J.; Schulte auf'm Erley, G. Does genotypic variation in nitrogen remobilisation efficiency contribute to nitrogen efficiency of winter oilseed-rape cultivars (Brassica napus L.)? Plant Soil 2013, 371, 463–471.
37. Girondé, A.; Poret, M.; Etienne, P.; Trouverie, J.; Bouchereau, A.; Le Cahérec, F.; Leport, L.; Orsel, M.; Niogret, M.-F.; Deleu, C.; et al. A profiling approach of the natural variability of foliar N remobilization at the rosette stage gives clues to understand the limiting processes involved in the low N use efficiency of winter oilseed rape. J. Exp. Bot.2015, 66, 2461–2473.
38. Zar, J.H. Biostatistical Analysis, 5th ed.; Prentice Hall: Upper Saddle River, NJ, USA, 2009.

CHAPTER 18

Potential Association between Soil and Leaf Chemical Properties, and Soybean Seed Composition

Luciano M. Jaureguy[1], Pengyin Chen[1*], Kristofor Brye[1], Derrick Oosterhuis[1], Andy Mauromoustakos[2], John R. Clark[3]

[1]*Department of Crop, Soil, and Environmental Sciences, University of Arkansas, Fayetteville, USA*

[2]*Agricultural Statistics Laboratory, University of Arkansas, Fayetteville, USA*

[3]*Department of Horticulture, University of Arkansas, Fayetteville, USA*

ABSTRACT

Maximizing the expression of seed quality traits by understanding how they are affected by environmental variables may help develop high quality nutritious soybeans. Eight specialty soybean breeding lines were grown at two Arkansas locations differing by soil texture, with three replications in 2011. Before the reproductive period, soil and leaf samples were collected from each plot. Soil samples were analyzed for particle size distribution, electrical conductivity, pH, and a set of extractable nutrients from the top 12.5 cm, while leaf samples were analyzed for the same set of nutrients as the soil samples. At maturity, seed samples were analyzed for the same set of nutrients as were leaf and soil samples, plus protein, oil, fatty acids, and sugars. All leaf elements were within the adequate levels for soybean production at both locations. Overall, seed composition of breeding lines did not differ among locations and no significant changes in breeding line ranking among locations were observed. Attempting to modify seed composition by nutrient fertilization may not be profitable, as no direct relationships were observed between leaf or soil chemical properties, and seed composition. These findings may provide a starting point for future studies on fertilization and management practices that improve soybean seed quality.

Keywords: Soybean, Seed Composition, Soil Properties

1. INTRODUCTION

Understanding the contribution of environmental variables to seed composition may help develop cultural and breeding strategies to produce high quality soybeans, especially due to the increasing global need of more nutritious food. Plant nutrient uptake is active during vegetative plant growth as soil extracted nutrients are incorporated into proteins to activate enzymes and contribute to the charge balance in the cell [1] . During the reproductive period, nutrients can be remobilized to seeds from vegetative tissue via the phloem, and to a lesser extent, newly uptaken by the roots via the xylem, transferred to phloem, and transported to the reproductive organs [2] [3] . Decreases in leaf element concentrations during the reproductive period have been well-documented, and have been used as evidence for nutrient remobilization in several crops [4] -[7] . Thus, factors affecting the availability of nutrients in the soil and the rate of uptake and remobilization of chemical elements by the plant have the potential to affect seed inorganic composition. Furthermore, because soil extracted nutrients are essential for plant metabolic processes, those factors also have the potential to significantly affect the organic composition of the seed [8] .

Bellaloui et al. [9] measured soil elemental concentration and soybean seed composition of field sections that differed in yield potential. Areas of low yield potential exhibited lower seed protein and oleic acid concentration than that in areas of medium and high yield potential. Areas of low yield potential had lower soil organic matter, and extractable sodium (Na), calcium (Ca), potassium (K), boron (B), and zinc (Zn) concentrations than areas of medium and high yield potential. Differences in seed protein concentration between areas were hypothesized to be caused by differences in soil Na, B, and Zn, which are important in enzymatic reactions involved in protein and nitrogen metabolism. In addition, differences in soil K between the areas studied were proposed to have determined differences in photosynthetic capacity and the consequent alteration of seed protein and oil concentrations. Similarly, Poutaraud and Girardin (2004) studied the correlation between soil chemical properties and the alkaloid concentration in meadow saffron (Crocus sativus) seeds in France. A significant positive linear relationship was observed between soil Ca and Ca + cobalt (Co) concentration and seed alkaloid concentration. The fact that the enzymes in the alkaloid synthesis pathways needed Co and that Ca prevented alkaloids from entering cell compartments, where otherwise they would be degraded, was proposed as the explanation for the observed results. No such soil-leaf-seed chemical composition studies have been conducted for soybeans grown in the mid-southern US. It is possible that soil chemical properties may have a significant effect on seed composition, as they greatly contribute to plant nutrient availability, their mobilization, and overall plant growth.

The objective of this study was to evaluate potential associations between soil properties and element concentrations in the leaf, with seed composition in four types of specialty soybeans: high protein, high oil, modified fatty acids, low phytate/low stachyose grown at two Arkansas locations in 2011.

2. MATERIALS AND METHODS

2.1. Field Experiment

Eight breeding lines with modified seed composition and a high-yield check were used for this study (Table 1). Breeding lines were arranged in a randomized complete block design with three blocks. Each plot was 5 meters long and 4 rows wide. Seeding rate was 30 seeds per meter. The experiment was planted at the Arkansas Agricultural Research and Extension Center in Fayetteville, AR (36.099, −94.178), and at the University of Arkansas Vegetable Research Station in Kibler, AR (35.398, −94.232) in 2011. The soil at the Fayetteville location is mapped as Captina silt loam (fine-silty, siliceous, active, mesic Typic Fragiudult) and described as a very deep, moderately well-drained soil developed on a thin mantle of silty material [10] . The previous crop in the Fayetteville field was wheat (Triticum aestivum). Kibler soil is mapped as Roxana silt loam [10] and described as a well-drained, moderately permeable soil formed in a stratified loamy alluvium [11] . The previous crop in the Kibler field was soybean. The row spacing was 100 cm and 90 cm and planting dates were June 2nd and May 31st for Fayetteville (planted on beds) and Kibler (planted on flat ground), respectively. Field plots were fully irrigated (overhead at Kibler; furrow irrigation at Fayetteville) and managed during the growing season using standard cultural practices adopted for full-season soybean production in Arkansas [12] . Cultural practices included tillage with chisel plow and disc, and fertilization based on soil test results and recommendation of University of Arkansas Cooperative Extension Service. Weed control was performed by applying pre-plant (e.g., glyphosate) and post-emergence herbicides at label rates. Average temperature and rainfall for these two locations during the 2011 season can be found in [15] , which is a larger experiment the present study was part of. At maturity, the two center rows of each plot were harvested using a plot combine. Seed was stored for phenotypic analysis.

Table 1. Seed traits of soybean lines used in this study.

Genotype†	Trait
R05-1772‡	high protein
R05-1415‡	high protein
R05-655	high oil
R02-6268F	high oil
R07-8292	high oleic acid
R05-5346	low linolenic acid
R06-814	low saturated fatty acids (stearic + palmitic acids)
R07-2000	high sucrose, low stachyose, high inorganic phosphorus
AG5605	high yield

†all lines, except AG5605, which is property of Asgrow Seed Company, were developed at the University of Arkansas soybean breeding program. ‡see Chen *et al.* [13] and Chen *et al.* [14] for more details on these lines.

2.2. Soil Sampling, Soil Chemical Element Composition and Soil Texture Analysis

Soil samples were collected from the top 12.5 cm of each plot about one week before the beginning of the reproductive period (flower initiation). Twelve soil cores (three per row) were extracted with a 2.5 cm diameter push probe, and mixed in a plastic bucket. Soil samples were dried at 50°C for 2 days and ground to pass a 2 mm mesh screen. Soil was extracted with Mehlich-3 extractant solution in a 1:10 soil/extractant solution ratio and analyzed for extractable nutrient concentration [phosphorus (P), K, Ca, magnesium (Mg), sulfur (S), iron (Fe), manganese (Mn), Zn, copper (Cu), and B] using a Spectro Arcos Inducted Coupled Plasma Mass Spectrometer (Spectro Analytical Instruments, Mahwah, NJ). Data were expressed in milligram of element per kilogram of dry soil. Soil pH and electric conductivity were determined with an electrode on a 1:2 soil/water paste. Sand, silt and clay were determined on dried ground soil using the 2 hours hydrometer method [16] .

2.3. Leaf Composition Analysis

Composite samples of the uppermost fully developed trifoliates were collected randomly from the four rows of each plot the same day soil samples were collected. Leaf samples were dried at 50°C for 2 d, and ground to pass a 0.84 mm mesh screen. Samples were digested using concentrated HNO_3 and 30% hydrogen peroxide on a heating block system (Environmental Express, Mt. Pleasant, SC) and analyzed using Spectro Arcos Inducted Coupled Plasma Mass Spectrometer (Spectro Analytical Instruments, Mahwah, New Jersey). Data were presented as percentage of dry tissue for P, K, Ca, Mg, S, and as milligram per kilogram of dry tissue for Fe, Mn, Zn and Cu.

2.4. Seed Composition Analysis

Protein and oil analysis. Samples containing 20 - 25 g of seed were sent to the USDA research facility at Peoria, IL or to the University of Missouri Delta Center at Portageville, MO for protein, oil and moisture analysis in a FOSS⍰ (Eden Prairie, MN) near infrared transmittance instrument. Protein and oil concentration were presented in milligrams per gram of seed in dry-weight basis.

Fatty acid analysis. Samples containing five seeds were sent the DNA facility at Iowa State University in Ames, IA. Fatty acid concentrations [oleic acid, linolenic acid, and saturated fatty acids (sats)] were determined by gas chromatography according to the methods developed by Hammond [17] . The fatty acid concentration was presented in milligrams per total grams of oil.

Sugar analysis. Each seed sample was processed separately for sugar extraction, sugar fractionation and identification, and sugar quantification by high-performance liquid chromatography (HPLC). For a detailed description on the HPLC methods see Jaureguy et al. [15] . Sugar data were presented in milligrams of sugar per gram of seed.

Inorganic phosphorus (Pi) analysis. Seed samples were ground in a coffee grinder (Krups?, Shelton, CT) and the powder was screened through a 450-μm sieve (VWR International, West Chester, PA). A 0.1 g of soybeanseed fine powder was weighed in a 2.0 ml centrifuge tube. One ml of extraction buffer (deionized water, 12.5% trichloroacetic acid, 1 M $MgCl_2$) was added to the tube, and the solution was vortexed until homogeneous. The solution was incubated overnight at 4°C, vortexed again, incubated at room temperature for 30 min, and centrifuged at 1400 g for 4 min. Samples were stored at 4°C until Pi analysis. For Pi analysis a 10 μl sample of each extract was placed in a well of a flat-bottom 96-well plate (Becton Dickinson, Franklin Lakes, NJ) containing 90 μl of deionized water. The plates included Pi standards consisting of proportional dilutions of K_2PO_4 in deionized water. Standards and samples were allowed to react with 100 μl of Chen's reagent (6 N sulfuric acid, 2.5% amomiummolybdate, 10% abscorbic acid, deionized water) for 1 h, and absorbance at 882 nm was measured in a plate reader (Biotek, Winooski, VT). Samples were analyzed in batches of 96, and all the samples were analyzed with the same batch of reagents. A sample of breeding line CX1834-1-6 (high Pi/low phytate) and cultivar Osage (low Pi/normal phytate) were also included in each batch of samples as controls for the extraction process. Inorganic P concentrations were determined by linear extrapolation using the formula μg Pi in sample = μg Pi in standard/curve slope. The conversion factor was calculated according to the changes in concentration that Pi undergoes during extraction procedures. Data were presented as micrograms of element per gram of seed.

Chemical element analysis. A sample of seed fine powder was digested using concentrated $HNO_3$and 30% hydrogen peroxide on a heating block system (Environmental Express, Mt. Pleasant, SC) and analyzed using Spectro Arcos Inducted Coupled Plasma Mass Spectrometer (Spectro Analytical Instruments, Mahwah, New Jersey). Data were presented as mg of chemical element per gram of seed for P, K, Ca, Mg, S, and as milligram of element per kilogram of seed for Fe, Mn, Zn and Cu.

2.5. Statistical Analysis

Soil, leaf and seed composition data were evaluated by analysis of variance (ANOVA) considering location and breeding line as fixed factors and block as a random factor using the PROC MIXED procedure in SAS [18] . Significant interactions were further investigated with the "slice" option, which allowed testing main effects or interactions at different levels of other main effects. Least-square means were compared using the "contrast" option.

The sum of squares for each of the factors of the model: soil element concentration + leaf element concentration + error = seed element concentration was calculated as a measure of the contribution of soil element concentration and leaf element concentration to the variability observed in the seed concentration. This analysis was performed for each location separately.

3. RESULTS AND DISCUSSION

3.1. Soil Characteristics

The soil particle size analysis showed that both soils used in this study were silt loams [10] [11] . However, the clay content was significantly greater at the Kibler site, whereas the sand content was significantly greater at the Fayetteville site (Table 2). Silt content did not differ between locations. Soil at Fayetteville exhibited lower pH than soil at Kibler and was slightly acidic. Although it is known to affect plant nutrient availability, soil pH at Fayetteville was adequate to not to limit plant growth. Soil electrical conductivity differed between locations, but both were below the range (>500 µmhos/cm) at which plant injury due to high soil salt concentration may be expected. Soil at Kibler had greater concentrations of all extractable soil chemical nutrients analyzed except for P and Fe, which did not differ between locations, and S, which was greater at Fayetteville (Table 2). Soil P concentrations were optimum at both locations, but soil K, Zn and Cu, and S concentrations were below optimum at Fayetteville and Kibler, respectively. Soil Ca, Mg, Mn, and B concentrations were above the optimum level at both locations. However, Kibler soil exhibited more properties in the optimum range and perhaps was a better overall soil environment for soybean growth (Table 2). The differences in soil chemistry between locations were possibly due to natural differences between soil parent materials and the cumulative effect of previous management practices (e.g. tillage, and fertilization) at the two locations.

3.2. Leaf Chemical Element Composition

None of the leaf chemical properties showed significant location by breeding line interaction ($P > 0.05$), which suggested that the leaf elemental concentration across all lines responded similarly to the environment and without significant changes in the ranking of the breeding lines between locations (Table 3). Thus, whether breeding lines had been selected for specific seed components did not influence the ability of the plants to take up inorganic nutrients from the soil and supply them to the metabolic systems during growth and development. However, leaf Mg, Fe, Mn, Zn, and B concentrations, averaged across breeding lines, differed between locations. Breeding lines had greater concentrations of leaf Mg, Fe, Zn, and B at Kibler but, a greater concentration of leaf M nat Fayetteville (Table 4). All lines at both locations had adequate leaf nutrient concentrations for normal growth, such that it would not have resulted in a recommendation for fertilization.

Table 2. Physical and chemical properties of the top 12.5 cm soil at the two Arkansas locations (Fayetteville and Kibler) where the study was conducted.

Soil properties†	Fayetteville	Kibler	Ratio‡	F test§	Low range¶
texture	Silt loam	Silt loam	-	-	-
clay %	9	23	2.6	***	-
sand %	38	24	0.6	***	-
silt %	52	53	1.0	NS	-
pH	5.3	7.1	1.4	***	-
EC (μmhos cm^{-1})	163	106	0.7	*	-
P	44	47	1.1	NS	16 - 25
K	81	152	1.9	**	61 - 90
Ca	682	1839	2.7	**	<400
Mg	51	382	7.5	**	<30
S	15	8	0.5	**	<10
Fe	142.2	129.5	0.91	NS	-
Mn	81.6	117.6	1.44	**	<40
Zn	0.8	2.7	3.44	**	<1.6 - 3
Cu	0.3	2.0	7.63	**	<1
B	0.3	0.6	2.02	**	<0.025#

†EC = electrical conductivity; all minerals expressed in mg·kg^{-1}. See Materials and Methods for details. ‡ratio between the seed component concentration observed at Kibler and the concentration observed at Fayetteville. §from the location effect in ANOVA. ¶the low range values were obtained, when available, from the University of Arkansas Soil Test report. NS not significant at 0.05 probability level. *significant at 0.05 level. **significant at 0.01 level. ***significant at 0.001 level. #data from Minarik and Shive [19].

3.3. Seed Organic and Inorganic Composition

Oleic and linolenic acids showed significant location by breeding line interaction, which indicated that the expression of these two traits depended on both breeding line and location (Table 5). Four breeding lines (R05- 1415, R05-1772, R05-655, R07-2000, R07-8292) had greater oleic acid concentration at Fayetteville than at Kibler, but only breeding line R07-8292 had greater oleic acid concentration at Kibler than at Fayetteville. In contrast, except for breeding line R07-8292, linolenic acid concentration of all breeding lines differed between locations. The different response of R07-8292 compared to the other lines may have caused the significant location by breeding line interaction for oleic and linolenic acid concentrations.

Differences in genetic background and breeding history of the breeding lines studied had a larger effect on seed organic components than on environment (locations) where the lines were evaluated. It is possible that the differences in soil properties and leaf inorganic nutrient concentration between locations did not contribute significantly to variation in organic seed composition. Maybe deficient or higher levels of soil nutrients were needed to detect considerable differences.

All seed chemical elements, except for Ca, showed no breeding line by location interaction (Table 5), which suggested that seed inorganic components responded to the effect of the environment, but without significant

Table 3. Analysis of variance of leaf chemical element concentrations for breeding lines with modified seed composition grown at two Arkansas locations (Fayetteville and Kibler) in 2011.

Source of effects	Dfn†	Dfd‡	P§	K	Ca	Mg	S	Fe	Mn	Zn	Cu	B
Location (Loc)	1	4	NS	NS	NS	***	NS	**	***	*	NS	***
Breeding line (Bl)	8	32	NS	NS	NS	NS	NS	NS	NS	NS	NS	NS
Loc x Bl	8	32	NS	NS	NS	NS	NS	NS	NS	NS	NS	NS

†degrees of freedom of numerator in F test. ‡degrees of freedom of denominator in F test. §P, K, Ca, Mg, S in %; Fe, Mn, Zn, Cu and B in mg·kg^{-1}. See Materials and Methods for details. NS not significant at 0.05 probability level. *significant at 0.05 level. **significant at 0.01 level. ***significant at 0.001 level.

Table 4. Leaf elemental concentrations averaged across breeding lines of soybeans with modified seed composition grown at two Arkansas locations (Fayetteville and Kibler) in 2011.

Mineral†	Fayetteville	Kibler	Ratio‡	F test§	Optimum range¶
P	0.3	0.4	1.14	NS	0.1 - 0.5
K	2.1	2.3	1.05	NS	0.5 - 0.6
Ca	1.1	1.1	0.95	NS	0.2 - 1
Mg	0.3	0.4	1.43	***	0.1 - 0.4
S	0.3	0.3	0.98	NS	0.25 - 0.3
Fe	126	179	1.42	**	50 - 250
Mn	61	40	0.66	**	20 - 500
Zn	34	43	1.27	*	25 - 150
Cu	10	9	0.99	NS	5 - 20
B	27	42	1.56	***	20 - 60

†P, K, Ca, Mg, S in %. Fe, Mn, Zn, Cu, B in mg·kg^{-1}. See Materials and Methods for details. ‡ratio between the seed component concentration observed at Kibler and the concentration observed at Fayetteville. §from the location effect in ANOVA. ¶from Marschner (1997). NS not significant at 0.05 probability level. *significant at 0.05 level. **significant at 0.01 level. ***significant at 0.001 level.

Table 5. Analysis of variance of seed organic and inorganic components for breeding lines with modified seed composition grown at two Arkansas locations (Fayetteville and Kibler) in 2011.

Source of effects†	Pro‡	Oil	Ole	Lin	Sats	Suc	Sta	Pi	P	K	Ca	Mg	S	Fe	Mn	Zn	Cu	B
Location (Loc)	NS	NS	**	***	NS	**	**	NS	**	***	***	***	**	NS	NS	**	NS	***
Breeding line (Bl)	***	***	***	***	***	***	***	***	***	***	***	***	***	**	NS	***	NS	*
Loc x Bl	NS	NS	***	***	NS	NS	NS	NS	NS	NS	**	NS	NS	NS	NS	NS	NS	NS

†degrees of freedom of numerator in F test: Loc = 1; Bl = 8, Loc x Bl = 8. Degrees of freedom of denominator in F test: Loc = 4; Bl = 25, Loc x Bl = 25. ‡Pro = protein concentration ($mg{\cdot}g^{-1}$); Oil concentration in $mg{\cdot}g^{-1}$; Ole = oleic acid concentration ($mg{\cdot}g^{-1}$); Lin = linolenic acid concentration ($mg{\cdot}g^{-1}$); Sats = saturated fatty acids (palmitic + stearic acids) ($mg{\cdot}g^{-1}$); Suc = sucrose concentration ($mg{\cdot}g^{-1}$); Sta = stachyose concentration ($mg{\cdot}g^{-1}$); Pi = inorganic phosphorus concentration ($\mu g{\cdot}g^{-1}$). From P to S in $mg{\cdot}g^{-1}$; from Fe to Zn in $mg{\cdot}g^{-1}$. See Materials and Methods for details. NS, not significant at 0.05 probability level. *significant at 0.05 level. **significant at 0.01 level. ***significant at 0.001 level.

changes in breeding line ranking. Alternatively, lack of genetic variation may also explain these differences. Overall, soybeans had greater linolenic acid, P, K, Ca, Mg, and B concentrations at Kibler, but greater oleic acid, sucrose, stachyose, S and Zn concentrations at Fayetteville (Table 6). Differences in leaf elemental composition between locations were not reflected in seed elemental composition. For example, leaf P, K, Ca, S, and Cu concentrations did differ between locations but, except for Cu, the same elements in the seed differed significantly between locations. Furthermore, Fe and Mn differed between locations in the leaf, but had similar concentrations in the seed. Seed-to-leaf enrichment ratios [20] showed that plants at Fayetteville had greater seed P and Mg relative to the leaf than plants grown at Kibler (Figure 1). Similarly, plants grown at Kibler had greater seed Ca relative to the leaf than plants grown at Fayetteville. Because leaf elemental concentrations were within the medium-optimum range at both locations, differences observed in seed enrichment ratios may have been caused by different rates of elemental remobilization/uptake between locations from the time. Plants were sampled until maturity. The lack of direct relationships could mean that the soil had no impact on leaf elemental composition and seed traits, or all soil properties measured were adequate, in other words, no deficiencies or excesses. Future research should involve experiments with minerals that showed the largest contrast between leaf and seed concentrations and applied at deficient, optimum, and excessive levels to the growing plant.

In general, except for in a few cases, soil and leaf variables were not correlated with organic seed components (data not shown). Variation in the concentration of inorganic seed composition was, in general, not explained by variation in soil or leaf elemental concentrations (Table 7). The only exception was seed Zn concentration at Fayetteville, in which 47% of the variation was explained by leaf Zn (Table 7). These results suggest that at Fayetteville, leaf Zn concentration may be an adequate predictor of seed Zn concentration.

Table 6. Seed elemental concentrations averaged across breeding lines with modified seed composition grown at two Arkansas locations (Fayetteville and Kibler) in 2011.

Seed component†	Fayetteville	Kibler	Ratio‡	F test§
Pro	362	369	1.02	NS
Oil	183	185	1.01	NS
Ole	255	235	0.92	*
Lin	41	61	1.48	*
Sats	138	137	1.00	NS
Suc	61	49	0.81	*
Sta	42	38	0.90	*
Pi	375	345	0.92	NS
P	6.7	7.0	1.06	*
K	19.2	20.7	1.08	*
Ca	2.2	3.0	1.36	*
Mg	2.4	2.7	1.11	*
S	3.6	3.4	0.94	*
Fe	55	56	1.02	NS
Mn	31	29	0.94	NS
Zn	46	38	0.83	*
Cu	15	16	1.07	NS
B	24	37	1.52	*

†Pro = protein concentration ($mg{\cdot}g^{-1}$); Oil concentration in $mg{\cdot}g^{-1}$; Ole = oleic acid concentration ($mg{\cdot}g^{-1}$); Lin = linolenic acid concentration ($mg{\cdot}g^{-1}$); Sats = saturated fatty acids (palmitic + stearic acids) ($mg{\cdot}g^{-1}$); Suc = sucrose concentration ($mg{\cdot}g^{-1}$); Sta = stachyose concentration ($mg{\cdot}g^{-1}$); Pi = inorganic phosphorus concentration ($\mu g{\cdot}g^{-1}$). From P to S in $mg{\cdot}g^{-1}$; from Fe to B in $mg{\cdot}kg^{-1}$. See Materials and Methods for details. ‡ratio between the seed component concentration observed at Kibler and the concentration observed at Fayetteville. §location effect from ANOVA. NS, not significant at 0.05 probability level. *significant at the 0.05 probability level.

Several studies have also failed to show significant association between soil or leaf elemental and seed organic compositions. Boswell and Worthington [21] studied the effects of varying amounts of soil B and Mn on seed protein, oil, and fatty acids in a field experiment. Soil Mn and soil B had only a slight effect on protein concentration and no effect on oil or fatty acids. Similarly, Heenan and Campbell [22] observed no effect of Mn in the growing medium on protein and oil of two cultivars grown in the greenhouse. Wilson [8] reported that deficient levels of soil Mn increased seed linoleic, linolenic, saturated fatty acids and protein, but decreased oil and oleic acid concentrations. Interestingly, these effects were absent when plants were above the deficiency levels (15 $mg{\cdot}kg^{-1}$). Ham et al. [23] studied the effect of S fertilizer on seed protein and fatty acids and reported no effect of S on these two seed components. Similar results were observed for protein and oil measured in soybeans grown with different S fertilizer rates in Missouri [24] , Kansas [25] and Iowa [26] , and increasing levels of macronutrients N, P, K, Ca, and Mg in soybeans grown in hydroponic conditions [27] .

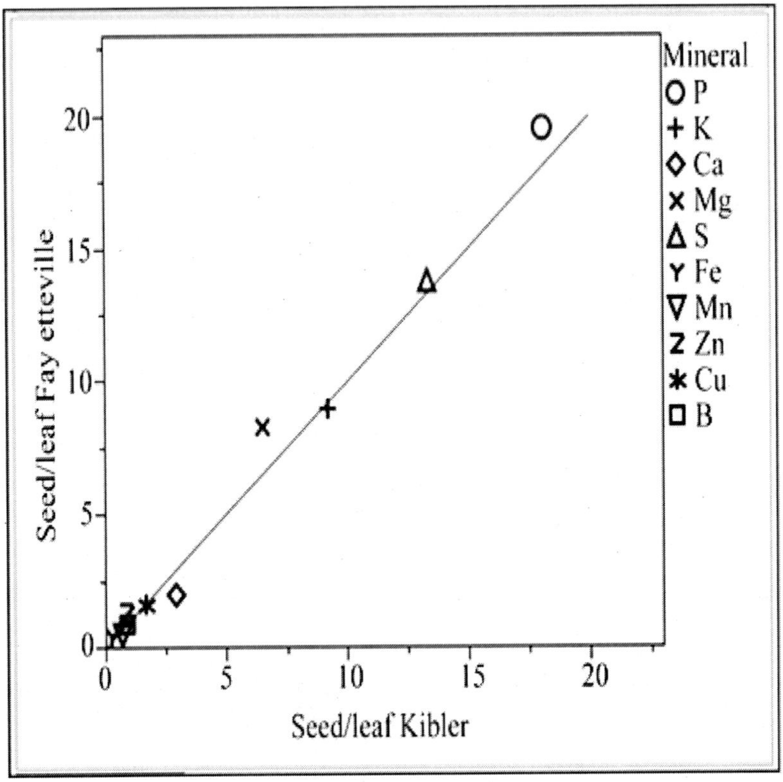

Figure 1. Seed enrichment ratios (seed chemical element concentration/leaf chemical element concentration) of breeding lines (combined) with modified seed composition grown at two Arkansas locations (Fayetteville and Kibler) in 2011. The line represents equal enrichment ratios at both locations. Elements above the diagonal have higher enrichments ratios at Fayetteville, whereas elements below the diagonal have higher enrichment ratios at Kibler.

Other studies have shown significant relationships between soil or leaf elemental and seed organic compositions. Gaudou and Arrivets [28] studied the effect of soil P, K, and N on soybean seed organic components and showed that P increased protein and oleic acid, whereas soil K increased oil and linolenic acid, but decreased protein concentrations. Bellaloui et al. [29] [30] studied the effect of foliar B application on soybean seed composition and demonstrated a significant increase in protein and oleic acid, but a decrease in oil and linolenic acid concentrations. In a different study, Bellaloui et al. [9] observed a significant correlation among soil K, B, Zn,

Table 7. Summary of the sum of squares for each of the factors of the model: soil element concentration (Soil) + leaf element concentration (Leaf) + error = seed element concentration, calculated as a measure of the contribution of soil element concentration and leaf element concentration to the variability observed in the seed element concentration of soybean breeding lines with modified seed composition grown at two Arkansas locations (Fayetteville and Kibler) in 2011.

	Seed P	Seed K	Seed Ca	Seed Mg	Seed S	Seed Fe	Seed Mn	Seed Zn	Seed Cu	Seed B
						Fayetteville				
Soil	0.63	0.07	0.08	0.01	0.20	4.42	18.14	27.34	0.11	7.60
Leaf	0.50	0.05	0.09	0.03	0.13	146.71	45.02	201.69	8.14	27.46
Error	5.66	18.86	2.03	0.81	0.77	317.77	424.30	194.76	50.51	151.70
Total	6.79	18.99	2.21	0.86	1.11	468.89	487.47	423.80	58.77	186.76
						Kibler				
Soil	0.06	1.10	0.14	0.00	0.00	4.91	2.08	105.06	0.03	4.49
Leaf	0.28	1.13	0.05	0.01	0.00	38.64	0.24	16.59	0.32	1.55
Error	8.25	45.26	6.18	1.07	0.97	745.97	172.41	331.63	99.06	188.79
Total	8.59	47.48	6.37	1.08	0.97	789.52	174.72	453.28	99.41	194.83

seed protein and oleic acid for a high-yield cultivar grown at one location for two consecutive years.

The lack of agreement among the studies described here suggests that the effects of cultivars (e.g. high yield, specialty), environments (temperature, moisture, soil type), and methods used (field, greenhouse) may have accounted for the observed variability in results. It appears that if producers want to attain different levels of organic seed components, fertilization with minerals may not be a recommended practice; alternative management strategies may be more adequate. For example, previous studies have shown planting early tends to increase oil and oleic acid concentration [31] [32] , whereas late planting tends to increase sucrose [33] .

In this study it was shown that although they had a different breeding history, all specialty soybean breeding lines were able to take up, transport, and increase leaf elemental concentrations to adequate levels before the re productive period. Differences in leaf minerals between locations were not translated to the seed. Other factors such as temperature during seed-fill may have overridden potential associations between leaf minerals and seed components. Altering fertilization practices may not result in changes in seed composition. Results presented in this study may be used as a starting point for research aimed to understand which mechanism plays a relevant role in seed composition, and ultimately develops fertilization and management practices that improve soybean seed quality [34] .

ACKNOWLEDGMENTS

Luciano M. Jaureguy, Pengyin Chen, Kristofor Brye, Derrick Oosterhuis, Andy Mauromoustakos, John R. Clark We wish to thank all the individuals of the University of Arkansas Soybean Breeding Program for their help with field and laboratory work. The research presented in this paper is part of LMJ PhD dissertation and supported by grants from the United Soybean Board and the Arkansas Soybean Promotion Board.

REFERENCES

1. Gardner, F.P., Pearce, R.B. and Mitchell, R.L. (2003) Mineral Nutrition. In: Gardner, F.P., Pearce, R.B. and Mitchell, R.L., Eds., Physiology of Crop Plants, Scientific Publishers, Jodhpur, 209-245.
2. Marschner, H. (1997) Mineral Nutrition of Higher Plants. Academic Press, London.
3. Grusak, M.A., Pearson, J.N. and Marentes, E. (1999) The Physiology of Micronutrient Homeostasis in Field Crops. Field Crops Research, 60, 41-56.
4. Pearson, J.N. and Rengel, Z. (1994) Distribution and Remobilization of Zn and Mn during Grain Development in Wheat. Journal of Experimental Botany, 45, 1829-1835.
5. Loneragan, J.F., Snowball, K. and Robson, A.D. (1980) Copper Supply in Relation to Content and Redistribution of Copper among Organs of the Wheat Plant. Annals of Botany, 45, 621-632.
6. Hocking, P.J. and Pate, J.S. (1977) Mobilization of Minerals to Developing Seeds of Legumes. Annals of Botany, 41, 1259-1278.
7. Waters, B.M. and Sankaran, R.P. (2011) Moving Micronutrients from the Soil to the Seeds: Genes and Physiological Processes from a Biofortification Perspective. Plant Science, 180, 562-574.
8. Wilson, D.O., Boswell, F.C., Ohki, K., Parker, M.B., Shuman, L.M. and Jellum, M.D. (1982) Changes in Soybean Seed Oil and Protein as Influenced by Manganese Nutrition. Crop Science, 22, 948-952.
9. Bellaloui, N., Hanks, J.E., Fisher, D.K. and Mengistu, A. (2009) Soybean Seed Composition Is Influenced by Within- Field Variability in Soil Nutrients. Crop Management, 8.
10. Soil Survey Staff (2011) Natural Resources Conservation Service, United States Department of Agriculture. Web Soil Survey. http://websoilsurvey.nrcs.usda.gov
11. Garner, B.A. and Cox, J.B. (1979) Soil Survey of Crawford County, Arkansas. Unites States Department of Agriculture, Soil Conservation Service and Forest Service, in Cooperation with the Arkansas Agricultural Experimental Station, 35.
12. Tacker, P. and Vories, E. (1998) Irrigation, Arkansas Soybean Handbook, University of Arkansas. Cooperative Extension Service, 42-49.
13. Chen, P., Sneller, C.H., Ishibashi, T. and Cornelious, B. (2008) Registration of High-Protein Soybean Germplasm Line R95-1705. Journal of Plant Registrations, 2, 58-59
14. Chen, P., Ishibashi, T., Dombek, D.G. and Rupe, J.C. (2011) Registration of R05-1415 and R05-1772 High-Protein Soybean Germplasm Lines. Journal of Plant Registrations, 5, 410-413.
15. Jaureguy, L.M., Ledesma Rodriguez, F., Zhang, L., Chen, P., Brye, K., Oosterhuis, D., Mauromoustakos, A. and Clark, J.R. (2013) Planting Date and Delayed Harvest Effects on Soybean Seed Composition. Crop Science, 53, 2162-2175.

16. Arshad, M.A., Lowery, B. and Grossman, B. (1996) Physical Tests for Monitoring Soil Quality. In: Doran, J.W. and Jones, A.J., Eds., Methods for Assessing Soil Quality, Soil Science Society of America, Madison, 123-142.
17. Hammond, E.G. (1991) Organization of Rapid Analysis of Lipids in Many Individual Plants. In: Linskens, H.F. and Jackson, J.F., Eds., Modern Methods of Plant Analysis, Vol. 12, Springer-Verlag, Berlin, 321-330.
18. SAS Institute (2002) SAS/STAT User's Guide, Version 9.2. SAS Institute Inc., Cary.
19. Minarik, C.E. and Shive, J.W. (1939) The Effect of Boron in the Substrate on Calcium Accumulation by Soybean Plants. American Journal of Botany, 26, 827-831.
20. Tyler, G. and Zohlen, A. (1998) Plant Seeds as Mineral Nutrient Resource for Seedlings—A Comparison of Plants from Calcareous and Silicate Soils. Annals of Botany, 81, 455-459.
21. Boswell, F.C. and Worthington, R.E. (1971) Boron and Manganese Effects on Protein, Oil, and Fatty Acid Composition of Oil in Soybeans. Journal of Agricultural and Food Chemistry, 19, 765-768.
22. Heenan, D.P. and Campbell, L.C. (1980) Growth, Yield Components and Seed Composition of Two Soybean Cultivars as Affected by Manganese Supply. Australian Journal of Agricultural Research, 31, 471-476.
23. Ham, G.E., Liener, I.E., Evans, S.D., Frazier, R.D. and Nelson, W.W. (1975) Yield and Composition of Soybean Seed as Affected by N and S Fertilization. Agronomy Journal, 67, 293-297.
24. Brown, J.R., Thom, W.O. and Wall Sr., L.L. (1981) Effects of Sulfur on Yield and Composition of Soybeans and Soil Sulfur. Communications in Soil Science and Plant Analysis, 12, 247-261.
25. Sweeney, D.W. and Granade, G.V. (1993) Yield, Nutrient, and Soil Sulfur Response to Ammonium Sulfate Fertilization of Soybean Cultivars. Journal of Plant Nutrition, 16, 1083-1098.
26. Haq, M.U. and Mallarino, A.P. (2005) Response of Soybean Grain Oil and Protein Concentrations to Foliar and Soil Fertilization. Agronomy Journal, 97, 910-918.
27. Harper, J. (1971) Seasonal Nutrient Uptake and Accumulation Patterns in Soybeans. Crop Science, 11, 347-350.
28. Gaydou, E.M. and Arrivets, J. (1983) Effects of Phosphorus, Potassium, Dolomite, and Nitrogen Fertilization on the Quality of Soybean Yields, Proteins, and Lipids. Journal of Agricultural and Food Chemistry, 31, 765-769.
29. Bellaloui, N., Abbas, H.K., Gillen, A.M. and Abel, C.A. (2009) Effect of Glyphosate-Boron Application on Seed Composition and Nitrogen Metabolism in Glyphosate-Resistant Soybean. Journal of Agricultural and Food Chemistry, 57, 9050-9056.
30. Bellaloui, N., Reddy, K.N., Gillen, A.M. and Abel, C.A. (2010) Nitrogen Metabolism and Seed Composition as Influenced by Foliar Boron Application in Soybean. Plant Soil, 336, 143-155.
31. Kane, M.V., Steele, C.C., Grabau, L.J., MacKown, C.T. and Hildebrand, D.F. (1997) Early-Maturing Soybean Cropping System: III. Protein and Oil Contents and Oil Composition. Agronomy Journal, 89, 464-469.
32. Wilcox, J.R. and Cavins, J.F. (1992) Normal and Low Linolenic Acid Soybean Strains: Response to Planting Date. Crop Science, 32, 1248-1251.
33. Bellaloui, N., Reddy, K.N., Gillen, A.M., Fisher, D.K. and Mengistu, A. (2011) Influence of Planting Date on Seed Protein, Oil, Sugars, Minerals, and Nitrogen Metabolism in Soybean Under Irrigated and Non-Irrigated Environments. American Journal of Plant Sciences, 2, 702-715.
34. Cakmak, I., Pfeiffer, W.H. and McClafferty, B. (2010) Biofortification of Durum Wheat with Zinc and Iron. Cereal Chemistry, 67, 10-20.

CHAPTER 19

Ultraviolet-B Radiation Alters Soybean Growth and Seed Quality

K. Raja Reddy[1], Hrusikesh Patro[2], Suresh Lokhande[3], Nacer Bellaloui[4], Wei Gao[5]

[1]*Department of Plant and Soil Sciences, Mississippi State University, Mississippi State, USA*

[2]*Department of Agronomy, Orissa University of Agriculture & Technology, Bhubaneswar, India*

[3]*Department of Sociology, Iowa State University, Ames, USA*

[4]*Crop Genetics Research Unit, USDA-ARS, Stoneville, USA*

[5]*USDA-UV-B Monitoring Network, Natural Resource Ecology Lab, Colorado State University, Fort Collins, USA*

ABSTRACT

Research on the effects of ultraviolet-B (UV-B) radiation on soybean seed quality is limited. The objective of this study was to quantify UV-B doses, 0, 5, 10 & 15 $kJ \cdot m^{-2} \cdot d^{-1}$, on soybean growth and seed quality. The experiment was conducted in the Soil-Plant-Atmosphere-Research (SPAR) facility. Chambers located at the R.R. Foil Plant Science Research Facility of Mississippi State University, Mississippi, USA, were used. Each SPAR chamber consists of a steel soil bin to accommodate the root system, a Plexiglas chamber to accommodate plant canopy and a heating, and cooling system connected to air ducts that pass conditioned air to cause leaf flutter through the plant canopy. The SPAR units, supported by an environmental monitoring and control systems, are networked to provide automatic acquisition and storage of the data, monitored every 10 seconds throughout the day and night. Soybean cultivar Pioneer 93Y92 (maturity group IV, Roundup Ready) was used in the study. The desired UV-B radiation was supplied by square-wave UV-B supplementation systems under near ambient PAR and delivered to plants for eight hours, each day, from 08:00 to 16:00 h by eight fluorescent UV-313 lamps. The results showed that increased UV-B did not influence many of the growth parameters because the treatments were imposed at mid-fruiting period. Seed quality parameters that are important for seed industry and human and animal nutrition were all affected by UV-B. Protein and palmitic and oleic acids declined linearly, while oil and linoleic and linolenic acid contents increased with increased UV-B. Sucrose, stachyose, and stearic acid contents showed quadratic trends, increased to about 4 - 5 kJ of UV-B and declined at higher doses. Thus, both current and projected UV-B radiation levels can modify soybean growth and seed quality. The functional algorithms developed in this study could be useful to develop UV-B- specific sub-models for soybean farm management and in policy decision areas.

Keywords: Ultraviolet Radiation, Biomass Parameter, Seed Quality, Soybean Nutrition, Seed Composition

1. INTRODUCTION

Soybean is an important source of protein, oil, carbohydrates, isoflavones, and other minerals [1] for humans and animal nutrition [2] . Seed quality is often determined by seed protein, oil, fatty acid, sugars and mineral content. Therefore, maintaining and improving soybean seed quality is a key to overall human and animal nutritional aspects. Projected changes in global climate include increasing atmospheric carbon dioxide concentration, temperatures, and ultraviolet (UV)-B radiation which have significant effects on plant vegetative and reproductive growth. Reductions in ozone column have led to substantial increases in UV-B radiation at the earth's surface with the amount and intensity dependent on atmospheric and geographic factors [3] . During the last few decades, the thinning of the stratospheric ozone has led to the enhanced UV-B radiation on the earth surface [4] . The current UV-B levels of 4 - 9 $kJ \cdot m^{-2} \cdot d^{-1}$ during June-August 2002 over USA and projected increases of 14% - 40% in near future will have profound effects on crop yield and quality. The relative plasticity of soybean growth and development to the current and projected UV-B radiation will play a major role in determining crops future as well the industry needs as quality will be dependent on changes projected in the climate. Numerous studies have shown that enhanced UV-B radiation can affect physiological and biochemical processes of many plant species, including altered plant photosynthesis [5] , changes in the carbon partitioning from growth pools to secondary metabolic pathways [6] , and thus changes in crop morphology, crop reproductive organ abortion and yield reduction [7] .

For the past 10 plus years, the effect of enhanced UV-B radiation on soybean yield has been extensively studied because of the importance of intensity and quality of solar radiation intercepted by the canopy in determining soybean yield and yield components [8] . It was suggested that enhanced UV-B radiation makes the soybean plants to be dwarf by shortening the internodal length [9] . In addition, enhanced UV-B radiation also caused decreased total biomass of soybean by 24% [10] . Pod number per plant was the most responsible component for yield change under UV-B radiation in the 2-year study. Seed number per pod was less affected by changes in UV-B light treatment, compared with the pod numbers per plant. However, enhanced UV-B radiation has been reported to have no significant effect on effective pod-filling period, but the seed size was negatively impacted and decreased 12% for three sensitive cultivars [11] . Reduction of seed size was mainly due to the decrease of cotyledon cell number [12] . Studies conducted in China revealed that the soybean crop under exposure to the treatment of weak UV-B radiation during the early stage of pod development greatly decreased seed number as well as seed weight per plant and seed yield [12] . It was reported that seed weight of the 15 out of 20 soybean cultivars decreased when exposed to UV-B radiation [13] .

Soybean is mainly produced for oil and soymeal, and the quality of oil and the soymeal depends on the composition of fatty acids and protein, respectively. Soybean seed protein concentration ranges from 341 to 568 $g·kg^{-1}$ of total seed weight, with a mean of 421 $g·kg^{-1}$. Oil concentration ranges from 83 $g·kg^{-1}$ to 279 $g·kg^{-1}$ with a mean of 195 $g·kg^{-1}$ [14] . Saturated fatty acids in soybean oil range from 100 $g·kg^{-1}$ to 120 $g·kg^{-1}$ for palmitic acid, and from 22 $g·kg^{-1}$ to 72 $g·kg^{-1}$ for stearic acid [15] . The mean concentration of unsaturated fatty acids is 240 $g·kg^{-1}$ for oleic acid, 540 $g·kg^{-1}$ for linoleic acid, and 80 $g·kg^{-1}$ for linolenic acid [16] . A negative correlation has been reported between protein and oil concentration and yield in many soybean cultivars [17] , and a negative correlation between protein and oil [18] . The average soybean seed contains 9% to 12% total soluble carbohydrates, of which 4% to 5% are sucrose ($C_{12}H_{22}O_{11}$), 1% to 2% are raffinose ($C_{18}H_{32}O_{16}$), and 3.5% to 4.5% stachyose ($C_{24}H_{42}O_{21}$) [11] . Raffinose and stachyose are undesirable components because they have detrimental effects on the nutritive value of the meal and are indigestible by human and animals, often causing flatulence and diarrhea in non-ruminants [18] . Therefore, soybean seed with low raffinose and stachyose is desirable for feeding non-ruminant animals to improve feed energy efficiency, increase mineral uptake, and reduce flatulence [17] . On the other hand, soybean seed with high sucrose concentration is desirable for the seed and food processing industry because it improves taste and flavor in tofu, soy milk and natto [2] . Seed composition constituents have also been reported to be genetically controlled [19] . However, seed composition has also shown to be affected by environment [20] , genotype, maturity, diseases [21] , temperature [20] , drought [22] , nutrients in soil and seed [23] .

Understanding soybean responses to the increasing intensities of UV-B radiation is much needed to develop suitable management and cultural practices for the present and future climates. Some studies have been carried out either under greenhouse or under unrealistic solar radiation environments and results may not be portable to field situations. Studies have been conducted on other crops such as wheat [24] and corn [25] . The effect of enhanced UV-B radiation on soybean physiology and yield has been studied extensively. However, information on the effects of enhanced UV-B radiation on seed quality characteristics of soybean is limited and little information is available concerning the UV-B radiation effects on the relationship among growth, seed yield and seed quality traits [26] . The objective of this study was to investigate the effect of UV-B radiation on soybean growth and seed quality as soybean seed nutrition is important for human and animal nutrition.

2. MATERIALS AND METHODS

2.1. Facilities and Growth Conditions

The Soil-Plant-Atmosphere-Research (SPAR) chambers located at the R.R. Foil Plant Science Research facility of Mississippi State University (33°28N, 88°47W), Mississippi, USA were used for this study. Briefly, each SPAR chamber consists of a steel soil bin (1 m deep by 2 m long by 0.5 m wide) to accommodate the root system, a Plexiglas chamber (2.5 m tall by 2 m long by 1.5 m wide) to accommodate plant canopy and a heating and cooling system connected to air ducts that pass conditioned air to cause leaf flutter through the plant canopy. The SPAR units supported by an environmental monitoring and control systems are networked to provide automatic acquisition and storage of the data, monitored every 10 s throughout the day and night. Many details on the operation and control of SPAR chambers were described by others [27] .

Soybean cultivar Pioneer 93Y92 (maturity group IV, Roundup Ready) seeds were planted on 5 July 2011 in pots (22 cm tall and 22 cm diameter) filled with fine sand and soil mixed in the ratio of 75:25. Seventy pots having planted with 3 seeds per pot were maintained in an outdoor environment. In each pot one healthy plant per pot was maintained a week after emergence. Twelve pots, 1 plant pot^{-1} and 3 plants row^{-1}, were arranged in 6 rows in each of the 4 SPAR chambers on 27 August 2011. Plants were watered three times a day with half- strength Hoagland's nutrient solution [28] , delivered at 0800, 1200, and 1600 h, to ensure favorable nutrient and water conditions for plant growth through an automated and computer-controlled drip system. Variable-density black shade cloth around the edges of plants was adjusted regularly to match plant height in order to simulate natural shading caused by presence of other plants. The treatments included 4 biologically effective UV-B radiation intensities of 0, 5, 10, and 15 $kJ \cdot m^{-2} \cdot d^{-1}$. The treatments were imposed on plants when they are at the middle of the pod-filling stage, 65 days after sowing.

The desired UV-B radiation was supplied by square-wave UV-B supplementation systems under near ambient PAR. The UV-B radiation was delivered to plants for eight hours, each day, from 08:00 to 16:00 h by eight fluorescent UV-313 lamps (Q-Panel Company, Cleveland, Ohio, USA) powered by 40 W variable dimming ballasts. The lamps were wrapped with pre-solarized 0.07 mm cellulose diacetate film to filter UV-C (<280 nm) radiation. The cellulose diacetate film was changed at 3 - 4 d intervals. The UV-B energy delivered at the top of the plant canopy was checked daily at 09:00 h with a UVX digital radiometer (UVP Inc., San Gabriel, CA, USA) calibrated against an Optronic Laboratory (Orlando, FL, USA) Model 754 Spectroradiometer, which was used initially to quantify lamp output. The lamp output was adjusted, as needed, to maintain the respective UV-B radiation levels. A distance of 0.5 m from lamps to the top of plants was always maintained throughout the experiment. The actual biologically effective UV-B radiation was measured during the crop growth period at six different locations in each SPAR unit corresponding to the pots arranged in rows. The weighted total biologically effective UV-B radiation at the top of the plant canopy during the treatment period is presented in Figure 1. The set UV-B conditions, average temperature, and measured chamber CO_2 from a typical day during the experimental period for each treatment is mentioned in Table 1.

Final harvesting was carried out on 13 October 2011 (100 DAS or 47 days after UV-B treatment). Plants were cut above ground and separated into roots, leaves, pods, and stems. Pods were kept outside under normal drying conditions while other plant parts were dried in oven at 75°C until it weighed constant during a period of 72 h. The biomass parameters such as dry weight of leaf, stem, root and total per plant, number of pods and seeds per plant, number of seeds per pod as well as seed weight per seed were measured.

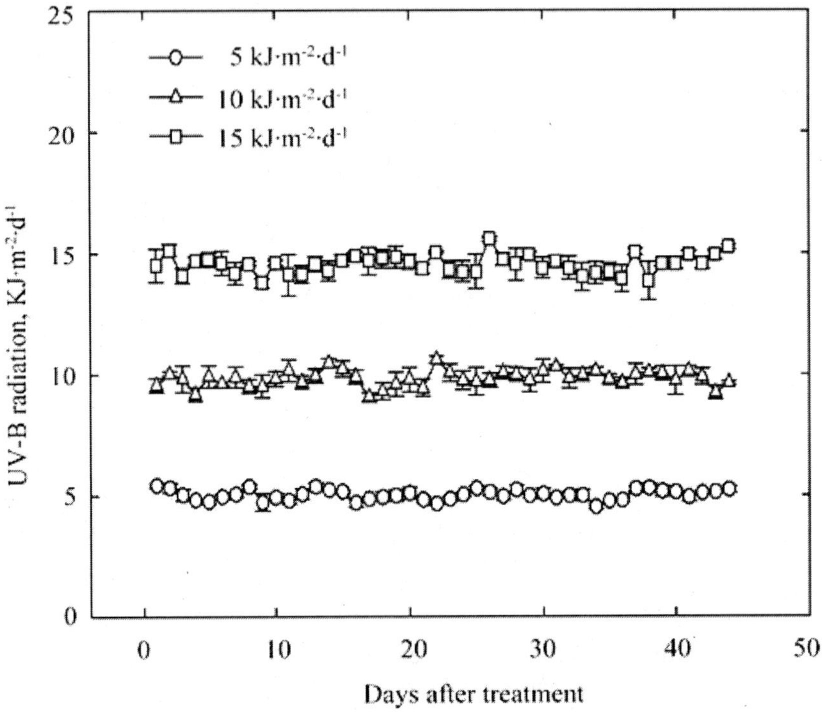

Figure 1. Daily ultraviolet-B (UV-B) radiation doses for three UV-B treatments measured at the top of the canopies, and lamps placed at 0.5 m above the canopy. Each data point is mean of three measurements measured at 4 different locations in each treatment and standard errors are shown if the values are larger than the symbols. The UV-B doses were measured daily at 10:00 h and adjusted to the target levels of 5, 10, or 15 $kJ \cdot m^{-2} \cdot d^{-1}$.

Table 1. The Set Ultraviolet (UV)-B conditions, average temperature, and measured chamber CO_2 from a typical day during the experimental period for each treatment.

Treatments	Measured variables	
UV-B ($kJ \cdot m^{-2} \cdot d^{-1}$)	Mean temperature (°C)	Chamber CO_2 ($\mu mol \cdot mol^{-1}$)
0	29.6 ± 0.04	409 ± 2.2
5	29.7 ± 0.03	407 ± 3.1
10	29.3 ± 0.05	406 ± 2.3
15	29.6 ± 0.03	404 ± 2.7

Each value represents the mean ± SE (standard error of the mean) for one typical day for [CO_2], and 27 August to 13 October 2011 for temperature.

Each value represents the mean ± SE (standard error of the mean) for one typical day for [CO_2], and 27 August to 13 October 2011 for temperature.

2.2. Seed Analysis for Protein, Oil, Fatty Acids, Sucrose, Raffinose, and Stachyose

Seed pooled from each row of plants were analyzed for protein, oil, fatty acids, sucrose, raffinose and stachyose. About 25 g of seed from each pot was ground using a Laboratory Mill 3600 (Perten, Springfield, IL). Analyses were conducted by near infrared reflectance [29] using a diode array feed analyzer AD 7200 (Perten). Calibrations were developed by the University of Minnesota, using Perten's Thermo Galactic Grams PLS IQ software. To improve accuracy, the calibration curve was updated as needed for oil, fatty acids, protein, sucrose, raffinose, and stachyose. Analyses of protein, oil, and sugars were performed based on a seed diameter basis [29] [30] . Fatty acids are analyzed on oil basis.

2.3. Statistical Analysis

To test the significance of UV-B radiation on growth and biomass components of soybean, analysis of variance was performed by using general linear model PROC GLM (SAS Institute Inc., Cary, NC, USA). Fisher protected LSD tests at $P = 0.05$ was used to determine significance of treatment effects. Regression analysis was performed between seed quality parameters and UV-B radiation using SAS (Sas Institute, Inc.) and SigmaPlot 11.0 (Systat Software Inc., San Jose, CA, USA) and best fit regression models were selected based R^2 values. The graphical analysis was carried out using SigmaPlot.

3. RESULTS AND DISCUSSION

3.1. Treatments and Growth Parameters

The ultraviolet-B radiation treatments were very close to the set points (Figure 1). The treatments represent the UV-B doses that are currently experienced for soybean and all other crops cultivated between 40°N and 40°S latitudes, currently at 2 to 10 $kJ \cdot m^{-2} \cdot d^{-1}$ depending on location and season [31] [32] and within the limits of projected UV-B in the 21st century [33] . Biomass parameters under UV-B radiation and natural light conditions obtained were summarized in Table 2. The effect of increasing levels of UV-B radiation at 5, 10 or 15 $kJ \cdot m^{-2} \cdot d^{-1}$ in respect of root and total dry weight per plant as well as number of seeds per pod over natural light (control, 0 UV-B radiation) was found non-significant. However, the total dry weight per plant decreased at all the UV-B radiation levels compared with the control, but the reduction was more at 10 $kJ \cdot m^{-2} \cdot d^{-1}$compared to either 5 or 15 $kJ \cdot m^{-2} \cdot d^{-1}$. This is in agreement with the study of Feng et al. [10] where enhanced UV-B radiation decreased total biomass soybean by 24%. Leaf dry weight had no distinct pattern with increasing levels of UV-B radiation. Stem dry weight, pod dry weight, number of pods, seeds per plant, and individual seed weight showed lower values at higher doses of UV-B compared to the control. However, the reduction in single seed weight was consistent with concomitant increase in UV-B radiation doses up to 15 $kJ \cdot m^{-2} \cdot d^{-1}$ (0.20 $g \cdot seed^{-1}$), which is significantly lower in comparison to control (0.22 $g \cdot seed^{-1}$).

Seed weight was negatively impacted by UV-B radiation. Chen et al. [13] reported similar findings in that seed weight of the 15 soybean cultivars out of 20 significantly decreased. Seed size was a function of the rate of seed growth and the duration of dry weight accumulation in the seed fraction and thereby the dry weight. Seed growth rate was shown to be a function of the cotyledon cell number and the supply of assimilate to the developing cotyledons [34] . Reduction of seed size was mainly due to the decrease of cotyledon cell number [11] .

Stem and pod dry weight per plant were reduced by UV-B radiation up to 10 $kJ \cdot m^{-2} \cdot d^{-1}$, while it enhanced at 15 $kJ \cdot m^{-2} \cdot d^{-1}$, but the enhancement remained lower than control. The significant reduction in stem and pod dry weight at 10 $kJ \cdot m^{-2} \cdot d^{-1}$ were of the order of 30 and 12% over control, respectively, while reductions were close and comparable with 5 $kJ \cdot m^{-2} \cdot d^{-1}$. Decreased stem dry weight might be due to reduction in main stem and branch elongation rates upon enhanced UV-B radiation resulting more compact and shorter plants [35] . It was suggested that enhanced UV-B radiation makes the soybean plant dwarf by shortening the internode length [9] . It was reported, working on pea plants, that low doses of ultraviolet-B or ultraviolet-C radiation had affected phytohormones, including indoleacetic acid (IAA) and abscisic acid (ABA) [36] . Mark and Tevini [37] speculated that the mechanism for reduced stem elongation by UV-B might be due to changes in the phytohormone levels, especially IAA which plays a role in stem elongation. It was also confirmed that UV-B radiation decreased plant height of two soybean cultivars 10% and 15%, and decreased dry weight of individual stem by 44% and 28% [38] .

Table 2. Biomass parameters of soybean as influenced by enhanced doses of UV-B radiation[m].

Biomass parameters	UV-B radiation, $kJ \cdot m^{-2} \cdot d^{-1}$				F test
	0	5	10	15	
Leaf dry weight, g	20.6bc	24.2ab	20.4c	24.4a	3.67
Stem dry weight, g	30.3a	29.5a	23.2b	30.1a	4.27
Root dry weight, g	18.4a	16.8a	17.7a	17.7a	ns
Pod dry weight, g	75.8a	68.1b	67.5b	69.3ab	7.51
Total dry weight, g	145.0a	138.5a	128.8a	141.5a	ns
Pods, no. $plant^{-1}$	109.9a	101.6b	105.7ab	107.8ab	6.57
Seeds, no. $plant^{-1}$	259.3a	237.7b	248.4ab	258.8a	16.19
Seeds, no. pod^{-1}	2.36a	2.34a	2.35a	2.41a	ns
Seed weight, $g \cdot seed^{-1}$	0.22a	0.21ab	0.21b	0.20b	0.013

[m]Values are means of four replicates; ns = Not significant at p = 5%; means within columns followed by the same letter are not significantly different at the 5% level.

Pod and seed number per plant were significantly decreased by UV-B radiation up to 5 $kJ \cdot m^{-2} \cdot d^{-1}$, exhibiting 8 and 9% reduction compared with control, respectively, where as it increased at 10 and 15 $kJ \cdot m^{-2} \cdot d^{-1}$, and the increases were maintained lower than control. UV-B radiation decreased average pod number per plant of three soybean cultivars by 34%. This suggests that UV-B radiation imposed during early flowering stage would change assimilates availability to the developing reproductive structures, influence flowering, and flower and pod abscission number with a resultant change in final pod number at harvest. Pod number per plant was the yield component most influenced by change in cultural and environmental conditions [39] . In our study, UV-B radiation had no significant effect on seed number per pod (Table 1). Herbert and Litchfield [40] and Board et al. [41] reported that seed number per pod is a minor component determining the yield of soybean. However, the small tendency of decreased seed number per pod under UV-B radiation indicated that seed number per pod is strongly determined by the internal genetic mechanism, and is less influenced by environment condition.

The attenuation of the stratospheric ozone has led to the enhanced UV-B radiation on the surface of land in recent decades [42] . The present study confirmed previous reports [9] [43] that UV-B radiation in the canopy changed soybean agronomic traits and decreased yield per plant. Furthermore solar UV-B radiation exclusion studies indicated that ambient levels of solar UV-B radiation reduce biomass accumulation and grain yield of soybean [44] . Few studies indicated a breakdown of IAA on exposure to UV-B radiation [45] . Crop biomass production in response to UV-B radiation was highly UV-B dosage dependent. The effects of enhanced UV-B radiation on plant growth and dry matter accumulation in growth chamber or greenhouse were usually much higher than those in field conditions [46] . In field conditions UV-B caused decreases in total above ground production of soybean when PAR and UV-A were reduced to less than half their flux in sunlight [47] . The UV-B radiation appears to have a strong ameliorating effect when PAR is not high.

3.2. Seed Quality Parameters

3.2.1. Seed Protein

Although crop yield and seed quality are important products of soybean crop, few studies addressed UV-B effects on seed quality parameters that are important to human and animal nutrition. Soybean seed protein reported in this study under various UV-B treatments, 407 to 440 g·kg^{-1}, is within the range, 340 - 570 g·kg^{-1}, reported in other studies (Wilson 2004). Seed protein content, however, declined linearly with increase in UV-B radiation, 2.21 g·kg^{-1} UV-B^{-1}(Figure 2). Since seed protein content is governed by nitrogen availability, UV-B might have affected both uptake and translocation of nutrients as reported in other studies under UV-B treatments in soybean and other plants under UV-B radiation [25] [48] [49].

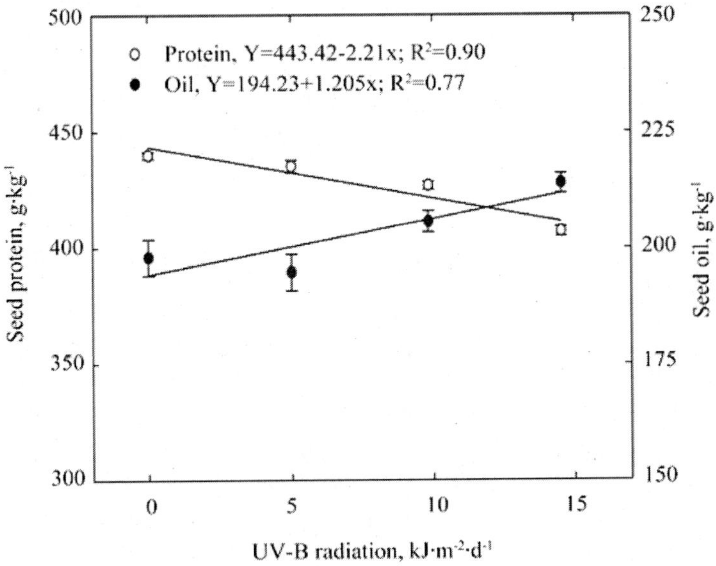

Figure 2. Ultraviolet (UV)-B radiation effects on soybean seed protein and oil content for plants harvested at 100 days after sowing. The UV-B treatments were imposed at early-pod-filling stage, 65 days after sowing. Each data point is mean of four replications and standard errors are shown if the values are larger than the symbols.

3.2.2. Seed Oil

Soybean seed oil is an important product with many industrial application. The seed oil content of found in this study ranges from 198 to 238 $g \cdot kg^{-1}$, are with the ranges reported in the literature of about 83 to 279 $g \cdot kg^{-1}$, depending on cultivar and growing season environment [14] . Unlike, seed protein content, soybean seed oil increased linearly with 1.205 $g \cdot kg^{-1}$ UV-B^{-1} (Figure 2). Alan et al. [50] reported similar declines in seed protein and oil content of two soybean cultivars with increased UV-B radiation. The seed protein content of the soybean cultivar Essex was decreased up to 5% while the seed oil content had no significant changes due to enhanced UV-B radiation. Reductions in seed protein content were observed in cultivar Williams in two of the years while it increased in two other years. Seed oils were also variously affected with reductions of 10% in one year and an increase of 8% in another year. However, no changes grain protein in maize and gluten concentration in wheat have been reported in non-leguminous plants [51] .

The inverse negative relationship between soybean seed protein and oil ($y = 752 - 1.6x$; $R^2 = 0.88$, where $y = g \cdot kg^{-1}$ protein and $x = g \cdot kg^{-1}$ oil) observed in our study when used all treatments and inverse relations with protein and yield [52] observed in many studies poses many challenges to breeders and farm managers to increase protein without reducing grain yield. Breeders have to consider UV-B into their grand scheme of things while increasing the soybean seed yield to meet the future human and animal nutrition with altered climate change.

3.2.3. Seed Fatty Acids

Soybean oils contain about 16% saturated, 23% monosaturated and 58% polysaturated fatty acids and these fatty acid composition play important roles in oil stability and human and animal nutrition [51] . The saturated fatty acids such as palmitic and stearic acid contents in seed oil showed linear and quadratic trends with increasing UV-B levels (Figure 3). The decline in pamitic acid content was 0.9459 $g \cdot kg^{-1}$ UV-B^{-1}. On the other hand, maximum stearic acid content of 40.5 $g \cdot kg^{-1}$ was observed at 4.94 kJ of UV-B and declined at higher and 0 UV-B levels. Plants grown at the highest UV-B had 28% less stearic acid content than the maximum levels observed at 4.94 kJ of UV-B. The most common monosaturated fatty acid in plants, oleic acid, declined linearly with increase in UV-B doses, 6.734 $g \cdot kg^{-1}$ UV-B^{-1} (Figure 3). The polysaturated fatty acids, linoleic and linolenic acid contents, on the other hand, increased with increase in UV-B doses (Figure 3). The increase was steeper for linolenic acid (69.45 $g \cdot kg^{-1}$ UB-B^{-1}) compared with increase in linoleic acid (0.1903 $g \cdot kg^{-1}$ UV-B^{-1}). High oleic acid and low linoleic and linolenic acids are desirable because they contribute to oil stability and human and animal nutrition [51] .

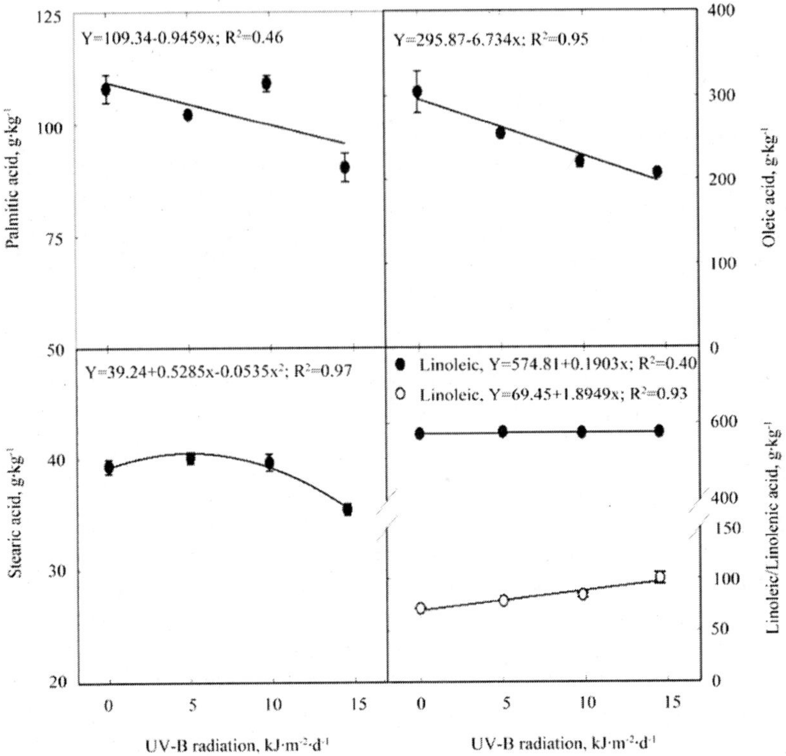

Figure 3. Ultraviolet (UV)-B radiation effects on soybean seed fatty acid content for plants harvested at 100 days after sowing. The UV-B treatments were imposed at early-pod-filling stage, 65 days after sowing. Each data point is mean of four replications and standard errors are shown if the values are larger than the symbols.

3.2.4. Seed Sugars

Growing season environment such as temperature [20] , genotypes, maturity and diseases [21] , drought [22] , and nutrients [23] affects soybean seed composition. Sucrose and stachyose contents showed quadratic trends with increasing UV-B doses, maximum sucrose of 38 mg·g^{-1}was found at 4.27 kJ of UV-B (Figure 4). Raffinose content, on the other hand, declined linearly with increase in UV-B doses, 0.013 mg·g^{-1} UV-B^{-1}. Experiments conducted by Gao et al. [25] also reported that maize seed sugar and starch contents decreased with enhanced UV-B irradiance similar to our results in soybean. Similarly, higher UV-B treatments showed decreased wheat grain sugar content [24] . Raffinose and stachyose are undesirable components because they have detrimental effects on the

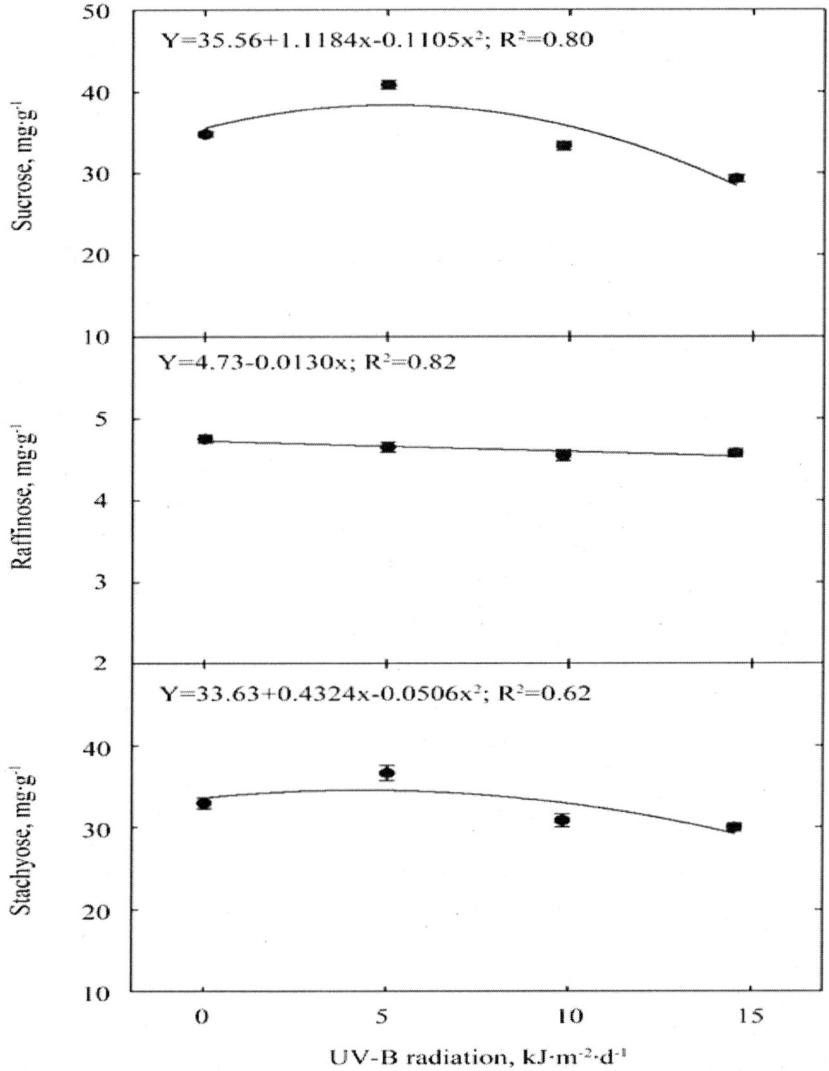

Figure 4. Ultraviolet (UV)-B radiation effects on soybean seed sugar contents for plants harvested at 100 days after sowing. The UV-B treatments were imposed at early-pod-filling stage, 65 days after sowing. Each data point is mean of four replications and standard errors are shown if the values are larger than the symbols.

nutritive value of the meal and are indigestible by human and animals, often causing flatulence and diarrhea in non-ruminants [18] . Therefore, soybean seed with low raffinose and stachyose is desirable for feeding non ruminant animals to improve feed energy efficiency, increase mineral uptake, and reduce flatulence [53] . On the other hand, soybean seed with high sucrose is desirable because it improves taste and flavor in tofu, soy milk and natto [2] .

4. CONCLUSION

This study reports the biomass and seed quality responses of soybean crop exposed to enhanced UV-B irradiance under sunlit controlled plant growth chamber conditions. Enhanced UV-B irradiance caused significant reductions in certain plant growth parameters, but the influence was not consistent as the treatments were imposed during mid-fruiting period. The protein, oil, fatty acid and sugar contents in soybean plants are critical to soybean seed industry and human and animal nutrition. Increased UV-B radiation altered soybean seed quality. Seed protein declined linearly while oil content increased with increase in UV-B radiation. Changes in these two important seed quality parameters have implications for seed industry and animal and human nutrition. The negative correlation between seed protein and oil content and their opposite responses to increasing UV-B doses poses challenges for breeders to develop soybean cultivars better suited for human and animal nutrition in a changing climate. The desirable fatty acids such as oleic acid declined while the linoleic and linolenic acids increased with increase in UV-B radiation. Palmitic and stearic acids showed linear and quadratic trends with increasing UV-B doses. While lower levels of desirable sugars such as raffinose declined linearly while stachyose increased up to 4 - 5 kJ and declined at higher UV-B treatments will have positive effects on improved feed energy efficiency in non-ruminant animals. The declined sucrose content at higher UV-B doses compared to the levels typically seen in soybean growing areas (~5 kJ), will have serious implications for human consumption. The functional relationships between UV-B radiation and soybean seed quality parameters will be useful to develop seed quality-specific sub-models under optimal temperature and nutrient conditions. Models equipped with seed quality would be useful not only for production optimization of natural resources such as water and nutrients, but also useful in assisting planting dates in the current environment and in policy decisions for the hypothesized changes in global environmental change on soybean production in the future.

ACKNOWLEDGEMENTS

We thank Mr. David Brand for their technical assistance. This article is a contribution from the Department of Plant and Soil Sciences, Mississippi State University, Mississippi Agricultural and Forestry Experiment Station, paper no. J-12658. Mention of trade names or commercial products in this publication is solely for the purpose of providing specific information and does not imply recommendation or endorsement by the Unites States Department of Agriculture (USDA). USDA is an equal opportunity provider and employer. This work was in part funded by the USDA-NIFA 2013-34263-20931, sub-award to Mississippi State University, G-7799-2. Funding for this research was also partially provided by USDA, Agricultural Research Service projects 6066-21220-012- 00D and 6066-21000-051-00D.

CITE THIS PAPER

K. Raja Reddy, Hrusikesh Patro, Suresh Lokhande, Nacer Bellaloui, Wei Gao (2016) Ultraviolet-B Radiation Alters Soybean Growth and Seed Quality. *Food and Nutrition Sciences*,**07**,55-66.

REFERENCES

1. 1. Grieshop, C.M. and Fahey, G.C. (2001) Comparison of Quality Characteristics of Soybeans from Brazil, China, and the United States. Journal of Agriculture and Food Chemistry, 49, 2669-2673.
2. Hou, A., Chen, P., Alloatti, J., Li, D., Mozzoni, L., Zhang, B. and Shi, A. (2009) Genetic Variability of Seed Sugar Content in Worldwide Soybean Germplasm Collections. Crop Science, 49, 903-912.
3. Madronich, S., McKenzie, R.L., Bjorn, O. and Caldwell, M.M. (1998) Changes in Biologically Active Ultraviolet Radiation Reaching the Earth Surface. Journal of Photochemistry and Photobiology B: Biology, 46, 5-19.
4. McKenzie, R.L., Conner, B. and Bodeker, G. (1999) Increased Summer Time UV Radiation in Four Aquatic Macrophytes. Chemosphere, 74, 642-647.
5. Reddy, K.R., Kakani, V.G., Zhao, D., Koti, S. and Gao, W. (2004) Interactive Effects of Ultraviolet-B Radiation and Temperature on Cotton Physiology, Growth, Development and Hyperspectral Reflectance. Photochemistry and Photobiology, 79, 416-427.
6. Bassman, J.H. (2004) Ecosystem Consequences of Enhanced Solar Ultraviolet Radiation: Secondary Plant Metabolites as Mediators of Multiple Trophic Interactions in Terrestrial Communities. Photochemistry and Photobiology, 79, 382-398.

7. Mohammed, A.R. and Tarpley, L. (2010) Differential Response of Southern US Rice (Oryzasativa L.) Cultivars to Ultraviolet-B Radiation. Journal of Agronomy and Crop Science, 196, 286-295.
8. Board, J.E. and Harville, B.G. (1996) Growth Dynamics during the Vegetative Period Affects Yield of Narrow-Row, Late-Planted Soybean. Agronomy Journal, 88, 567-572.
9. Barnes, P.W., Flint, S.D. and Caldwell, M.M. (1990) Morphological Response of Crop and Weed Species of Different Growth Forms to UV-B Radiation. American Journal of Botany, 77, 1354-1360.
10. Feng, H.Y., Chen, T., Xu, S.J., An, L.Z., Qiang, W.Y., Zhang, M.X. and Wang, X.L. (2001) Effect of Enhanced UV-B Radiation on Growth, Yield and Stable Carbon Isotope Composition in Glycine max Cultivars. Acta Botanica Sinca, 43, 709-713.
11. Kim, E.H., Seguin, P., Lee, J.E., Yoon, C.G., Song, H.K., Ahn, J.K. and Chung, I.M. (2011) Elevated Ultraviolet-B Radiation Reduces Concentrations of Isoflavones and Phenolic Compounds in Soybean Seeds. Journal of Agro Crop Science, 197, 75-80.
12. Liu, X.B., Li, Y.S. and Herbert, S.J. (2004) Effects of Enhanced UV-B Radiation on Seed Growth Characteristics and Yield Components in Soybean. Field Crops Research, 154, 158-163.
13. Chen, J.J., Zu, Y.Q., Chen, H.Y. and Li, Y. (2004) Influence of Enhanced UV-B Radiation on Growth and Biomass Allocation of Twenty Soybean Cultivars. Journal of Agriculture and Environmental Sciences, 23, 29-33.
14. Wilson, R.F. (2004) Seed Composition. In: Boerma, H. and Specht, J.E., Eds., Soybeans: Improvement, Production, and Uses, 3rd Edition, ASA, CSSA, and SSSA, Madison, 621-668.
15. Cherry, J.H., Bishop, L., Hasegawa, P.M. and Leffler, H.R. (1985) Differences in the Fatty Acid Composition of Soybean Seed Produced in Northern and Southern Areas of the U.S.A. Phytochemistry, 24, 237-241.
16. Schnebly, S.R. and Fehr, W.R. (1993) Effect of Years and Planting Dates on Fatty Acid Composition of Soybean Genotypes. Crop Science, 33, 716-719.
17. Burton, J.W. (1984) Breeding Soybean for Improved Protein Quantity and Quality. In: Shibles, R., Ed., Proceedings of World Soybean Research Conference, III, Ames, Westview Press, Boulder, 361-367.
18. Liu, K. (1997) Soybeans Chemistry, Technology, and Utilization. Chapman & Hall, New York.
19. Hartwig, E.E. and Kilen, T.C. (1991) Yield and Composition of Soybean Seed from Parents with Different Protein, Similar Yield. Crop Science, 31, 290-292.
20. Dardanelli, J.L., Balzarini, Martinez, J., Cuniberti, M., Resnik, S., Ramunda, S.F., Herrero, R. and Baigorri, H. (2006) Soybean Maturity Groups, Environments, and Their Interaction Define Mega-Environments for Seed Composition in Argentina. Crop Science, 46, 1939-1947.
21. Bellaloui, N., Smith, J.R., Ray, J.D. and Gillen, A.M. (2009) Effect of Maturity on Seed Composition in the Early Soybean Production System as Measured on Near-Isogenic Soybean Lines. Crop Science, 49, 608-620.
22. Bellaloui, N., Mengistu, A., Fisher, D.K. and Abel, C.A. (20112) Soybean Seed Composition Constituents as Affected by Drought and Phomopsisin Phomopsis Susceptible and Resistant Genotypes. Journal of Crop Improvement, 26, 428-453.
23. Bellaloui, N., Hanks, J.E., Fisher, D.K. and Mengistu, A. (2009) Soybean Seed Composition Is Influenced by Within-Field Variability in Soil Nutrients. Crop Management, 8.
24. Zu, Y. (2004) Intraspecific Responses in Grain Quality of 10 Wheat Cultivars to Enhanced UV-B Radiation under Field Conditions. Journal of Photochemistry and Photobiology, 74, 95-100.
25. Gao, W., Zheng, Y., Slusser, J.R., Heisler, G.M., Grant, R.H., Xu, J. and He, D. (2004) Effects of Supplementary Ultraviolet-B Irradiance on Maize Yield and Qualities. A Field Experiment. Photochemistry and Photobiology, 80, 127-131.

26. Kakani, V.G., Reddy, K.R., Zhao, D. and Sailaja, K. (2003) Field Crop Responses to Ultraviolet-B Radiation: A Review. Agriculture and Forest Meteorology, 120, 191-218.
27. Reddy, K.R., Hodges, H.F., Read, J.J., McKinion, J.M., Baker, J.T., Tarpley, L. and Reddy, V.R. (2001) Soil-Plant-Atmosphere-Research (SPAR) Facility: A Tool for Plant Research and Modeling. Biotronics, 30, 27-50.
28. Hewitt, E.J. (1952) Sand and Water Culture Methods Used in the Study of Plant Nutrition. Technical Communication No. 22, Farmham Royal Commonwealth Agriculture Bureaux, Bucks, 187-190.
29. Wilcox, J.R. and Shibles, R.M. (2001) Interrelationships among Seed Quality Attributes in Soybean. Crop Science, 41, 11-14.
30. Boydak, E., Alpaslan, M., Hayta, M., Gercek, S. and Simsek, M. (2002) Seed Composition of Soybeans Grown in the Harran Region of Turkey as Affected by Row Spacing and Irrigation. Journal of Agriculture and Food Chemistry, 50, 4718-4720.
31. Gao, W., Zheng, Y., Slusser, J.R., Heisler, G.M., Grant, R.H., Xu, J. and He, D. (2004) Effects of Supplementary Ultraviolet-B Irradiance on Corn Yield and Qualities: A Field Experiment. Photochemistry and Photobiology, 80, 127-131.
32. McKenzie, R.L., Aucamp, P.J., Bais, A.F., Björn, L.O. and Ilyas, M. (2007) Changes in Biologically Active Ultraviolet Radiation Reaching the Earth's Surface. Photochemistry and Photobiology Sciences, 6, 218-231.
33. WMO (2006) Scientific Assessment of Ozone Depletion. Global Ozone Research and Monitoring Project—Report No. 50. WMO, Geneva.
34. Egli, D.B., Ramseur, E.L., Yu, Z.W. and Sullivan, C.H. (1989) Source-Sink Alterations Affect the Number of Cells in Soybean Cotyledons. Crop Science, 29, 732-735.
35. Zhao, D., Reddy, K.R., Kakani, V.G., Read, J.J. and Sullivan, J.H. (2003) Growth and Physiological Responses of Cotton (Gossypium hirsutum L.) to Elevated Carbon Dioxide and Ultraviolet-B Radiation under Controlled Environmental Conditions. Plant, Cell & Environment, 26, 771-782.
36. Katerova, Z., Ivanov, S., Prinsen, E., Van Onckelen, H., Alexieva, H. and Azmi, A. (2009) Low Doses of Ultraviolet-B or Ultraviolet-C Radiation Affect Phytohormones in Young Pea Plants. Biologia Planturum, 53, 365-368.
37. Mark, U. and Tevini, M. (1996) Combination Effects of UV-B Radiation and Temperature on Sunflower (Helianthus annuus L., cv. Polstar) and Maize (Zea mays L, cv. Zenit) Seedlings. Journal of Plant Physiology, 148, 49-56.
38. Mathew, J.P., Herbert, S.J., Zhang, S.H., Rautenkranz, A.A.F. and Litchfield, G.V. (2000) Differential Response of Soybean Yield Components to the Timing of Light Enrichment. Agronomy Journal, 92, 1156-1161.
39. Herbert, S.J. and Litchfield, G.V. (1982) Partitioning Soybean Yield Components. Crop Science, 22, 1074-1079.
40. Board, J.E., Kamal, M. and Harville, B.G. (1992) Temporal Importance of Greater Light Interception to Increased Yield in Narrow-Row Soybean. Agronomy Journal, 84, 575-579.
41. Schrope, M. (2000) Successes in Fight to Save Ozone Layer Could Close Holes by 2050. Nature, 408, 627.
42. Qiang, W.Y., Yang, H., Chen, T., An, L.Z. and Wang, X.L. (2004) Effect of the Combination of Cadmium and UV-B Radiation on Soybean Growth. Chinese Journal of Applied Ecology, 15, 697-700.
43. Mazza, A.C., Boccalandro, H.E., Giordano, C.V., Battista, D., Scopel, A.L. and Ballare, C.L. (2000) Functional Significance and Induction by Solar Radiation of Ultraviolet-Absorbing Sunscreens in Field-Grown Soybean Crops. Plant Physiology, 122, 117-126.
44. Ros, J. and Tevini, M. (1995) UV-B Radiation and IAA Interaction during Growth of Seedlings and Hypocotyl Segments of Sunflower. Journal of Plant Physiology, 146, 295-305.
45. Caldwell, M.M., Flint, S.D. and Searles, P.S. (1994) Spectral Balance and UV-B Sensitivity of Soybean: A Field Experiment. Plant, Cell & Environment, 17, 267-276.

46. Murali, N.S. and Teramura, A.H. (1985) Effects of Ultraviolet-B Irradiance on Soybean.VII. Biomass and Concentration and Uptake of Nutrients at Varying P Supply. Journal of Plant Nutrition, 8, 177-192.

47. Ros, J. (1995) Synergistic and Antagonistic Effects of Enhanced UV-B Radiation at Various Nitrogen Levels in Crop Plants. In: Bauer, H. and Nolan, C., Eds., Proceedings of the First European Symposium on the Effects of Environmental UV-B Radiation, European Commission, Luxemburg, 245-252.

48. lan, H., Teramura, N. and Sullivan, J.H. (1994) Effects of UV-B Radiation on Photosynthesis and Growth of Terrestrial Plants. Photosynthesis Research, 39, 463-473.

49. Calderini, D.F., Lizana, X.C., Hess, S., Jobet, C.R. and Zuniga, J.A. (2008) Grain Yield and Quality of Wheat under Increased Ultraviolet Radiation (UV-B) at Later Stages of the Crop Cycle. Journal of Agriculture Science, 146, 456-367.

50. Obendorf, R.L., Horbowicz, M., Dickerman, A.M., Brenac, P. and Smith, M.E. (1998) Soluble Oligosaccharides and Galactosylcyclitols in Maturing Soybean Seeds in Planta and in Vitro. Crop Science, 38, 78-84.

51. Gao, W., Zheng, Y.F., Slusser J.R. and Heisler, G.M. (2003) Impact of Enhanced Ultraviolet-B Irradiance on Cotton Growth, Development, Yield, and Qualities under Field Conditions. Agricultural and Forest Meteorology, 120, 241-248.

52. Rotundo, J.L. (2009) Physiological Bases of Environmental and Genotypic Effects on Soybean Seed Composition. Paper 12113, Graduate Theses and Dissertations, Iowa State University Ames, Ames.

53. Sherif, M. (2013) Soybean, Nutrition and Health. In: El-Shemy, H., Ed., Soybean—Bio-Active Compounds, InTech, Rijeka, 453-473. http://www.intechopen.com /books/soybean-bio-active-compounds/soybean-nutrition-and-health

Index